The Devil's Blood
Expanded from Season of the Vigilante

To Alta —
Keep a step ahead of the devil!
Your friend,
Kirby Jonas 11/25/02

Books by Kirby Jonas

Season of the Vigilante, Book One: The Bloody Season
Season of the Vigilante, Book Two: Season's End
The Dansing Star
Death of an Eagle
Legend of the Tumbleweed
Lady Winchester
The Devil's Blood

Books on audio tape read by James Drury, "The Virginian" (Available at Books in Motion or through the author at www.kirbyjonas.com) .

The Dansing Star
Death of an Eagle
Legend of the Tumbleweed
Lady Winchester
The Devil's Blood (pending)

The Devil's Blood

Kirby Jonas
Cover art by Author

Howling Wolf Publishing
Pocatello, Idaho

Copyright ©2003 by Kirby F. Jonas. All rights reserved. No part of this book may be reproduced or transmitted in any form or by any means, electronic or mechanical, including photocopying, recording, or by any information storage and retrieval system, without permission from the publisher. Request for such permission should be addressed to:

Howling Wolf Publishing
P.O. Box 1045
Pocatello ID 83204

This is a work of fiction. Names, characters, places and incidents portrayed in this novel are either fictitious or are used fictitiously.

For more information about Kirby's books, including his novel-in-progress with actor Clint Walker, check out:

Kirby Jonas
kirby@kirbyjonas.com
www.kirbyjonas.com

James Drury, "The Virginian"
www.thevirginian.net

Manufactured in the United States of America

First Edition

Jonas, Kirby, 1965 —
    The devil's blood / by Kirby Jonas. -- 1st ed.
      p.   cm
    ISBN 1-891423-07-X
    1. Title
      Library of Congress Control Number: 2002105618 "

                                                                                                     CIP

Acknowledgments

Out of an entire book, this little piece is probably the hardest to write. It seems inevitably I leave out the name of a friend or colleague or acquaintance who, if not instrumental in the research of the book, was monumental in at least one aspect of it. So, here I go again, and if you are one of those few who escapes mention, please accept my sincerest apologies. I promise you it was done out of forgetfulness, not spite.

To start at the beginning, I have to thank Mom and Daddy for putting up with cowboy dolls all over the house, the seeds I was planting that would grow eventually into (thus far) more than eighty book plots. While my folks worked the land and took care of the details of creating a secure home, my brothers and I rode our dusty mounts through the canyons and over the prairies of childhood fantasy. Without that freedom, this book and many of my others would never have existed.

Thanks go to "Miss Hone," alias Janice Vasas, my third grade teacher and lifetime favorite. She understood the magic it took to build a child's dreams; Reed Regan, for introducing me to the enchantment of Louis L'Amour; and Jarvis Anderson, for making me understand the joy of putting my own stories to the printed page. Leslie Kidman, Ted Potter, and Von Mortensen, and Mrs. Arnold, for unending support. Stuart Fredrickson and Richard Hobbs and Shawn Curtis, for giving me local heroes.

Richie Smith, Russell Smith, Mark Jones and Loui Novak, who took their valuable time to read and critique. And Scott Darger and his entire family, who took in a young rebel long enough for him to scour the Arizona desert and adopt it as his second home.

Dean and Nancy Hoch and Mary Croney, whose original faith got this work of fiction off the ground, and Steve Medellin, who kept the dream rolling. Kelly Lance and the Naylors and Dan Gilbert, for helping to fill in the blanks on horsemanship and the frailties of an animal that once seemed invincible to me. The boys from CAS-1, too numerous to mention by name, who faithfully fed me information on guns that I might never have found elsewhere, and that certainly would never have been delivered with more enthusiasm. James Drury and Gordon Perry, who found mistakes that were invisible to everyone else and reported them dutifully. And James again, for giving me the thrill of hearing "the Virginian" relate every single instance in the book that gave him chills. Clint Walker, for being that reachable star on the screen that even a child could tell was real. Peter Breck and

VI The Devil's Blood

Lee Majors, James Arness and Robert Conrad, Buck Taylor, Dobe Carey and Ben Johnson, without whom life would have lost a lot of color. Chris Taft, Jose Olano, Todd Shank, Gunn Mosqueda, Dell Mangum, Ron Ciancutti, Dave Lundy, Gene O'Quinn—my brothers that should have been. The Bauers and the Kelseys and Uncle Berdett, and the spirit of Grandpa Arch Hess, who would have loved to put away his sheriff's guns and sit in a quiet moment to read this book if he had had the chance.

There are too many friends who support me to list them all. You know who you are. Thank you from the bottom of my heart. Perhaps I would be better served by listing all of those people in my life to whom this book is *not* indebted. If no one in the world read my books, I would still write them. But knowing they are enjoyed by others is a reward that cannot be matched.

Without brother Jamie no dream this big could have been kept alive. Thanks must go to him for all his support and understanding at the same time that my apologies fly to him for the long wait on the book we are writing together.

To Cheyenne, Jacob, Clay and Matthew, who with hats firmly planted on their heads and guns on their hips are keeping the flame of the West alive.

And to Debbie, my shadow, my friend, my partner, my lover, my wife. I've tried in all of my books to get across what you mean to me. No words can say it. I hope you can see it in my eyes.

Author's Note

The following is an excerpt from the original version of this book, which was published in two volumes. It was known then as *Season of the Vigilante, Book One: The Bloody Season* and *Book Two: Season's End*:

This book began in 1978 as "The Vigilante." Its main character, Tappan Kittery, was then a huge, bearded brute of a man who killed mercilessly anyone he suspected of a crime or anyone who got in his way. But my heart wasn't in it, and it died with Chapter One.

Then, in 1982, new life was breathed into The Vigilante. *Although stuck in Idaho, I dreamed of a faraway place on an encyclopedia map of Arizona: The Baboquivari Mountains (pronounced Bob-o-KEE-vuh-ree). I dreamed up a town that was an oasis near there, south of Tucson. The town was named Castor, the Spanish word for beaver. And into this fictional town rode Captain Tappan Kittery, now a much gentler, deeper character.*

In 1982-83, I wrote in pencil in three spiral notebooks, and soon I had finished my first complete novel, The Vigilante, *over three hundred pages long. With a borrowed electric typewriter, I then typed up the book after proofreading it myself. I typed it again in 1984, completely, because I had no computer.*

In 1985 and 1986, I proofread that typed copy while living in France, and in the absence of a typewriter, rewrote it with a ballpoint pen.

At last, in 1986 and 1987, I had the chance to move to Arizona with my friend Scott Darger and his family. I didn't hesitate. I moved to Mesa, Arizona, a short *drive from Tucson. While in Mesa, I made many weekend trips to Tucson and south of it, to the setting of* The Vigilante. *I rode the hills where the book takes place. I scoured them on foot and by automobile. I mapped and planned and dreamed. And in that time period Tappan Kittery, to me, became a real being, and somehow like a close friend.*

I returned to Idaho in June of 1987, bought a word processor, and typed my book again, this time as Season of the Vigilante. *Now it seemed real, because I knew all the places I wrote about. I again honed it and reprinted it in 1989. By then, I had met and married my wife, Debra Chatterton. It is because of her that you hold this book in your hands now. She read the book, she loved it, and it was she who pushed for it to be published.*

Debbie and her mother met local publishers, Dean and Nancy Hoch. They said they would read my book and give me a critique and possibly some help finding a market. Instead, after Nancy read the book, she asked to be the publisher. Needless to say, I was elated.

I retyped Season of the Vigilante *completely on an Apple computer in 1990. I then reworked it again, and it took me three eight-hour days to print it out, in 1991. It was while working on the computers at Idaho State University that I met Chris Taft, and it is due in great part to his friendship and his immense help with the computer that my final draft was able to be printed.*

The decision was made by the publishers and myself to divide Season of the Vigilante in half. Thus, it became Book One: The Bloody Season, *and* Book Two: Season's End. *I hope you enjoy them both, and enjoy the story as much as I enjoyed writing it."*

Seven years have passed since I wrote those words—a lot of water under the bridge. And I've thought of many things I wish I had written in my author's note back then. With your kind permission (and I guess without it), I would like to write them now.

The number one most important point I'd like to bring out is how my admiration for one Western actor who was nearly larger than the silver screen he filled became the inspiration for my main character, Captain Tappan Kittery. If you see a little of Cheyenne Bodie—of Clint Walker—in the broad shoulders and the dark hair, and the full chest of Tappan Kittery, it is not by mistake. When I was young, he became the icon I strove to emulate, both on screen and in real life. It was almost as if I had no choice but to model my first protagonist, who was also bigger than life, after Clint Walker, the man who captured my dreams and those of so many other Americans.

Secondly, I'd be remiss in not thanking the Louis Marx Toy company for being indirectly responsible for the creation of the story of *The Vigilante*. For those of you who don't have the fond memories that thousands of us do of Marx's Johnny West dolls, let me paint you a picture. Johnny West rode into the homes of tens of thousands of children from 1965 to 1975, a light brown plastic-clothed man with reddish brown hair and a penchant for adventure. He stood eleven and one half inches tall and came housed in a thin cardboard box, its lid taped once on each side. (This stands out in my mind, as my brother and I had to sneak a pocketknife into King's toy store every year when Christmas drew near to cut this tape and pick the best of the dolls!)

VIII   The Devil's Blood

Johnny West had family, too, and friends. Not the least of these was Captain Tom Maddox. To make a long story short, over several Christmases and birthdays, and through the shopping trips provided by a strenuous winter of shoveling neighborhood sidewalks, the collection of Marx dolls owned by my brothers and me became quite impressive. It was natural that this should carry over into my love of Western literature.

And so, in the summer after my junior year of high school, I began to fashion a western town, complete with a working gallows, wagons whose wheels I carved out of solid pieces of pine, a water tower and a cantina with swinging doors that really swung. I dressed Johnny West up in clothes I sewed myself, glued real hair from wigs, our dogs, and my own head, to their heads and faces. The Captain Tom Maddox doll became Captain Tappan Kittery. And then, borrowing Mom's '64 Chevy pickup, I headed out into the mountains to create a town. Along Cedar Creek, up the Blackfoot River, my brother and I built the town of Castor, replete with corrals, a sign welcoming visitors to Castor, and enough odd characters like monks and fancy gamblers and freighters to give the place some cultural diversity. We even found a cave, which became the Desperado Den of the book.

I spent a week creating intricate scenes in which I used cotton balls to form gunsmoke, fishing line to hold up falling guns, hats, or men who were supposed to appear as if they were in mid-air. I even went so far as to drag a sharp stick through the street to create wagon tracks. Yep. This, folks, is how Castor and the *Season of the Vigilante* were born. And in the eight days I spent up Cedar Creek, I was only caught once, by some folks going fishing. They had the audacity of commenting of me and my brother, "Aren't they a little old to be playing with dolls?" I guess we were. But this was *art!*

Taking all the slides gleaned from this trip, I wrote my first draft of the book, after going through them trying to make sense of the week's work. I still own the slide show—a prized possession.

And so the games of childhood managed to create a bridge between youth and a shaky manhood my wife claims I still haven't made it to. If I have my way I never will.

Then came the perils of having the books published. As I said in my original "author's note," the publishers and I made the decision to cut the book in half. As for my part, I made that decision because my other choice was to cut it in thirds. But cutting it in half still became one of the most disastrous publishing choices I can imagine. We waited and waited, far too long, and the second half of the book came out over a year after the first, which left my readers hanging. Many of them forgot all about me by then. Consequently, it took four years to sell all of those books.

I didn't want to let Tappan Kittery and Castor die, but I never wanted to see a divided book again. And so, in the last two years, I made the decision to rejoin the books, the way I always knew they were meant to be. I couldn't name it *Season of the Vigilante*. It has been recreated, added to, refined. You'll find in this book a story more the way I would have written it had I done so today, although the basic plot is the same. And so I named it after one of the several songs I wrote for the book. *The Devil's Trail.* Then, shortly before publication, I discovered that a friend of mine, another author, had taken the title for his newly published book that winter. So *The Devil's Trail* became *The Devil's Blood.* Here, my friends, is that story.

Kirby Jonas   IX

To Clint Walker, the man who brought Yellowstone Kelly to life on the big screen, rode into the living room every week as Cheyenne Bodie, and drove the dreams of a boy growing up in the West. To Clint Walker, the inspiration for Captain Tappan Kittery.

Cast of Characters

<u>The Vigilantes</u>

Tappan Kittery

Lafayette Bacon—cowboy on the Double F
Cotton Baine—Kittery's friend
Adam Beck—Kittery's friend
Durnam Beehan—one-time outlaw, older brother of the twins
Edward Beehan—a Beehan twin
Linus Beehan—one of the Beehan twins, of the JP Ranch
Tick Bell—miner
Mort Dansing—ex-deputy sheriff of Pima County
Johnny Hedrick—gunman
John Laynee—gunman
Jake Long—cowboy on the Double F
Randall McBride—banker
Zeff Perry rancher—owner of the Double F
Jed Polk—owner of the JP Ranch
James Price—federal hangman, vigilante captain
Black Ramblene—owner of the Empire Mountain Ranch
Jed Reilly—co-owner of the R Slash R
Joe Reilly—co-owner of the R Slash R
Frederick Ridgon—miner
Bison Sabala—blacksmith
Tule Simpton—freighter for Tully, Ochoa and Company, in Tucson
Miles Tarandon—deputy sheriff of Pima County
Dabney Trull—grave digger

<u>The Outlaws</u>

"Savage Diablo" Baraga
Major Morgan Dixon—Desperado Two
Samuel Colt Bishop—Desperado Three
"Bloody Walt" Doolin—Desperado Four

X   The Devil's Blood

"Big Samson" Rico Wells—Desperado Five
Noble "Silverbeard" Sloan—Desperado Six
Crow Denton—Desperado Seven
"Slicker Sam" Malone—Desperado Eight
Pete Bandy—horse thief
Ned Crawford—Kittery's initial target
Gustavo Ferrar—brutal Mexican ringleader
Paddles-On-The-River "Paddlon"—Baraga's guard
Shorty Randall—Baraga informant
Amayo Varandez—Mexican outlaw
Tawn Wespin, "The Skunk"—Baraga henchman

Townfolk

Sheriff William Barden—historical character, representation of Sheriff William Oury
Mario Cardona—owner of the cantina
Nelson Chace—government agent
Judd Creech "Creature"—lonesome trapper
Marshal Francis H. Goodwin—historical character
Emilita Greene
Maria Greene
Miguel Greene—Maria Greene's baby
Thaddeus Greene
Albert Hagar—owner of the warehouse and mercantile
Doctor Adcock Hale
Solomon Hart—warehouse worker
Jarob Hawkins—livery stable owner
Jose—Kittery's young helper number two
Bartholomew Kingsley—owner of the cafe
Tania McBride
Raoul Mendoza—friend of Gustav Ferrar
Pepe—Kittery's young helper
Lieutenant Babar "Bab" Peppridge—army officer hunting down the Desperados
Marshal Joe Raines
Adam Rolph—government agent unjustly killed by the vigilantes
Governor Anson Safford—historical character
Marshal Wiley P. Standefer—historical character
Juan Torres—friend of Gustav Ferrar
Efraín Valesquez—Mexican sheepherder
Beth Vancouver
Sheriff Luke Vancouver

A Note from Clint Walker

In 1999 I was introduced to Kirby Jonas by my friend, James Drury—"The Virginian"—who narrated four of Kirby's books on audio tape. After listening to three of those audio books I, too, became a fan of Kirby Jonas.

I was so impressed with his work that I shared with him a story idea of my own and asked if he was interested in collaborating on the story. He agreed, and the story is now near completion. We are very pleased with the result, and that Kirby Jonas magic is bringing the story to life.

This is not a run-of-the-mill Western. It has some unique twists—believable but full of surprises—and it will capture your imagination.

Kirby and I are excited about *Yaqui Gold,* and we think you will be too.

Clint Walker

XII The Devil's Blood

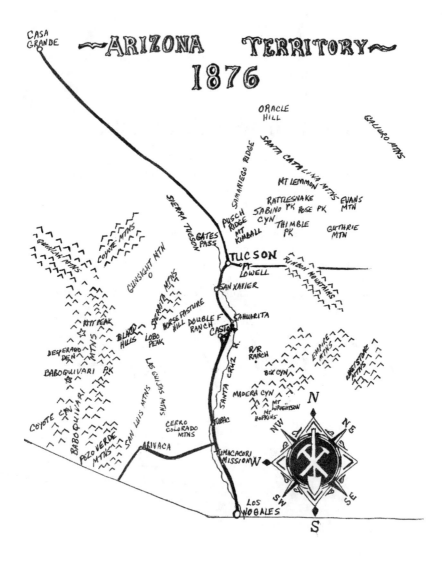

Part One
Season of Gloom

Chapter One
The Desperado Den

Mercy sings a hopeless song the ears of a killer cannot hear.

In the shadows of the blue-black hole eating into the side of the mountain, at the foot of the cliff where Indian gods made their home, merciless men waited in the dark shadow of death. The Desperados Eight were led by one man. He was called Savage Diablo Baraga.

Longhaired little Blue Bell Smith waited with them. It was like waiting for a breeze to blow by—not just any breeze, but that one in particular that every man would recognize when he smelled it, when he felt it on his skin, when its acrid musk tainted his tongue. Blue Bell Smith didn't know why they waited. He only knew the shadow of death lurked there for him.

Seven of the men in this hole in the rock they called the Desperado Den belonged there. Blue Bell Smith did not. The little man rode with Diablo Baraga when they let him, but he was nothing more than a cur to Baraga. He met neither his standards nor the standards it took to gain fame in Arizona newspapers.

"The Desperados Eight" was the romantic name with which the *Tucson Citizen* had dubbed the gang and which had been spread across the continent by the hot butter knife of the press. They were *Los Desperados Ocho* in the little border towns and in the *villas* below the border in Sonora. To the world they were evil in the flesh.

Today in the Desperado Den the only air that moved was a black wind. It moaned through Blue Bell Smith and called his name. It was that breeze from which they awaited a sign, guided by the bizarre protocol these men required before they made their move. Smith knew he should have left here in the night. He should have left after he awoke from the dream that left him cold and shaking, sweating and nearly in tears in the dark. He should have left, but he stayed.

There was no talk, no laughter among the Desperados Eight, or rather, among the seven present. With cold eyes, they waited in restless silence, avoiding the little man's glance. It was a spiteful thing. He would look at one of them, and they would look away. Maybe he had a guilty conscience. But his memory of speaking with the marshal—what was his name? Joe Raines . . . That memory was so vivid it gave him the eerie feeling Baraga could read his mind.

Smith ran slim graceful fingers over his mustache and through his shoulder-length blond hair. He pulled out his Army Colt pistol and used the oil of his hair to smooth its red walnut grip. He slipped the gun back into the cross-draw holster on his left hip. He thought of his mother, of his younger brother dying of scarlet fever years ago when he was seven and his brother was only four. The doctor himself had succumbed the same morning. No one had been able to save even him—a man of medicine. Smith thought of his father crying, when men weren't supposed to cry. He thought of Christmas in the hardwoods of upstate Michigan, of snows ten feet high on the flat. He thought of home and family . . . and he waited to die.

Blue Bell Smith looked at each of the Desperados. He tried to get one of them to meet his gaze. There was Savage Diablo Baraga. Satan incarnate. Savage Diablo—Savage Devil. No one knew his real name, and no one knew for certain where this one came from. But it fit him like a well-tailored coat.

In spite of everything else impressive about Baraga, one feature grabbed a man first. Baraga surely knew that, and maybe it was why he was so full of hate. His right arm was missing at the elbow; all of the right sleeves of his garments were sewn up on the ends so he wasn't obliged to tuck or roll them. He'd lost the arm on Cemetery Ridge, at Gettysburg, where he had ridden under the command of General George Pickett.

Six pistols festooned Baraga's frame, strapped to him, slung from him, stuffed behind his belt. A one-armed man must assure easy access to a loaded gun. Most of his weapons were Colt .45's, one on his left hip, butt to the rear, one at the right in a cross draw holster. Another hung across his chest from a leather and chain rig of his own invention. He wore two converted .36 caliber Wells Fargo Colts, in holsters sewn on the outside of each boot, and when he rode another short-barreled Peacemaker rested against the inside of the buckle of his gun belt. He kept the .45's to avoid the confusion of loading several pistols of different caliber during a heated battle, should he ever unload all six pistols and have time to reload.

Baraga sat massaging the stub of his arm. He gazed out into the sunshine. Even seated, Baraga cut an imposing figure. His big-boned, cruelly handsome face with its eagle nose commanded admiration. It was framed by short, golden, brown-streaked hair that swept back from a broad forehead. Piercing, ice blue eyes gazed from beneath thick brows, and a dark, closely trimmed beard surrounded his mouth. A man couldn't have looked once at Savage Diablo Baraga without turning to look again.

Major Morgan Dixon was Baraga's second in command. He sat near Baraga studying a book of war that spoke nothing of the guerilla warfare that was Baraga's hallmark. Dixon had the kind of looks envied by men, desired by women. But from head to toe he was characterized by darkness. Black hair slicked down with bear grease reared back from an olive forehead, and a trim beard accentuated his jaw. He wore dark clothing, and his hat was gray, but in spite of all that darkness Smith couldn't imagine it shaded the darkness of his soul.

There were times the Major sported a tattered gray campaign coat—his badge of honor. He had survived the harsh conditions of a northern prison camp alongside men like Baraga, and now he, too, hated the Union and anything that represented it. Other than Baraga and Bishop, he didn't care much for *anything*. Or any*one*. His indifference toward the lives of others showed in the weapon he favored, a twelve-gauge shotgun with sawed-off barrels.

The third Desperado, Samuel Colt Bishop, lay stretched full-length across a blanket near the grotto's entrance. His gaze roamed the sunny canyon wall three hundred yards away. That outlaw was a puzzle Smith had never solved. Bishop wasn't like the other outlaws. For one thing, he was one of only two who hadn't fought in the war. They said he'd stayed out west, though he'd been twenty-one years old when the war began—and primed for a fight. For another thing, Colt Bishop was a gentleman, and as such he was out of place in the Desperado Den. He always treated Smith with kindness, when he treated him at all.

Bishop's raw-boned face was whiskered, and a thick dark mustache drooped over his lips. Facial hair was a sign of the times and seemed a requirement to ride with the Desperados. Roughly cut brown hair etched with gray scratched a haphazard line across his forehead and half-covered his ears. Though of average height, and not uncommonly muscular, Colt Bishop enjoyed immense respect wherever he went due to his reputation with the Colt pistol he wore. Blue Bell Smith had never seen the gunman touch liquor.

As Blue Bell's anxiety grew, his eyes ticked quicker and quicker across the rest of the gang. Bloody Walt Doolin was a big, black-haired man, broad across his chest and narrow between his slanted dark eyes. Silverbeard Sloan's nickname was a fit enough description of his outward appearance, and then there was Crow Denton, the man they called the Breed, and Slicker Sam Malone, who seemed as much an outcast as Blue Bell himself. These were men who had always scorned Blue Bell Smith but who had paid him well to be a spy. A list of men who would, without twitching, watch Blue Bell Smith be slain . . .

A molten-yellow sun poured down on the vast, rolling, rock-strewn hills. In that inferno of sand and rock, curled dead grass and thorn, nothing weak had hope for survival. Even the strong died horrible deaths unless fortune chose their side. The timid, and some not so timid, saved their travels for the light of the stars to avoid the lethal temperatures and the bandits and Apaches who might raid this land without as much warning as a javelina's charge.

Among the rocks and cactus and thornbrush of that land, on a light chestnut gelding with a glistening white mane and tail, rode a giant of a man by the name of Rico Wells. Today he had a shadow, a little man named Shorty Randall. Shorty's mane, in color, mimicked Wells's chestnut, and he rode a white horse that shone for miles across the desert—visible to any wandering Apache or Mexican outlaw or cutthroat who wanted a stake or just had a thirst for blood. Shorty was unwise to the ways of the wild country. And Rico Wells was too big and mean and reckless to care if the little man's horse drew attention to them.

The government of Arizona Territory called Rico Wells a menace to society. The newspapers called him, in kinder moments, Desperado number five. Or Big Samson. Mexicans called him *Sierra Grande*—Big Mountain.

Like sun-darkened strands of corn silk, Big Samson's fine, flowing hair lifted as a breeze swirled past his plodding chestnut, making a strand of it brush his cheek. He raised a hand and twirled a finger absently into the stray lock of hair as he kept his wary blue eyes moving, like two pieces of sky roving in a wall of sandstone. Twisting and twirling his hair with his finger was a habit Shorty had noticed whenever any loose hair tickled the big man's skin. Whatever the cause of the hair-twirling habit, Big Samson did it enough that the tips, at least around his face, had taken on a definite curl. If it had been any other man but Wells, Shorty might have called it a nervous habit, and he might have laughed. But a man did not laugh at Rico Wells.

Wells sat a Texas saddle with the three-banded barrel of a Springfield rifle balanced across its flat-topped horn. They could have named him Goliath, this Rico Wells; only the fifteen hundred-pound horse he rode could dwarf him. His head was large, and his hips were not narrow, but the enormous set of his shoulders made his head seem small and his hips the hips of a long distance runner. The relative thinness of his jaw was disguised behind a bush of dark beard, contrasting the yellow hair flowing past his shoulders.

Rico Wells had the tastes of an Indian, with beads on his moccasins, knife sheath and holster. He wore a cotton shirt that hung untucked, sashed at the waist, and a breechclout of shimmering golden satin over soiled buckskin trousers. The beads on his buckskin were drab in their color, and the only thing about Rico Wells that could have been called gaudy was the golden clout. Shorty Randall rode in awe beside this monster. Shorty was a few inches below average height; Wells was a good foot beyond.

Shorty pulled his furtive eyes away from Big Samson and studied the land. Mountain ranges, hazy and indistinct on the horizon, sawed against a heat-bleached, cloudless sky. To the northeast, the Sierritas; to the east, the Cerro Colorados; to the southeast, the San Luis range—all melded together, giving the appearance of one long series of ragged hills.

Nearer still rose the Baboquivaris.

This twenty-mile backbone jutted from rocky earth. Its granite rims braced the skyline, flanked by a narrow band of grassed-over foothills. Mesquite, with its wiry limbs and shriveled pods of beans, and umbrella-like paloverde—"green stick"—hunkered in washes and on the lower hills. Ironwood, bur sage, cholla, ocotillo and the occasional saguaro cactus, massive, spiny arms curved upward, broke the monotony of rock and sand.

As Wells and Shorty climbed into the foothills, the vegetation thickened, mixed now with evergreen oak and taller shrubs, purple prickly pear and Spanish bayonet. Before them, less than three miles away, the granite broke loose from the soil and heaved into the sky. The high, steep, rounded dome of Baboquivari Peak dominated the Baboquivaris, looking like the bald head of some giant whose body must have been the size of Arizona.

This was Indian ground, the land of the Papago—the *Tohono O'odham*. The Tohono O'odham were a peaceful people, for the most part friendly to whites. But to them Baboquivari was a special place, a land of mysterious gods. They resented the whites' encroachment here. According to their legends, an ancient god, *I'itoi*, had taken refuge here from his enemies in a labyrinthine cavern deep in the base of the peak. They expected him to return one day to purge his lair. Shorty was no man of superstitions, but he dared not entirely discount the myth. He had been to the base of Baboquivari. He'd felt its mystery and its astonishing power. The first afternoon he ever drew near the peak, he'd heard the wind rush over its granite face like some mighty waterfall. Yet on that particular day there had been no other sign of wind . . .

The Tohono O'odham weren't the only inhabitants of the Baboquivaris, of course. In spite of the quiet of the land during the day, the nighttime hills teemed with life. Mule deer, and whitetails the size of dogs; cottontail and jackrabbits;

javelinas and desert sheep all browsed the brush and trees. Coyotes and Mexican wolves sang lonely songs to the stars and fed on the less fortunate of the plant eaters. There were also the big cats, the preeminent cougar and the elusive jaguar. These desert dwellers and the Papagos knew of the cavern deep in these mountains toward which Wells and Shorty rode. But the Papagos never wandered near. In that hidden cavern, only death could await them.

They came in sight of the cavern from over the rim of the canyon. It was a path only Rico Wells liked to use, a trail running deep into a jungle of thorns that always left his face and the gold-toned hide of his horse bleeding from many tiny wounds. The canyon opened out before them, four hundred yards at its widest. Green marked the surfacing of a spring, and among the oaks and juniper Shorty could see horses foraging.

Across from them, the solid limestone of the western canyon wall gave way to a gaping black hole partly concealed by gnarled oaks. The cave bore deep into the mountain, harsh and forbidding against the flat tan of the stone. Shorty had been to the Desperado Den before, but only twice. He remembered the cave floor, thick with cool, powdery sand, reaching back sixty feet before coming up against a solid wall of rock. It then thrust up to a ceiling twenty feet high in some places. But most of all he remembered the hard men who found respite from the sun and planned their plundering raids from that cave. And he remembered the pale deadly eyes of Baraga.

Blue Bell Smith's teeth hurt from clenching his jaw. He wanted to walk outside, but somehow he knew he would be overstepping his bounds. He wanted a cigarette, but his papers were gone, and he wouldn't on his bravest day have dared ask any of the Desperados for a loan—not even Colt Bishop, not on a day like today. He walked to the rear of the grotto, then started back toward the front.

Blue Bell Smith became aware of the sounds of horses moving up through the brush in the bottom of the canyon. It was the sound not of horses drifting in their quest for nourishment, but the sound of ridden animals. He tensed, and sweat formed little beads on his cheeks and forehead. With his hands shaking, he stopped at the big wooden chest at rest in the middle of the floor. He picked his hat off the chest, clamping it down to his ears. With outward calm, he looked at the hot ashes of the fire. He cursed himself for coming back here—for staying after his nightmare bade him run.

He waited, he sweated. He didn't believe in God, but he prayed.

And then Big Samson Wells stood framed in the sunlight, towering like his namesake. Shorty Randall moved in beside him, insignificant as a hare next to this bison. Smith's heart jumped when he saw the little man. He knew he lived in Castor. He might have seen Blue Bell with the marshal!

Smith remained still. Wells didn't look at anyone but Smith, and the little man tried to meet his eyes. But he couldn't. He tipped back his hat and placed a boot on the wooden chest. He rested his elbow on the upraised knee. The sweat beads melded and began to roll down his cheeks.

Wells stared with wolfen gaze. His right hand held his long-barreled Springfield rifle. "Shorty and me talked about you," Wells said, his voice a deep purr Smith had found strangely lulling in the past. But lulling was not the word for it now.

Smith shifted his weight and dropped his foot from the wood chest, straightening.

Without thinking, he dried the palms of his hands on his shirt and glanced quickly around him, then back at Big Samson. "You talked about *me?* Anythin' I might be interested in?"

Wells moved only his lips. "Somethin' you should be."

"Well, try me." Smith tried to smile, but he was aware of how he failed. He edged his trembling fingers closer to the achingly small Army Colt on his hip. It seemed less threatening than David's slingshot.

"The story's on your face, Smith," said Wells, raising his voice. "Don't play games—you sold us out."

Smith tried to reply, but his mouth wouldn't open. He felt nerves twitch along his jaw like flies dancing in a line. He found his voice at last. "Sold you out? What're you talkin' about?"

"Shut your mouth," Wells said, his voice lowering. "I wanna know one thing: how much did they pay you?"

"Pay me for what?"

"To die."

The room was silent, and everyone watched. Samuel Colt Bishop had stood from his blanketed, sandy bed, and he watched both men. Sam Malone, the man the government called Desperado Eight, watched Smith. His eyes nervous, he wiped his palms against his shirt and looked from Smith to Rico Wells, swallowing hard. He was an underdog, too. He could feel the fear of any man who had to face Rico Wells.

Blue Bell Smith twisted up his face. "I—I don't wanna call anybody a liar, but . . . Well, *somebody's* been fillin' yer head with trash."

"You make a bad liar. Shorty don't make up stories. He saw you and the sheriff of Castor together—plain as I see you now. You spilled your guts. I hope you got to spend the money."

Blue Bell Smith looked outside, feeling cold all over. Many eyes drilled through him. Murderous eyes. He *hadn't* spent the money. He had tasted bourbon in Castor—good bourbon. But it hadn't lasted long enough, not to be standing here. There were no women in Castor—not the kind he wanted. And so he had gold coin in his pocket. Two double eagles—and the promise of more. But what did a dead man need with gold? It couldn't buy him a place in Heaven.

His eyes narrowed, and he felt an odd sensation—the cold walnut grip of his 1860 Army Colt folding inside his hand. It was odd because he didn't realize he had reached for it.

Grunting, Wells lurched close, swinging with the butt of his Springfield. The sharp edge of the butt plate caught Smith in the teeth, and as he went down his pistol flew wide. He landed on his back but saw the pistol from the corner of his eye. With the heels of his boots he thrust himself backward, clawing for the butt of the gun. Wells used his rifle like a shovel, thrusting it into the little man's ribs. The crackling snap of bone and cartilage was loud between the close rock walls.

With his mouth set, Wells held to his rifle barrel and raised the butt high. It came down against Smith's left temple with all the smooth, easy strength Wells had in his hands and long arms. The little man's eyes rammed shut, and his hands tightened as Wells stepped back. A quiver passed through Smith's face and frame and hands. Then he was still, and his blue eyes regarded the ceiling with utter indifference.

Wells watched for a long moment and no one spoke. At last, the big man pivoted, his moccasined heels shifting the deep sand on the floor. He sought out Baraga, and their eyes met. "He sold us out."

Baraga nodded, his eyes almost bored. He glanced toward Smith, then back at Wells. "So I heard. That explains the way he's been acting since you left. Like the devil was prodding his rump with a hay fork." Baraga took a step closer and stared down at Smith. At last, he let out a huff and frowned thoughtfully, turning to Sam Malone.

"Grab yourself a plate, then go out and tell the Injun to come and have his share while you take his watch. Rico, drag this coyote out of here and down the canyon three hundred yards or so. Throw him over the side for the buzzards."

Wells nodded, then bent and slung Smith to one shoulder like a sack of grain and tramped out of the grotto. Sam Malone picked up his rifle and disappeared into the sunshine, and after several minutes another man appeared. He was an old Indian Baraga's bunch had named Paddlon—an unceremonious abridgment of his Papago name, Paddles-On-The-River. No one knew Paddlon's age, but his face was gaunt and wrinkled like a sun-dried apple, and his shoulder-length hair was thin and bleached white. In his eyes was the old man's only visible strength. They snapped when he felt put-upon and gleamed when he thought of his family, whom Apaches had killed many years ago. Paddlon's people detested the Spanish-given name Papago, and referred to themselves as either Tohono O'odham or as *parientes*—kinsmen. But the Desperados knew him and every redman by one term: Injuns.

When Paddlon paused at the warm coals of the fire pit, Baraga threw him a preoccupied glance. "Eat up, Paddlon. And take a canteen back up with you. It's getting hot."

They all sat down to fresh-boiled rice and red peppers, stewed jackrabbit and steamed cornbread. Crow Denton was Desperado Seven—a half-breed Yaqui Indian whose walrus mustache made the wrinkles alongside his mouth appear to go clear to his jaw. Denton, with Bishop, finished up the pair who had taken no part in the big war. This meal was his doing, and so the others ate with satisfaction. His cooking was a match for his marksmanship, and no one could best Crow Denton with a rifle.

Rico Wells whispered back in after several minutes and took up a plate, seating himself Indian-style in the sand. He ate almost ravenously, catching up with the others and surpassing them, emptying his plate first. He pressed the last of his cornbread against his tin plate to embed the remaining rice in it before popping it in his mouth. As he chewed, he looked up at Savage Diablo Baraga and studied him. Finally, he said, "Shorty says Mouse is gettin' hungry again. He wants fifty percent more than he's bein' paid."

Baraga stabbed his knife into the floor and threw a harsh look at Shorty Randall. "Who is this man who thinks he's worth so much? I'll wager he comes from the north. Union man. Greedy as a man can come, but he probably ranks with gutter trash."

The little blond man swallowed hard, his eyes flickering between Baraga and the floor. He must have thought Baraga expected no response, for he remained silent.

"Shorty, I'll tell you right now—Mouse's cut goes no higher than it is now. You tell him. If he isn't satisfied, we'll have to kill him," Baraga said. "Tell him *that*. As far as I'm concerned, we'll pay you to do the job. You're not stupid. You're well aware who's buttering your bread. Say I'm wrong if you believe it."

Shorty shook his head, aware of every eye on him. "There wouldn't be no money without you. I know that."

Baraga gave a sharp nod. "Right. Now what about the payroll? The payroll to the Dolce Vita mine? Any word?"

Shorty swallowed his last bite. "Not yet. But next week we'll know. There was a delay in Frisco."

"All right," Baraga said. "Sloan will be in to meet you." He glanced over at Silverbeard Sloan, the Desperado Six of journalism fame. "By then, I hope there's word on the payroll. But either way, you'll have spoken with Mouse about his cut. He can be replaced, in which case he'll be dead. I want him to be aware of that with every breath he takes. He stays with the present deal or he dies," Baraga said flatly.

"Shoulda killed 'im a long time ago," Silverbeard Sloan growled. "He's comin' out of this whole deal sweeter'n I ever dreamed. An' he don't even risk his life for it."

"He risks it to us," Rico Wells put in.

The others nodded agreement, and Major Morgan Dixon spoke for the first time. "If he's a Federal man, he'll stay with his cut and keep his mouth shut. Craven Union trash wouldn't stand up to any one of us."

When the meal was finished, the Indian, Paddlon, returned to his rock perch above the cavern, and Malone came in to stretch out on the sandy floor, careful to avoid the spots of blood in the sand where Blue Bell Smith had died.

Samuel Colt Bishop sat at the rear wall of the cave with his eyes half-closed. But he was very aware of those around him, what they were doing, what they were saying—sometimes what they were thinking. He was a perceptive man, a man who felt things others didn't, saw more than others saw because he breathed in his surroundings. His speed with a gun dictated he never let down his guard because there would always be someone who wanted to kill him.

Colt Bishop's father had named him after the inventor of the Colt revolver when he was born not long after the appearance in Texas of the Patterson revolver. His father's love of guns had rubbed off on him. But nothing else had. His father had been a wife and child beater, and Bishop had used his namesake revolver to kill him when he was only fifteen years old.

Bishop hired out as a mine payroll guard at sixteen. After that he'd been a wrangler, a cowhand, and a foreman on a sizable California ranch. He'd worked as a barkeep, too, in a rowdy border town, before killing a U.S. marshal in a drunken fight and finding his calling in the gun. For a little while he'd been a marshal himself, but there was no money in that. He'd fought private wars—against Indians and Mexicans, usually, but he drew no border. He was not one to choose between colors of skin. A Colt .45 would kill a white man as easily as a red or brown or black. And when he found no work, he made his own, robbing the occasional bank or stagecoach. He was a gunman firmly planted on the south side of the law, and all money, like all men, to him was the same color.

Silverbeard Sloan left his plate on the floor when he was done eating and flopped down near the grotto's entrance, just out of the punishing sunlight. He set a dented oilcan nearby and pulled a greasy rag from the pocket of his gray wool vest. Then he reached with a practiced hand and withdrew from his right holster a silver-plated Colt Peacemaker with a four and three quarter-inch barrel.

For a long moment, he tilted the revolver back and forth, letting it catch and reflect the blue sky and tawny limestone against its intricately engraved frame and barrel. The pistol, like its twin in his second holster, had cost him twenty-eight dollars brand new. It had smooth ivory grips that fit his delicate hand and caressed his long, slender fingers like silk.

Noble was the outlaw's given name, a name so unbefitting a man of his character as to sound ludicrous, for he was a born killer—far from noble. A woman or a child in his path was as likely to be trampled into the dust as a dog or a possum. A barbarian void of morals, he derived his greatest pleasure from shooting his Colts, especially if his targets were alive when the lead began to fly. It was by that act he could add to an already deadly reputation. It was time—time and the romantic journalists of the nineteenth century—who had done away with the ill-fitting name of Noble. Time, although he was not yet forty, had given a silver sheen to his thick head of hair and distinguished looking beard. Newspapers had renamed him for it. Most people in Arizona Territory couldn't have said what his real name was. Silverbeard Sloan was a name of legend.

Sloan cleaned his already flawless Colt. He spun the cylinder and listened to its revolutions. He cocked it and felt the buttery flow of the hammer beneath his thumb. The best gunsmith in New Mexico had seen to the action, changing it from a standard factory issue Colt to this instrument of perfection.

Sloan found himself looking over at Colt Bishop every now and then. He didn't like the man—never had. The newspapers insisted on saying Bishop was the faster of them, and that was not true. But there was only one way to prove he was faster . . .

He knew he could outdraw Colt Bishop, and in his mind he had done so many times. He gunned him down, laughing as he fell. He had imagined it so often it had begun to take place in an exact succession. They faced each other. Bishop cowered but tried to defend his reputation. The twin Colts came to Sloan's palms like they had lives of their own—although in real life he would never have tried to use them simultaneously in a fight. They bucked against his palms. He watched the agonized twist of Bishop's face, the blood on his shirt, his fall to the ground. Sloan stood over him, blowing smoke from the barrels of his Colts.

But of course he couldn't face him. Not that he couldn't beat him. He had no doubt he could. In spite of what everyone said, he knew he was the faster of them—by far. That would one day be a well-known fact. But for now . . . well, it would be foolish for two men in the same gang to fight each other. He would let them all think Bishop was faster, and he would bide his time and practice with his Colts.

Silverbeard caught himself staring at Bishop's walnut-handled Colt and blinked his eyes. Blast it, he was the fastest of them. He knew he was. Someday. Someday they would all see . . .

Chapter Two
Captain Tappan Kittery

There was about to be a lynching. Then Captain Tappan Kittery came along, and the odds of the necktie party's honored guest leaped a hundredfold.

Following the sudden, distant click of metal on stone, an abrupt silence descended on the shaded, rocky wash. Even the breeze paused in its springtime dance. The grass waited, the fresh new leaves of the oak and sycamore and ash trees hovered in silence. A lizard sunned itself on a granite boulder, unmoving and wary. Death often came to the first one who moved.

The wash was dry, where at times the snow waters crashed and roiled and sheared off hunks of stone, sweeping trees and boulders along like a freshet moves weeds and pebbles. At the head of the draw it was deep and dark and full of upslung granite boulders. The heavy old gray-trunked trees spread their branches over it and wove them together until it resembled an isolated hunk of rain forest cast away and lost in the middle of the Arizona desert. The rock-scarred roots of those trees gnawed and curled their way out of the steep banks and from between the rocks, and slithered along the rim of the wash among dead white branches and tall green grass.

The lizard sat on its rock, its black eyes staring up the draw. It extended and flexed its legs, making it rise, then drop. The paper-thin skin of its abdomen stretched with a breath, relaxed, stretched again. No one could spot the tiny, dust-colored body against its boulder perch. And no one could catch it if they did.

Then, from the deepest shadows of the draw, a rider appeared atop a giant black horse. As if sighing in relief, the breeze picked up again. The grass and leaves began once more to whisper, and a corner of the horseman's yellow scarf fluttered loosely.

The big rider swayed in the saddle, his eyes watchful for any movement. From beneath the wide black brim of his hat, sweat made glistening trails down through the dust on his tanned cheeks, running into a week's growth of heavy dark beard. The rider had been away from the comforts of men, it was plain. Salt stains splotched his faded blue shirt, and cooking grease marred the white and gray stripes of his trousers. His hat and his boots, even his horse and saddle, wore a veneer of white dust.

His name was Tappan Kittery. He was a handsome man, but handsome in a way that aped the angular and ragged mountains of Arizona, not the lush green grass and the nodding array of flowers that garnished the spring countryside. His blue eyes were constant-roving beneath his hat brim. His full lips lifted in a natural smile, and straight, almost-black hair ended above his collar line, cropped short against the heat of the Arizona desert.

The big man's eyes missed little. They even saw the brown lizard that seemed to sense his gaze. The creature pushed away and skittered into a crevice in the rocks. Then only the man and the horse and the wind in the draw remained.

They made their way up out of the wash at the first shallow place. There was a cattle trail there, and the black made its way along it. Minutes passed now with only the sound of the saddle creaking and the black's big hooves clopping against the sun-bleached trail. The late afternoon sun, although sapped of its noontime ferocity, still sat harsh and unforgiving above the horizon as Kittery reluctantly guided the black horse into it.

The sound of voices reached them through the trees, and Kittery drew the horse in. After a moment or two, he had heard what he wanted, and he prodded the horse on at a slow walk. They stepped into the grass, and now they seemed to glide forward with no sound.

On the other side of a little rise appeared two men.

Kittery sat his horse and studied them. They faced each other in knee-high green grass shaded by gnarled oaks and sycamore. One of them, a Mexican, was short, stocky, with long hair and a mustache and VanDyke beard. He stood with his hands over his head, backed up to a massive sycamore. The other, a lanky, sandy-haired man in the garb of a cowpuncher, held a pistol in his right hand, a coiled reata in the left. A bay horse stood ground-reined next to him.

By the sound of bleating back in the trees, and the occasional glimpse of a light gray form, Kittery figured the Mexican was a sheepherder. By the way the other one dressed, it was a cinch he wasn't. From the puncher's voice and ungainly movements, Kittery guessed he had been drinking.

Anger had burned in Kittery's heart all day. The sins of a lawless killer by the name of Ned Crawford, as elusive as he was murderous, filled his thoughts. Kittery was in no mood to be civil, but he had always been a rational man. He decided to take the smoothest road he could into the confrontation between the herder and the puncher. This proclivity was innate to Kittery's nature, and had been honed by mimicking the behavior of his older brother, Jacob, in their childhood. Jacob had always been a placid youth who made his decisions carefully and stuck by them with honor to the bitter end yet did all he could short of betraying his principles to make no man his enemy.

The cowboy took two steps closer to the sheepherder, brandishing his pistol. Kittery let the black walk in to twenty yards away, its hooves swishing in the long grass.

"Nice day for a ride."

The puncher wheeled around, swaying as his momentum tried to carry him farther than he had planned. He turned his Remington Frontier pistol on Kittery.

"How do." Kittery smiled, hiding his irritation at being called upon to act as guardian over a flock of range maggots.

"Howdy. Who're you?" The puncher squinted his eyes suspiciously.

"Name's Tappan Kittery."

The puncher dipped his chin and spat a dark stream into the grass. "Well, I'm Jed Reilly." He straightened himself up and squared his shoulders, puffing out his chest like a Bantam rooster.

Without any undue notice of the name, Kittery inclined his chin toward the Mexican. "Who's your friend there?"

Jed Reilly returned a wry smile. "*Friend?* That's a mutton puncher."

"So I figured. What do you plan to do with him?"

"Doesn't matter so much—maybe nothin'. You John Law?"

Kittery forced a chuckle. "No, I told you: I'm Tappan Kittery. Just a public-minded fool."

"That so? *Puh- blic-* minded," he cut the word in pieces to exaggerate it. "Then you'll be wantin' to give me a hand here." Reilly's throat erupted with a bray of laughter.

Kittery had allowed the black to walk in closer, until now they were ten feet away from the puncher's horse. "No-o-o," he soothed. "May as well put up the gun, friend. I don't think you really wanna hurt anyone today."

As he finished speaking, Kittery kicked his right leg up over the neck of the horse and slipped from the saddle, pausing to crouch and stretch his legs. He watched the puncher's face for any signs of alarm as he straightened up and walked closer to him. Reilly's only change of expression was a flickering of the eyes, but he backed up three steps. He shot a glance at the sheepherder, back at Kittery, and then to his own horse. His eyes rested at last on Kittery.

"Maybe, maybe not."

From the trees, a lamb blatted. The cowboy's eyes widened, and he gritted his teeth. His Remington, which had started to lower, swung up to bear on Kittery's chest. "No, mister, I guess yer wrong. I guess I do wanna hurt somebody t'day—this pepper belly an' anybody that tries to git in my way." He hefted the coiled reata toward Kittery. "Here, take this. Yer gonna help me swing that piece o' sheep meat."

"It won't change anything. You kill him, five more'll replace 'im. And they'll each have a thousand more sheep. It's a matter o' time."

Reilly shrugged, the corner of his mouth twisting upward. He dropped the coil of reata to his side. "Sure, but I'll feel a whole lot better about myself in the meantime."

Kittery grunted at that inane rationale. "I doubt that." His eyes bore into Reilly though he tried not to look threatening, a pretty big order for a man who stood well over six feet and outweighed the puncher by a good seventy pounds of hard flesh. The only thing that helped Kittery appear non-menacing was the natural upturn of his lip corners. In his past, men had made the mistake of believing that was a sign of perpetual good nature.

Tappan Kittery liked to avoid violence if he could, again a creed he had adopted from his brother Jake. He collected bounty on a criminal or two now and then, between more stable jobs, but the law in southern Arizona had come to know him as the one who brought them in alive—unharmed, if they were willing to come in without a fight. But Jed Reilly, with his liquored belly, wasn't in the same peaceable mood Kittery was. He was going to make a fight of it. Kittery had no use for the sheepherder either. He hated the sound of sheep on an otherwise peaceful mountain morning. He detested the stench they left hanging like a disease in the air. But he did have a liking for mutton and lamb chops, and, after all, the army-issue shirt he was wearing was made of wool, and so was the vest in his saddlebags. He couldn't say he had no use for *dead* sheep, and to have dead ones there had to be live ones at some time. Besides, the Mexican had as much right to a living as anybody else. He might have chosen a less menial calling if it had been placed before him.

When Reilly held out the reata again, shaking it insistently, Kittery reached for it with his right hand. Then, with no warning in his eyes, he grabbed the puncher's gun arm with his other hand, shoving it skyward. He jerked the reata out of the shocked puncher's hand and swung with it, striking him on the jaw. Reilly's finger jerked, sending a .44 caliber round into the sky. Kittery shifted his hand from Reilly's forearm to the pistol and wrenched it free. He slammed Reilly in the collarbone with the side of the gun as his hand came down.

Stunned, Reilly fell back against the tree, clutching at his collar. His knees sagged, and Kittery reached out with his right hand—freed by letting loose of the reata—and grabbed him by the shirt. He flattened the cowboy back against the tree and held him upright.

Shoving against Reilly's chest, Kittery stepped back, his hand dropping toward his own Remington Army that rode high on his right hip. He held the puncher's pistol loose along his thigh, forgotten.

Reilly leaned against the tree for a few more moments. He raised his hand to rub his jaw after the pained look on his face went away. He looked up at Kittery and tried to straighten up, but he fell back, slamming his eyes shut and grabbing at his collarbone again. It took him half a minute to ease away from the tree. Other than his horse stamping, Reilly was first to break the silence. He glared from the sheepherder to Kittery.

"Looks like you won yourself a greaser, mister. Yer lucky day."

Kittery's lip corners bent a little more upward. "Reckon so." His eyes swung to the Mexican. "What's your name?"

A look of suspicion still clouded the Mexican's eyes. He gawked at Kittery for a few moments, looking up and down his frame like he was some monster, stepped out of a cave. At last, he raised his eyes to meet the bigger man's.

"*Me llamo* Efraín Valesquez. I would wish to thank you for your help. *Gracias.*"

"Don't mention it. But look at it all this way: you'll have an exciting tale now to tell your family about."

Valesquez pondered that for a moment, then smiled, showing broken yellow teeth, the combined curse of bad nutrition and worse care. "You are smart, *señor. Inteligente.* A man must look at the bright side, yes?"

"Can't hurt. Say, how far to a town called Castor?"

"Not far. Nine, ten mile perhaps. But you should be warned—there is nothing there in this Castor. Very small, not grand—like Tucson."

"Thanks for the advice," Kittery replied with an amused nod. "But I have some business to tend to there. Besides, I'm tired of 'grand like Tucson.' Maybe we'll meet again, friend."

He held out his hand, and Valesquez shook it hesitantly. Then Kittery turned back to the puncher. "Reilly, I'll turn your pistol over at the sheriff's office in

Castor. You can keep the rope. I'll leave you enough to hang yourself."

"Thanks," said Reilly with a voice full of venom and liquor. He took the reata Kittery passed him and went to his bay. He struggled onto its back and turned again to Kittery. "Maybe you'll meet me again, too . . . *friend.* Me an' my brother Joe. That oughtta even the odds up a bit."

With that, he turned and spurred the bay, galloping west down the mountain slope toward the same road Kittery would soon have to ride. The pair stood and watched the rider's dust settle like a blanket back onto the vegetation and rocks.

Kittery climbed into the saddle and looked down at the Mexican, touching the brim of his black hat. "So long, Valesquez."

"*Adios,* Señor Kittery," the Mexican said with a smile.

Kittery twitched the reins and touched the black with his boot heels. The big horse moved west at a long walk under the shadows of the trees. Forty minutes of healthy trotting later, they had left the Santa Rita Mountains far behind them and came to a road running north and south, a good-sized road rutted with horse and mule and oxen tracks and the grooves of many wheels. This was the King's Road, the Royal Road, more commonly known by the Mexicans' name—El Camino Real. This same road led from deep in Old Mexico, up through Tucson, and on to San Diego.

After a glance to left and right, Kittery turned south. There was no sign anywhere of Jed Reilly.

The sun bedded and dusk came to the desert. Kittery pulled up in the twilight, conscious of the hissing, scratchy cry of nighthawks overhead, and the harsh *chug, chug, chug* of a wren's call back in heavy mesquite. Somewhere a poor-will made its lonesome song, and for a moment Kittery's thoughts turned melancholy, and he was back home in the Smoky Mountains of Carolina. There, the bird he and Jacob or his little sister Annie listened to from the porch in the gloaming was the whip-poor-will, and in the early spring the fireflies danced and flickered in the dusky air. In his dream, many loved ones, all his sisters, he and his three brothers, gathered around a table of hand-hewn slabs, and his mother said grace. Hickory smoke puffed occasionally from the fireplace, where a pot of fresh corn simmered next to a browning haunch of venison and a cast iron oven full of biscuits. From the west, a pumpkin-colored sun stole through the window, painting a soft glow on everything. Nudging nostalgia aside, Kittery brought his thoughts back to the present.

Castor wasn't much farther. But he wasn't in any hurry to arrive tonight. The town would be closed up, except perhaps for the local watering hole. He was weary from his day in the saddle, and though he would have liked to shake the dust from his clothes and wash it from his sweat-tainted body, that could wait another day.

Kittery sat there for some time in the saddle, watching along his backtrail. There was no sign of another human being. And it wasn't just Reilly he looked for. This was the land of the Chiricahua Apache, the most adept camouflage expert and guerilla fighter in the world. And although the great chief, Cochise, had died a couple of years ago, there were others. Victorio, Nana, Juh . . . and an upstart shaman by the name of Geronimo. That man had a nasty habit of jumping the reservation to plunder and kill. There was no telling where he might be at any given time, and though Kittery should have heard word in Tucson of any uprising, it wouldn't do to trust his life to that assumption.

To make matters worse, the border of Arizona and New Mexico territories with Old Mexico was infested with bandits of all skin colors, men with hearts numbed to killing by war of one kind or another. Kittery's outfit would make a good prize—for a man big enough to take it.

A small, sandy cove nestled in the rocks beside the road. When Kittery had satisfied himself that he was alone, he pulled into this shelter. There was plenty of room here for both him and his horse and patches of spring-green grass for the horse to eat, and for him to lie on.

There was no water here but that in his canteen—enough for the stallion and perhaps a cup for himself. His throat was parched so he could hardly swallow—Arizona did that to a man—but he had been through worse. As for coffee, he would do without. Deprival had trained him against that craving years ago.

Pouring some water into his hat, he let the horse drink while he removed the forty-pound Denver saddle and eased it upside down onto the ground. "Ain't fair you have to carry that and my fat carcass too, is it, boy?" He patted the animal's broad, damp back, but he would wait to rub it down. It wasn't good to curry a horse's hair until after its sweat had dried.

Kittery called the horse Satan, not because the horse had some evil inclination, but because he was big and powerful and dark. Back only a generation in his family tree ran the royal blood of the Friesian breed, accounting for both his size and the extra hair built up over his hooves. His feet weren't gigantic, just large enough to keep them from sinking too far in desert sand. His color was jet black, blacker than any night, rescued from total blackness only by the large blaze down his nose and four matched socks that ended at his knees. He was a proud horse, too, displaying the tools of a sire and the muscular, curved neck, long, wavy forelock and massive chest of a knightly steed.

Kittery picketed him between himself and the road. The horse would sound a warning if anything or anyone drew near in the night. He sank onto his stretched-out bedroll and chewed on a stale biscuit and some half-burned bacon he'd commandeered from the Shoo Fly Restaurant in Tucson too many mornings before. It wasn't much of a meal, but his needs were simple. And tomorrow he'd eat in a cafe; that was consolation enough.

Sitting there letting the cool air drift down and circle around him, Kittery listened to the stallion cropping grass. He listened to the birds and the other sounds of the night. The wren was gone. The fiddle of crickets and screech of nighthawks, the occasional far-off whistle of the poor-will, the kingly *hoo-hoo* of an owl—these night sounds and the scents of the desert were all that was left to Kittery now, these and a sky full of stars and stardust that must hang brighter than anywhere else in the world.

He listened to the whisper of wind through the grass, and among the branches of mesquite and paloverde and ironwood. Downing the last of his water, he laid the canteen beside his bedroll. Still sitting up, he started to doze, and then when he jerked awake he heard the crinkle of paper in the pocket of his gray-striped jeans. Sleep flew away from him, along with the pleasures of the night. Replacing the peace of solitude was the thought of the newspaper clipping he carried that told the story of Ned Crawford, of how he had raped Sarah Dodge and killed the entire Dodge family in cold blood.

These killings and nothing else had made Tappan Kittery quit his job driving freight for Tully, Ochoa and Company. This Ned Crawford was a stench in the air of a ruggedly innocent land. If the Apaches killed brutally, at least they had a cause. Ned Crawford couldn't say that.

Tappan Kittery had learned to kill in the war. And he had killed plenty. At one time or another he had killed Yank and Reb alike. He didn't want to kill anymore. It had left a sour taste in his mouth, and a heavy load in his heart. But Sarah Dodge had been a kind young woman and a lady, her youngest daughter no more than two. Tappan Kittery meant to bring justice to Ned Crawford if he burned every last short dime in the accomplishment. That was one man even Tappan Kittery would kill at the slightest provocation. And his brother Jacob would have approved.

Chapter Three
Castor, Arizona Territory

With dawn, the stars faded one by one, blinking out. A chill tinged the air, and the caliche was as cold as ice in comparison to the furnace blast of the previous afternoon. The diluted purple of the eastern sky foretold the waking of the sun and begged the big man to rise. Pushing up onto his elbows, he gazed over the spot he had chosen to make his camp. Beyond the cove and across the road sprouted a garden of mesquite, saguaro, bur sage, cholla and ocotillo. With God as its caretaker, it had flourished. Accenting the wild beauty were the hues of spring flowers: the lavender of owl clover, yellow of brittlebush in bloom, blue of lupine, orange of Mexican poppies.

Kittery rose and readied Satan, ignoring the hunger that gnawed at his stomach. There couldn't have been much more than a strip of jerky or a biscuit left anyway. He would just suck on a chunk or two of horehound candy and drive his weakness away.

Before mounting, Kittery slid his single-shot Springfield carbine out of its boot and blew the dust from its working parts. He did the same with the Remington Army revolver he wore on his hip, and a smile came to his face as he turned it over in his hand. He had owned the pistol since buying it new back in sixty-four. The only change to it was the numerous nicks and scratches and the dull, blackened look his sweat had given its walnut grips. Its method of loading was cumbersome for a man who lived dangerously, but maybe Tappan Kittery liked living in the past. Maybe he was a relic himself.

As for the carbine, it was not outdated. It was only two years old, an army-issue piece chosen for its supposed reliability in the most adverse conditions. And as long as he loaded it with the new cartridges made with brass casings rather than the older ones of copper, which seemed to get too easily jammed inside, it had treated him well. Yet Winchester offered two good rifles, each with the firepower to wipe out a gang. The '73 model was a reliable saddle gun, packing fifteen or sixteen .44 cartridges, each with a walloping two hundred grains of lead. The brand-new '76 model was even bigger—big enough for a buffalo or a horse, if that was called for. Its twelve .45-75 caliber cartridges could be an effective argument against violence.

But Tappan Kittery wasn't ready for change. He wore the same sweat-stained hat he had bought three years ago. It was one hundred percent beaver and tough as leather. He wore the same cavalry boots he had worn when he rode away from the army three years before, and he had owned them for three years before that. They'd been re-soled three times, but it was still the same leather. Change was for a man who had more money than brains. What served Kittery well, whether tangible or intangible, he saw no reason to do away with.

Kittery moved the big black out with crisp air brushing against his darkly whiskered cheeks. As the road wore on, the vegetation changed. Creosote, whose notorious poisonous roots prevented other plants from thriving close by, dominated the land, interspersed with an array of cactus and the mesquite, whose roots stretched far beneath the top soil to tap the ground water below. Now the sandy soil stretched for miles to the south, and the rugged ridges of mountains dominated the skyline in the east and west. One brushy rise provided a stage for two coyotes saying hello to the morning.

The sun cast a golden-orange glow across everything. After the cold night, the sky seemed bluer, the flowers brighter, the trees and cactus greener; and perhaps they were, before the heat of the day that promised to be.

Four miles farther, Kittery became more and more aware of a long ridge of light colored clay that heaved up on the horizon to the west. He discerned what appeared to be a smattering of earth-colored structures along the top of that ridge, and some white ones built along several roads that twisted the length of its tawny flank. Soon, a sign inviting with a painted white arrow told him to turn off on a side-road if Castor was his destination. Here he left the Camino Real.

To Kittery's left, in a rocky gulch, a stream gurgled along from its beginnings in the Santa Rita Mountains, and a grove of small cultivated pecan trees displayed their deep green leaves not far from its banks. Several beaver dams hampered the flow of the stream though most were long forgotten. But at one Kittery was surprised to see one of the plump denizens making its way across a deep, murky pool. It seemed as if it had no natural fear of man. The thought crossed Kittery's mind that maybe the beaver had sanctuary near Castor, for Castor itself meant beaver, in the Spanish tongue.

While these idle thoughts occupied his mind, Kittery rode down through a sandy wash and up over a rise. Topping out, he saw with relief that the dusty road descended into another, shallower wash, then leveled out, fading into a scattering of buildings.

Castor looked like the type of town where a man would stop for supplies before hurrying on—clean enough, but void of the luxuries enjoyed in larger cities, and even in many small towns in more hospitable environments. Few trees, other than those stunted ones the Sonoran desert normally grew on its own, and very few cultivated flowers. Barren except for the usual mixture of cactus and thorn bush growing along the ridge. The only place that really showed the kind of green that welcomed a man home was several more little patchy orchards of pecan trees that grew off to the left, and a field where grape vines crawled tentatively up some four-foot stakes set in neat rows.

Beyond the greenery squatted an assortment of faded privies and broken-down, slatternly adobe huts flanked by brush ramadas. Strings of red peppers hung from many of the protruding rafters, laundry was slung to dry from dirty windowsills, and almost all of the doorways were served by drooping blankets or cowhides in place of wooden doors.

Here arose the noises of many people, all of them he could distinguish speaking Spanish, and all of them sounding carefree. Little dark children darted to and fro, and one brown dog barked a warning toward him while another white one with black spots sat and licked itself, pausing only long enough to watch him and Satan pass. His little black eyes were keen and intelligent, and Kittery wondered what he was thinking, for he seemed to be contemplating the appearance of the man and the horse.

Kittery circumnavigated this section of town on a little dusty road plaited with cart and donkey tracks, passing a man dressed in a white outfit with an over-large straw sombrero. He pushed a cart that carried buckets of fresh water. A brown dog with stubby legs walked beneath the cart, taking advantage of its shade. The entire scene recalled the remnant of Old Mexico Kittery would always love, and he paused long enough to take a drink from the dipper the man offered him and to hand him down a dime.

Kittery had to ride past the rear of the business section of town to clear the last of the adobe houses, and as he came back into the main road and turned left into it he smiled at the orderliness of the little settlement. Two pine trees had been cut, probably from the Santa Rita Mountains, and been planted here at the entrance to the town to support a long bar between them, from which a sign hung by rusty lengths of chain. It read *CASTOR, ARIZONA*, and below, in much smaller letters, *Founded 1862*.

The main street of Castor was wide. Its wheel-rutted length ran straight until reaching the loading dock of a huge gray warehouse, then branched, becoming half its original size in the transfer, until it rejoined on the other side, widened out and continued on its desert-wending route back to the Camino Real.

Wood made up the majority of the buildings on Main Street, an oddity in a southwestern community. Perhaps that was another reason for the town fathers having named the town after the beaver. On the east side of the street, to Kittery's left, were a mercantile, combination barbershop and doctor's office, hotel, cantina and café. To his right he could see a dry goods store (which by its signage also served as hardware and grocery store), bank, sheriff's office and jail, blacksmith shop and a huge livery stable. Only the sheriff's office and jail and the blacksmith's, all strung together, were of the customary adobe. Pine boards housed everything else. Even the imposing warehouse was of wood, and situated width-wise at the end of the street it gave Main Street the appearance of a three-sided box. It seemed to be Castor's answer for the customary central plaza of most southwestern towns. The only masses of color offered by the entire town were the blue of the hotel and the faded green of the dry goods store, with its white signboard on which was emblazoned the fitting jade-colored legend, GREENE'S DRY GOODS.

Besides the two divisions of town Kittery had studied up close were the structures he had observed from a distance. Rising up with the long ridge, the better off seemed to make their nests. These homes, like the ones on the east side of town, were of the customary adobe—sensible if one wanted to remain cool in that often furnace-like environment—but most of these bore new whitewash, and they appeared orderly, with larger yards, oaken doors, and many with picket fences. Some of these adobes seemed precarious in their perch on the steep side of the ridge, but the ones he could observe in any detail were shored up with stones.

Kittery rode nearly the entire length of the street and pulled up before the livery stable. A leather-faced man with a peg leg, long, thin, snowy white hair and a smattering of hoarfrost for whiskers stepped out to meet him. Deep wrinkles that might have been from laughing surrounded the old man's mouth and eyes, but as he tugged the pipe from his lips he greeted Kittery with little expression, leaning a gnarled right hand on his cane.

"Ther's an empty stall in the back."

Kittery nodded and dismounted in one easy, flowing motion. He led Satan back into the shadows, unsaddled and released him in the farthest stall. When he returned, the old man looked him up and down before speaking. "You're a sizeable one." Without awaiting a reply, he continued, "Costs two bits a day fer the stall. I'll handle yer horse like he's my own. Corn an' grain cost fifteen cents extry, and it's worth it."

Kittery nodded again, glancing over the town. "Yeah, it *is* worth it. Give him the both. That animal means more to me than this entire ramshackle town—and everyone in it."

The old man flicked his eyes up at Kittery's face, then allowed them to sweep the town. A grin revealed the tobacco-stained stems of his teeth, and he chuckled, poking the pipe back in his mouth.

"Never seen a horse quite like him. What is he?"

"Part Friesian," Kittery replied. "Friesian and saddlebred."

The old man nodded and glanced back in the horse's direction.

"Do me a favor and let him roll here in a few minutes," said Kittery. "And can you tell me where I'll find Sheriff Vancouver?"

"Who're you, stranger?" The old man seemed to ignore his question.

"Tappan Kittery's the name."

"Wahl, I'm Jarob Hawkins. An' yer 'Sheriff' Vancouver is yonder." Without turning, he thrust a thumb over his shoulder. A man wearing a Pima Country deputy sheriff's badge had just leaned up against the top pole of Hawkins's corral. When Kittery's eyes fell on him, the lawman let a smile cross his face.

"Sheriff?"

"That's what they call me here—sort of an honorary title. Head *deputy* sheriff, actually."

"I'm Kittery."

"I heard. And I figured you were. I've heard about you before." The lawman smiled again, stepping close. "*Captain* Kittery?"

"Tappan," corrected Hawkins, off to the side.

"Well, Captain, too. Another of those 'honorary' titles."

Hawkins gave a sheepish shrug, and he turned and went on inside.

"Tappan, is it? Well, I received your letter. Why don't you come to my office—we'll talk a bit."

"Only over breakfast, I'm afraid." Kittery patted his lean stomach. "I haven't had a decent meal since Thursday mornin'."

"I have just the remedy for that. Come on up to the place and eat with me and my wife. She's a good cook, and a picture, too, if I can boast a little. You won't be sorry."

"You haven't eaten?" Kittery raised his brows in mild surprise.

"As a matter of fact, I worked night shift, so I'm just now headed home for the day. You might say it's my suppertime."

"Well, if you're offerin'. But first, I have somethin' for you." He drew Jed Reilly's pistol from behind his belt and handed it butt-first to Vancouver with a brief explanation of how he had come by it.

Vancouver shook his head and sighed. "Sounds like Jed. He's a hothead when he's been drinking—but a good boy when he's sober. I'll talk to him next time he comes to town. Thanks for your help—you saved the boy from his own stupidity."

"He didn't seem stupid," Kittery countered. "Just drunk."

Vancouver grunted. He turned to go, but something caught his eye, and he turned back. "Looks like we may have another guest for breakfast."

Kittery followed the invisible line drawn by Vancouver's eyes to a lone horseman on a long-legged dark bay. Badge agleam on his vest, the man let the horse plod past Kingsley's Cafe.

"Joe Raines," Kittery said quietly.

Vancouver looked around. "You know him?"

"All too well. We've rode together a time or two."

By this time, Raines had reached the pair, and he climbed down in front of Kittery, smiling with surprise. "Tap!" The marshal thrust out his hand and shook Kittery's, grasping his forearm with his other hand.

The marshal was in his mid-forties, with an aquiline nose, cheeks as dark and full whiskered as Kittery's and a mustache with elegantly curved ends. He wore a brown, flat-crowned hat tipped high on his forehead and dark hair thrown to one side, with silver-edged sideburns grown to the ridge of a rock-hard jaw.

"How are you, Joe?" Kittery beamed.

"Seeing you, Tap? I'm doing real well now. I didn't expect to see you so soon."

Kittery shrugged. "In my line of work, I could turn up anywhere. I meant to say g'bye before you left Tucson—what happened?"

"I got a tip. I had to get down here as quick as I could. They told me you were in to the office the day before I left. Sorry I missed you."

"Such is life," said Kittery with a grin.

Joe Raines smiled at Vancouver. "Howdy, Luke. Didn't mean to ignore you." His eyes swung back to Kittery. "What brings you down? Or should I say *who?* You're not still following that blasted Ned Crawford."

"He's not dead, is he?"

"No, I guess not. I sure wish you'd use your talents differently."

"Behind a badge?"

"You'd make a good lawman."

Kittery smiled again. "Too much responsibility. I answer t' no one."

"Yeah, well, a badge is nothing. One of these days some lady's going to catch you, and you'll be tied down like you can't imagine. Just wait."

"I'm waitin'."

"So you've met Luke," Joe Raines changed the subject abruptly.

"Luke? Oh! Yeah, sort of." Kittery turned to the sheriff. They hadn't shaken hands before, but they did now, when Vancouver offered, and Kittery looked him over anew. Raines seemed to like him; that meant he was all right.

Vancouver was a handsome man by anyone's standards, with slightly thinning light brown hair and clean-cut features and a good-humored glint to his bright blue eyes. He appeared to be in his mid-thirties, a little older than Kittery. He stood six inches shorter, but his lean frame and work-hardened muscles gave him the appearance of a taller man.

"Well, Joe." Vancouver turned his eyes to Raines. "Mr. Kittery and I were going to take the wagon up to the house to talk—over 'supper.' You're welcome to come along."

"Sounds fine, Luke, except it'll be breakfast for me."

Kittery said, "Me, too, but he's a late-nighter."

Raines chuckled "If you'll let me put up my horse, I'll be right with you."

Raines joined Kittery and Vancouver in the sheriff's office several minutes later. They were thumbing through some wanted posters, and Vancouver took several of them in hand.

"My rig's out back."

Stepping once more into the gathering heat, the trio climbed onto a buckboard that stood in the alleyway between the sheriff's office and the bank, which gleamed brilliant white in the sun, set off by a blue door and window frames. Vancouver started the horses moving with a twitch of the reins.

As they rolled away, Kittery glanced over a permanent gallows rising up behind the jail and a spacious boarded arena that faced it. They started up a winding, dusty road, and in minutes pulled to a stop in a quiet yard dotted with shrubs and cactus plump with winter's moisture.

Kittery surveyed the cottage with approval. A comfortable looking home, it was surrounded by a covered porch and bore a fresh coat of whitewash, a guard against the battering Arizona sun. A picket fence surrounded the yard and a new garden of beans, potatoes, corn and carrots. Flowers growing in a meticulously weeded bed painted the front wall in all their glorious array of color.

Climbing from the buckboard and kicking dust off their boots on the edge of the porch, the trio stepped through the front door, greeted by a spacious front room and the smell of bacon frying. Kittery glanced about, appreciating the quiet, domestic charm of the place, the soft cowhide chairs and a hand-made dining table, graced by a vase of lupine and daisies. Photographs lined the right wall, shelves laden with books the left. Soft blue curtains diffused the sunlight from the small window that faced the road, but on the north the curtains hung open on a larger window to reveal the hills and the creosote flats that stretched away toward Tucson. A braid rug softened the puncheon floor, and a long couch faced them against the far wall. Behind one of the closed doors leading from the room, between the shelves of books, they could hear the voice of a woman singing. It ceased when Vancouver called out.

"Company, Beth."

His hand had grasped the doorknob when the door opened, allowing the passage of a handsome woman whose gold hair was piled in braids above her head.

"This is Beth, boys."

The sheriff presented Kittery and Raines, and they shook the hand that, though completely feminine, was strong, the way a woman's should be. Kittery appraised Mrs. Vancouver but only nodded greeting. He had never been one for talk—in feminine company, anyway. He had grown up among rough-natured men, back home in the Great Smoky Mountains of Carolina. There had been little time for socializing or calling on the womenfolk, so the only women he spoke to were his sisters and his mother. It had been a hard life, with little or no time for the soft, pretty things that many set such store by. He had spent most of his growing years behind a plow or in the wooded hills with brother Jake, a long rifle in his hands, stalking a black bear or a white-tailed deer. But he knew true grace when it passed, and Beth Vancouver was a beautiful woman, a perfect match for the sheriff. Jake would have known how to act around this woman. He had always been the smooth boy with the easy, friendly character. That was one thing he hadn't passed on to his little brother.

A smile crossed Vancouver's face as he caught the admiration in Kittery's glance for his wife. "I'm going to feed the horses. You all get acquainted." With that, he strode outside, leaving momentary silence in the room.

"I've heard quite a bit about you, Marshal." Beth smiled at Raines. "It's good to make your acquaintance at last. Luke has had nothing but good to say about you."

Raines smiled back. "That was decent of him. Unfortunately, he never mentioned you to me."

"Oh, really? Perhaps he's ashamed of me."

"I seriously doubt that, ma'am."

"Well, I'll take that as a compliment."

"Do so, by all means."

A blush colored Beth's cheeks, but Kittery had no doubt she was used to praise. She turned her attention then to him. "My husband called you 'Captain.' You're an army man, Captain Kittery?"

"No, ma'am. Not anymore, anyway. I was a captain in the army during the war. Sergeant for eight years after. That captain bit just hung on."

"It suits you, however. The title 'captain', that is."

Kittery chuckled, then turned sober. "Sometimes it puts a man too much in mind of his past."

"That war was an awful thing," Beth agreed with a sigh, its memory washing through sky-blue eyes that took on a distant look as she gazed past the two men, out the window toward Tucson.

"Yes it was. I knew it a little from both sides."

"From both sides?" Beth returned curious eyes to Kittery.

"Everyone I knew, even my own family, was joining the Confederacy, so I did, too. I made lieutenant right off—because of my size, I suppose. After a year, I got to studyin' on things, and I decided I didn't wanna be any part of this country bein' torn in half. So I came over to the Union and became a captain before the end."

"There aren't many who can boast of something like that," Beth said. "And what about you, Marshal? You were in the War, too, I suppose."

"No, ma'am. By 1860 I was a deputy sheriff in El Paso. Frankly, by the time the war broke out I considered myself a soldier already. A lawman leads his own kind of war, ma'am—a war that never ends. I guess I'm making excuses for myself, but . . . I'm forty-two years old now and still fighting my wars every day. One of these days I'll come up against someone who's tougher, meaner and faster than

I am. I guess that, ma'am, is when my war will end."

Beth offered an understanding smile, letting her eyes turn for a moment in the direction her husband had gone. There was sudden sadness in the blue eyes, and she didn't look back at the marshal when she spoke. "Yes, I know how you feel, Marshal. I know all too well."

She cleared her throat, turning to Kittery with a forced smile. "And what is it you do now, Captain?"

The front door swinging open cut off Kittery's answer before he could begin, and Luke Vancouver stepped inside. "Let's eat, folks. It's gettin' cold."

Kittery and Raines seated themselves, and Luke and Beth set the table together, then brought out platters of food—flapjacks, fried eggs, bacon, sausage and a bowl of fresh-churned butter. Vancouver's helping his wife impressed Kittery, and he smiled at the lawman as he returned to the table for the last time. It took a big man to let a stranger see him doing what many considered woman's work.

Luke pulled a chair out for Beth, then seated himself, glancing at Raines, then Kittery. Last, his eyes flickered toward Beth. "I hope you boys won't mind my saying grace."

Kittery shrugged, and Raines said, "Please do," and they all bowed their heads while Vancouver uttered words of gratitude that came from his heart, mentioning the good food and the pleasant company. The words moved Kittery inexplicably.

With "amen", Vancouver took a knife in one hand and a fork in the other. He smiled and said, "Boys, eat hearty."

They did, too, for it was prime fare. Kittery couldn't help but take a second helping of flapjacks, and his eyes flickered to Beth. "Ma'am, these have to be the best slapjacks I've ate in . . . well, I don't know how long. But they couldn't be better."

"Thank you, Captain." Beth looked at her husband, then back down at the fork poised in her hand. "It's an old West Virginia recipe. Mountain folk food. I only wish we had blueberry syrup to go with it."

Kittery smiled and nodded at her. "I don't know what this syrup is, but it calls to mind somethin' my mama used to make, and it suits these cakes fine."

"That's just corn cob syrup," Beth said with an embarrassed laugh. "Do you really like it?"

Kittery smiled. "It's home fixin's, ma'am. It doesn't come any better."

While Tappan Kittery ate his food, Beth found her eyes drawn back frequently to him, in spite of herself. He hadn't had the chance to answer her earlier question, so she still wondered. But she didn't know if he wanted to answer it. What *did* he do for a living? He seemed to be an intelligent man, and graceful in his movements. But he was quiet. A distant sort, it seemed. A loner. She noticed him look out the window to the north a time or two, and a few times to the east. He had the look of a man who didn't like to remain in one place very long, she decided.

But in spite of the captain's reserve, Beth found herself drawn to him by something indefinable. There was something special about him, about the sunburned broadness of his face, about his steady blue eyes. She glanced down to cut off a triangle of flapjack and to move her fixed gaze from the big man's face. While she chewed the bite and listened to her husband and Marshal Raines talk, she couldn't help analyzing the way the Captain had affected her. She would never be unfaithful to Luke, even in her mind. As far as she was concerned, there was no man more handsome than Luke, and none could be more attentive and affectionate. But the Captain had a way of making her feel safe. No doubt it was partly his size, for he was uncommonly large and obviously very fit. He owned the broadest shoulders and back she had ever seen on a man with a slender waist, and his arms bulged with power even at the simple task of raising a fork to his lips. His broad, black-haired hands looked like they could crush a squash with little effort. But it wasn't only his physical appearance. It was his demeanor. He seemed shy, yes, but Beth felt that would change in the face of danger. It was only the timidity of a man who didn't spend much time in feminine company.

The captain hadn't shaved in a few days, and that gave his face, from low down his neck almost up past his broad cheekbones, a black, coal-dusted look. And there was a musky man-scent about him that made it plain he hadn't bathed in a while, which wasn't uncommon to earthy folk. She was used to men. There wasn't much about them that bothered her. In fact, it was the kind that tried too hard to be clean and proper that gave her cause to turn up her nose. As for Kittery, there was something particular beneath the dust, the man-scent, and the beard and sunburn—a sense of gallantry, she supposed. And an obscure gentleness.

Glancing around the table at her husband, at Joe Raines, and at big Tappan Kittery, she guessed she couldn't be much safer with any three men who had ever lived.

Vancouver paused to swallow a bite and looked at Raines. "So how did your ride go, Joe? Anything promising down that way?"

Raines pondered the question a moment as he chewed, looking down at his plate. He swallowed and touched a napkin to the corner of his mouth, then raised his eyes to meet Vancouver's. "Baraga's in there, Luke. I have no doubt of it. Your Blue-Bell Smith didn't seem to know how much time the gang spent in there, but it appeared to be considerable, by the worn-down trail I found going in. I put my faith in Smith. He may not be the best of characters, but he has some reason for squealing now. And if they ever learn he spoke to you, his life is worth little more than Confederate paper."

"You're right. So . . . tell me more. What'd you see out there?"

"I followed the trail a ways into the mountains—well up into the oaks. There was only one evident water source on the way in, at least anywhere near the trail. But good water. And there are a number of other trails that join the main one along the way, mostly before it leaves the foothills. After that there aren't many places a man would want to break off. That thorn jungle would rip the clothes right off you.

"That's limestone country mixed with granite. Limestone's prone to caves. I must have seen twenty or so, a couple of them good-sized. I have no doubt there's at least one big one back there that could house that entire gang. The question is, can we reach it? Even without that Indian guard Smith talked about, it sounded like a fortress. Might as well have a moat around it."

Vancouver nodded. "Well, I think you're right in your judgment of Blue-Bell Smith. He may be a weasel, but at least for now he's our weasel. And you're right about the Desperado Den, too. That Baraga's no fool. It'll take an army to bring them out. Or ten good men who don't care much for their own safety." His eyes flicked to Kittery, then back to Raines.

Kittery had followed the conversation of the two lawmen. They spoke of men and of places everyone in Arizona Territory was familiar with, after reading about Savage Diablo Baraga and his "Desperados Eight" for the past six or seven years in every paper in existence.

The Desperados Eight (every time Kittery heard that ridiculous sounding name some newspaperman had come up with he had to scoff) had become by far the most notorious band of outlaws in the southwest, a gang of ex-confederates and degenerates with no regard for life, not even their own, judging by some of the outrageous raids they undertook. The Eight had been numbered by local newspapers in order of their worth to the law, with Baraga, of course, as number one. Falling in place behind him were Major Morgan Dixon; Samuel Colt Bishop; "Bloody Walt" Doolin; Rico "Big Samson" Wells; Noble "Silverbeard" Sloan; Crow Denton; and "Slicker Sam" Malone.

Diablo Baraga was missing an arm—his right, if Kittery remembered correctly. From newspaper accounts—whether they were fact or not no one could prove—everyone knew the story. Baraga, Confederate colonel at a green age, serving under George Pickett, the pretty boy general, was wounded on Cemetery Ridge, at Gettysburg. Baraga had been taken prisoner after the horrible battle. Believing his arm could be healed, he had been forced, fighting like a crazed badger, to watch the doctors at the prison camp cut it off and throw it on the reeking heap with hundreds of other limbs.

They said Baraga changed after that. An already burning hatred for the Union grew, extending in time to the entire human race. When the war ended, he stole a pistol, and by long hours of practice he taught himself to handle a handgun better with his left hand than he ever had with the right. So began his spree of plunder and murder, landing him here, leader of the most feared band of killers in the entire southwest.

In some of his more desperate moments, Kittery had contemplated the feat of going after the Desperados Eight. The man or men who brought them down would be heroes. But the chance they would live through the encounter was slim. And Captain Tappan Kittery was a contented man at heart. He was seldom desperate enough to consider such an undertaking for long. Not when there were so many smaller fish in the sea such as Ned Crawford.

"Speaking of the Desperados . . ." Vancouver stood up as he spoke, walking around the table to the window that overlooked the town. "Well." He turned to look at Raines, who was stepping over to join him. "You ought to see this. I wanted to point out Shorty Randall's horse to you—so you'd recognize it if you ever saw it away from town. But you get a bonus. Shorty's down there, too."

Kittery stood and followed the other two to the window. The street below was quiet, but among the few horses tied along its length the bold white of one stood out at the cantina's hitching rail. Even at three hundred yards distance, Kittery could see the long flaxen hair of the man who stood beside the mount, tugging on its latigo. The man's head barely came to the shoulder of the horse. He looked both ways along the street before clambering aboard the saddle and trotting off toward the Camino Real.

"Shorty Randall," Raines said with a grunt. "The man of infamy." He looked over at Kittery. "Shorty's been pegged a spy, of sorts. He often rides toward the Baboquivaris, sometimes for days at a time. Actually, we think he's a mediator between Baraga and someone either in Castor or in Tucson. Someone that knows an awful lot about financial affairs in this territory."

"Why doesn't somebody follow him to the Desperado Den? Clean up that place?"

"Oh, that'll come," Vancouver spoke up. "But not yet. We're hoping he'll lead us to his informant first, if there is one. Baraga, he'll always be there. He's the known factor. I'd like to find out the unknown one before we tip our hand to Shorty."

"He's headed the right way," Raines mused. Kittery looked back to see Shorty Randall turning off the Castor road in a westerly direction, toward the Baboquivaris. Raines smiled grimly. "Looks like Baraga's about to find out I'm on his tail. I hope Blue Bell is standing clear."

When the three men were seated once more, Vancouver spoke to Raines. "Is there anything you'd like to know about the Desperados, Joe? Outside of what the records say?"

"Well, outside what the records say, I don't know much."

Vancouver chuckled. "The records tell most of it, I guess. But Blue Bell Smith and I have talked over dinner a half dozen times, so I think I'm one up on the government when it comes to knowledge of Baraga's bunch."

"There's one thing I'm curious about," said Raines. "Colt Bishop—is he as fast as folks want to make out?"

"Faster. Or so Smith tells me. He watched Bishop work with his gun every day—without fail. Even when most of the others were too tired to move. Said you couldn't even see his hand move. But you oughtta know, he says Silverbeard is near as good as Bishop—and with both hands. I guess that would make him twice as good."

"A nice tidbit to know, I guess," said Raines with a nod. "But it won't make much difference. I never claimed to be any quick-draw artist. It wouldn't take very much to beat me on the draw. I plan to have that cavern surrounded by so many rifles Baraga doesn't have time to blink."

"You're smart there," Luke said. "Pistols are a close-up weapon, for a fact. Dangerous enough, but not like a rifle. Unfortunately, Crow Denton's supposed to be one of the best shots there is with a rifle, so he'll be one to watch. And Big Samson's no slacker, either. He used to hunt buffalo for a living, I hear. So they can both take you out as far away as you can them. As far as being most dangerous, I'd almost have to cast my lot with Bloody Walt Doolin. He's vicious and unpredictable. He'll kill you for no reason and without any warning. Well, him and Wells both. That story about Doolin cutting his initials in men's flesh, Smith said it's true. Said he saw him do it once, down in Sonora."

Kittery glanced over at Beth, who shuddered and looked away. He had heard that story before. It was legend in Arizona, legend like the one about Doolin owning only two shirts, both red longjohns he redyed in the blood of his victims. He didn't believe that one, but he'd seen some awful things in his travels, and it sounded as if maybe Bloody Walt was indeed that twisted.

"Is it true Ned Crawford rides with Baraga now and then?" Kittery asked.

"That's the rumor," replied the sheriff. "You'll find many do, time to time. They seem to find security with the Desperados, and no wonder. Take a look at these flyers I brought up. Four of these are supposed to have worked with Baraga, to my knowledge. Crawford's another, and you have the poster on him. He's a henchman, you might say."

Kittery paused and nudged his empty plate away. "Your wire said you knew where Crawford is."

"Yes, I know where he is—or was. Up in the northern end of the Baboquivaris, laying low—as far as we know. They call that stretch of the range the Quinlans. Like your newspaper clipping said, Crawford's a backshooter and a coward of the blackest stripe. He doesn't care who he kills—or how, or why. And he's not alone, either. There are a couple hardcases with him—Amayo Varandez, for one. He's as bad as Crawford and worth a hundred or two more to the government."

Kittery nodded. "Well, I'll watch myself—when the time comes."

Beth Vancouver took in a deep breath and let her eyes settle on the captain's, which watched her husband so intently. So, that was it. Whatever the captain normally did to pay his way in life, he had forsaken it to come to Castor. He had come here to hunt down Ned Crawford. It sent chills down her spine to think of Crawford, for in spite of her over-protective husband's attempts to shelter her, she'd heard the stories. Crawford was a killer not only of men, but more often of women, and even children. And he didn't just kill them. He did things to them—brutal, inhuman things that even devils shouldn't do to one another. Beth didn't know if she approved of a man who went after other men without the proper authority of a badge, just for the money. It was so . . . mercenary. But Luke didn't like it either. He'd made that plain in the past. And yet he seemed to like the captain. Considering the brutality of the man he was attempting to bring to justice, Beth would have to forgive Tappan Kittery, too.

Chapter Four
Stepping on Toes

It seemed little time before the breakfast dishes were cleared away. Vancouver, Kittery and Raines stood at the door, hats in hand. Vancouver had decided to sacrifice some of his valuable sleep to drive Beth into town for supplies, and they waited now for her to ready herself.

The trio walked outside, and on the porch Kittery felt the heat press down. An oriole flitted off into the catclaw, flashing bright yellow and black plumage. It sang a song much like that of the meadowlark, full of cheer. When it dropped from its perch and sailed down the ridge, the silence resumed, and the yellow sun raged on.

"Quite a view from here," Raines commented.

They gazed over the desert floor, where Castor showed the only signs of habitation. In the not-so-distant east, the ridges of the Santa Rita Mountains appeared purple in the heat haze, and a soft, golden light played upon some of their ancient, wind-smoothed surfaces. To the north, they could see the Sierra Tucson, the Santa Catalinas and the Rincons, all blended together at that distance like one vast range, a far-off promise of greenery and abundant water, though the Tucsons offered precious little of the latter. In the foreground, the valley stretched away, rich with the succulent vegetation that gave the Sonoran Desert a look like no other desert in the world.

Beth came to join them wearing a blue cotton blouse that accented her eyes, a gray wool riding skirt and a navy blue cornette tied beneath her chin. Joe Raines let his eyes pass over her in frank appraisal. "You'd make anything look in fashion, Mrs. Vancouver. If you don't mind my saying it, you're truly a handsome woman."

A blush crossed her face for the second time that day. "Why thank you, Marshal." As if she couldn't help it, her eyes touched Kittery, but he merely smiled and turned away.

As they talked, Kittery looked over the sheriff's horses, now hitched to a heavy farm wagon. Both bays, they were well-bred animals and well cared-for. Indicating the larger of the two, Kittery commented on his fine breeding.

Vancouver answered with a pleased smile. "Thanks. That's Spade. I saw the one you rode in on yourself. You have an eye for good horseflesh. Tap, one of these evenings you oughtta go riding with Beth and me."

Kittery glanced at the sheriff with surprise. Nothing about Vancouver's demeanor indicated to Kittery that he had given any particular thought to the abbreviation of his Christian name. If another man had done the same, Kittery might have taken exception to the familiarity, but something about Vancouver set him apart, and instead it warmed his heart. Kittery had already begun to look at the lawman as a friend.

"We'll do that, Luke."

Vancouver nodded. "Good. Well, we may as well head down the hill."

With that, he turned to give Beth a hand onto the wagon seat and then climbed on himself. Kittery and Raines piled on back, and the rig wheeled around and rumbled down the twisting ridge road. The wagon slowed as it pulled into Main Street, leaving a filmy white cloud of dust in its wake. Vancouver tugged the horses to a halt in front of the green frame building that was Greene's Dry Goods.

A young girl, perhaps twelve or thirteen years of age, was sweeping the boardwalk. She looked up, and her large, very dark eyes caught Kittery's. She was a pretty girl, with clean features and a fine, straight nose. She had golden blond hair, but he guessed by the cast of her skin that she was a half-breed Mexican. She smiled, showing a perfect row of teeth and making her eyes crinkle up at the corners. Kittery smiled back, thinking she was going to make some boy very happy one day. When Kittery tipped his hat to her, she smiled bashfully and looked back down at her broom.

Leaving Beth to make out her order, Luke Vancouver invited himself to join Kittery and Raines in washing the dust from their throats at the cantina, farther down and across the street. As Kittery stepped up to the swinging doors of the cantina, in front of Raines and Vancouver, a heavy-jowled man staggered out. He careened into Kittery and knocked him off balance. Instinctively, Kittery reached out and shoved the other man away.

"Watch yourself, mister. This street's plenty big enough for two of us," Kittery said in a chiding voice.

The fat man's was not a pleasant face, but a cruel-looking one with small gray eyes set far back beneath shaggy brows and sunken behind corpulent cheeks. He appeared to be in his late forties or early fifties, and though several inches shorter than Kittery he easily outweighed him by forty pounds, with a huge belly pooching out his suspenders.

Kittery's mild demeanor disappeared in an instant when the fat man pulled back a fist. He stopped himself for no apparent reason other than perhaps Kittery's size and ready stance. "Keep out of my way, then!" the man bellowed. He turned and staggered off toward the dry goods store.

Vancouver laughed with little humor when the man was gone. "Tappan, meet Thaddeus Greene. The *gracious* Thaddeus Greene. He owns the dry goods store. Don't worry about him—he treats most people that way when he's drunk—and he's drunk half the week."

Kittery chuckled as he forced himself to take a calming breath. They entered the cantina, swallowed by its murky shadows and lingering, frowsty odor of stale whisky and tobacco smoke. Turning to the right, they went to lean up against the scarred plank bar. "Greene appears to be all mouth and no gumption," Raines observed as they settled themselves down on their elbows.

"Could be true," Vancouver replied. "But he's a big man in more ways than one in this part of the territory. Thaddeus Greene's a wealthy man, though you wouldn't guess it at a glance. He has more than his share of political pull."

Vancouver took a long black cigar from a tray on the bar and offered one to the others, which they turned down. Biting off the tip of the cigar, Vancouver touched a match to the other end. He dropped the match into a dish as Raines pulled out a pipe and began to tamp it. A short, mustachioed Mexican appeared from a door at the end of the bar. After greeting Vancouver and Raines, he picked up a dark brown bottle and, coming near, tipped it to fill three glasses.

The man smiled at Kittery, showing broken yellow teeth. "You are a stranger here, no? Me, I am Mario Cardona."

Kittery shook Cardona's outstretched hand. "Tappan Kittery."

The Mexican moved on to a newly arrived customer, and Kittery turned to the lawmen. They spoke of the town, for a time, of events along the Camino Real, and of politics in Tucson. Francis Goodwin was making a disaster of his position as United States marshal, Raines said. He was suspected of padding his own bank account with already insufficient government funds, and came nowhere near providing the Territory of Arizona with adequate protection from the lawless elements. Governor Anson Safford continued to perform excellently for the people of Arizona, yet he had been in office over seven years and wasn't expected to hold the position much longer. "Jason the Coachman," also known as Jason Hall, was scheduled to hang at the end of the week on Castor's gallows. Though he'd been a holdup man for ten years, he had never killed a man until three months previous. Then he had shot down a preacher and a deputy sheriff in one day, and now his career would come to a close at the end of a rope. Such were the affairs of Tucson and Castor.

In the broader picture, newspapers were filled with tales of George Armstrong Custer and the Seventh cavalry, making plans to deliver one final, decisive blow that would land the rebellious Sioux, Cheyenne and Arapaho on their reservations forever. And here in Arizona and New Mexico, lands famous for their military campaigns against the Chiricahua and Western Apaches, a temporary lull in the fighting continued but wasn't expected to go beyond the next reservation jump. With Nana, Juh and Geronimo alive, that wouldn't be long.

The three talked for a while of times gone by, and of the Desperados Eight and Ned Crawford. Then once more Mario Cardona stood before them with a bottle. "More wheesky?"

Vancouver raised a hand in refusal and stood up, snuffing out his cigar in the clay dish and placing a silver coin on the bar to pay for the drink and the smoke. "None for me, thanks. It's time I went to help Beth. Can't leave her out there alone with Thaddeus Greene all day. Nobody deserves that."

The lawman said goodbye and parted, his boot heels making clopping sounds as he moved down the porch.

Kittery, too, refused a refill. "I'm lookin' for a place to stay, *amigo*. You think there's any rooms left in the hotel?"

"That I do not know. But I know there ees men there who make much trouble. Bad men. You go look. You don' find room, I have one in the back of the cantina. You can stay there for . . . for feefty cents. *Bueno?*"

"Sounds good," Kittery said with a smile. "I'll try the hotel an' maybe be back."

Raines stood away from the bar. "I'll go along. After sleeping on rocks and scorpions all night I could stand a soft bed," he said with a laugh.

Paying for their drinks, the two stepped next door to the blue-painted hotel. They had to shoulder their way through an unyielding handful of whiskered Mexicans lounging in the doorway. The Mexicans glared at them but made no move of aggression.

The spacious lobby, lit only by one small front window that was shuttered against the sun, smelled of an unlikely mixture of scents—that of peppermint and fried beans. A paisley wool carpet traced the main walkways, the same carpet which clad a narrow staircase on the right, leading up to the second story. Its humble banister was worn almost bare of paint, and three or four balusters were completely missing.

Behind a desk at the left wall sat a middle-aged Mexican woman, her jet-black hair piled above her head. The style accentuated a slender neck and wide cheekbones and made large dark pools of eyes almost as black as her hair. She looked up as Kittery and Raines entered, and her full, almost sanguine lips parted.

Raines doffed his hat and held it at his side. "Ma'am. Per chance, do you have two empty rooms?"

"I am sorry. There are none."

"Oh. All right then." Raines turned to give Kittery a disappointed glance. "Thanks anyway," he said, and they turned away together.

They had made it halfway across the floor before her voice stopped them, not much more than a loud whisper.

"Wait."

Kittery and the marshal turned back. The woman had stood up, revealing a swollen abdomen to which her red calico dress clung. Her eyes touched Kittery and took him in at a glance, and then she turned to the marshal, her eyes lingering on his badge.

"If you would like, you cou' talk to some of the others. Perhaps they would move out an' leave their rooms for you." Her voice held a cryptic plea Kittery caught but that seemed to go unnoticed by his friend.

"No, ma'am, we'll try elsewhere. I wouldn't want to put anyone out if they're set already."

"Please."

The word caught the pair turning around, and they faced back around. The woman's eyes were on Kittery now, beseeching him.

"You havin' some kind o' trouble?" the big man asked.

Raines cut in. "You have the right to evict anyone from your place, if that's your wish."

"Yes, please," she repeated.

She had looked several times toward the group of Mexicans in the doorway, and Kittery waved toward them. "Is it them you want out?"

She nodded vigorously. "Sí, sí. They do not pay most of the time. An' they fight an' break everything. They are bad men."

"All of 'em? You want 'em all out?"

"Sí."

Raines nodded. "Well, that's good enough." He and Kittery looked at each other, and together they strode toward the door.

The biggest of the Mexicans, a man in a woven blue serape, stood in the center of the doorway, his broad back toward them. When Raines laid a hand on his shoulder, he whirled about. "You touchin' me?"

Raines felt no need to respond to the obvious. "If you have belongings in here it's time to move them out. And no trouble, *amigos*. No trouble."

A defiant look came into the man's eyes, and he turned to look at his compadres. Then he squared himself before the two would-be evictors. "Why?"

"We've a request to clear this hotel of destructive boarders," Raines replied.

"What?" The Mexican cast a malevolent look toward the lady proprietor. "Oh, now I understan'. So what we need to do?"

"You heard him," cut in Kittery. "You move out."

The Mexican turned to Kittery with a sneer, looking him up and down. "I hear heem—but I'm no listen."

Raines gave Kittery a cautioning glance. "You'd do well to listen . . . *señor.* And move out."

"Why?" said the Mexican with a sneer, his rotted yellow teeth flashing through a field of thatched whiskers. "We won' make no more trouble."

"What's your name?" asked Raines.

"I am Gustavo Ferrar. You remember eet."

"I'll remember. And you remember me. My name is Josiah Raines, and I'm a United States marshal."

Tappan Kittery wasn't a prejudiced man. He had nothing against the Mexican race or against any other. But he had dealt with Ferrar's type before, and all they seemed to understand was force. Mexican, Indian, Chinese, black or white, hardcases were all alike. He had listened to this Mexican's mouth long enough. Just looking for any length of time at his belligerent leer was a challenge, and his anger was building to the breaking point. Had he been his older brother Jake, he might have been more patient and tried harder to talk the Mexican into leaving peaceably. Unfortunately, Ferrar failed to see his growing anger. He glared at the badge on Raines's vest and sneered. "A piece of tin. Ees that s'posed to mean sometheen, you son of a devil?"

"Just move out, Ferrar," Raines said, shifting into a position of advantage. "I told you I didn't want any trouble with you."

"Oh, sí! No trouble! You move us out on the street, an' you weesh no trouble. I take you an' cut off your ears. Then I take that gringo-loving *puta* and show her—"

Those words rang in Kittery's ears as he lurched forward and backhanded Ferrar across the lips. As Raines ducked out of the way, Kittery followed the Mexican onto the porch and threw a left cross which took him between the eyes. Jake had been a peaceable man, but he had also trained himself to be a pugilist, and he had taught his little brother to throw that wicked left. Ferrar, trying to catch his balance, stepped off the porch and landed on his backside in the thick dust.

The Mexican boiled to his feet. His compadres scrambled out of the way, and Raines stayed where he was. He had seen Kittery in fights before, and it was unwise to interfere.

Ferrar was game—more so than Kittery had figured. Though a big man among his friends, he weighed an easy thirty pounds less than Kittery and stood four inches shorter, along with the arms to match. But without hesitating he came at Kittery swinging, and Kittery absorbed two blows to the face before coiling to strike back. With every ounce of power, he drove up hard and fast to the bottom of Ferrar's chin, sprawling him flat out into the street this time.

Once more, Ferrar managed to reach his feet, this time shaking his head and trying to clear his eyes by rapid blinking. With a snarl, he charged in, his arms flailing. Kittery stepped back and blocked a roundhouse right, then popped a left jab to Ferrar's forehead, snapping his head back.

When the Mexican swung again and missed, Kittery smashed him in the solar plexus, striving to push it through to his chest. Gasping for air, Ferrar stumbled off the porch, barely managing to keep his feet. His eyes and mouth were open wide.

The Mexican was defenseless, but Kittery was in no mood to let him recover. He followed him from the porch and stepped around to his left side. Face void of emotion, Kittery cocked his right fist and grabbed Ferrar by the collar to hold him steady. He hammered him twice in the ear, splitting it down the top. On the second blow, he let go of Ferrar's collar, then slammed him on the side of the head with the palm of his hand. Ferrar swayed and fell to the street, where he lay groaning and bleeding. He made no move to rise.

Kittery stood over Ferrar, his hands poised. When he saw the Mexican wouldn't get up, his eyes raised to the others. There were five of them, gathered around Ferrar in a crescent shape. They eyed him with the hating looks of men trying hard to cover up their fear. One of them finally made a move, and that was to crouch near Ferrar to help him to his feet.

One of the others looked at Kittery. "It is all right we go find our things?"

Kittery shrugged, taking a breath to calm himself. "Help yourself."

The Mexicans managed to get the now docile Ferrar to his feet, and with two of them steadying him, they filed through the hotel doorway past Joe Raines and ascended the stairs.

As the last one topped out at the landing, Raines turned to Kittery with a shake of his head. "You've picked yourself some trouble this time, Tap. That's a bad one, there."

"Yeah, I know. You fight one bean, sooner or later you'll have to fight the whole bowl," Kittery said dryly. He stepped inside the hotel and walked over to the desk. The Mexican woman had returned to it after watching the combat.

"How much for a room, ma'am?" he asked, as halcyon as if there had been no disturbance.

The woman stared at him, her eyes large. She finally found her voice. "Two dollars."

"How 'bout two dollars for a pair of 'em? Facin' the street. That's the price at the Cosmopolitan, up in Tucson."

The woman's mouth drew tight. But a second later she let a smile touch her lips, continuing to study Kittery's face. "All right. Then one dollar each. But only for you an' the sheriff."

Kittery and the marshal exchanged their dollars for keys and went up the stairs. One of the Mexicans was picking up his belongings in Kittery's room. He hurried out, avoiding Kittery's eyes to hide the fear and odium in his own.

Kittery let his eyes scan the room and frowned. He didn't like to swear, but he did now. It was plain why the lady had wanted the Mexicans out. This room had been lain squalid—ruined. Someone had scribbled words on the stained walls, and from what he understood of Spanish they didn't come more vulgar. Bits of food and clumps of mud littered the floor, tracked from the doorway to the bed. Crusted blood spotted the floor and sheets, as if the occupant had suffered from frequent nosebleeds. And he appeared to have a fondness for knives, for he had carved the bedposts and woodwork with words and intricate designs. The last thing Kittery noticed was several bullet holes in the floor, affording a glimpse of light from the room below. That proved the man who had slept here was not only a slob, but also a menace.

Like the rest of the room, it was plain the woman had given up on the bedsheets. They didn't appear to have been washed in months and were nearer black than white. He was half-afraid of what minuscule wildlife lurked there. Holding his breath, he grabbed a corner of the sheets and tore them and the blankets from the mattress, tossing them in a reeking heap into the middle of the hall. Going to the dirty, streaked window, he removed a curtain, rod and all, and let it fall to the floor among the dust and the dead flies collected there.

The sun blasted off the rooftops and a huge water tower across the street, next to Greene's Dry Goods. It wouldn't cool it off in the room to open the window, but it might move some of the stale air. Kittery struggled with the stiff window until he managed to force it open a foot. It wouldn't go any farther.

Outside, the town was beginning to grow quiet, as the heat drove people into their homes. He envied those with adobe houses, especially the ones with the foresight to construct their windows small and high in the wall, according to custom. Those people, although their homes might sometimes be dirtier than those of wood, could at least enjoy a measure of comfort against the onslaught of the sun. Kittery had no such comfort. He took off his hat, and his hair was plastered back with sweat, the same sweat that started to bead up on his cheeks and trickle in little rivulets down his chest and back.

Inside the huge doorway of the warehouse, he could see people lounging. As the Mexicans left the hotel, he saw some of them drift toward the warehouse and join the others inside. On the dock of the warehouse, awaiting delivery or to be moved inside the building, were stacked burlap sacks containing grains, cotton ones full of flour and sugar, barrels of unknown substances, and crates that held anything from rifles or chickens to melons and yard goods. There was also a load of mining gear there—shovels, picks, jacks, drills and lanterns. Mining was the number one industry around Castor and the surrounding settlements.

Alongside the warehouse loomed huge stacks of lumber, not only timbers for the mines but materials for building in the "New Southwest", as some merchants were beginning to call it. Like the designers of this hotel and most of Castor, they had begun to forsake the comfortable atmosphere of adobe for the cleanliness and style of wood. Prescott, the territory's first capitol and a very attractive little burg, was constructed mostly of wood. It was responsible in large part for the transition from adobe to wood, and even though Tucson, a predominately adobe city, was now the capitol, wood's popularity continued to grow. Its biggest drawback was the expense of freighting it, but much of the wood in Castor would be from the Santa Rita Mountains, only a matter of miles to the east.

At that hour Greene's Dry Goods enjoyed the most business, with the mercantile across the street a close second. Kittery could see Luke Vancouver loading sacks from Greene's into his wagon while Beth watched from inside the store. Toward evening the cantina would be the busiest spot. In such a community, home of miners, cowboys and drifters, the local watering hole was the hub of society, a place for refreshment, excitement and camaraderie, and a place to catch up on news.

Kittery wiped his forehead with his hand and dried his hand on his pants, then replaced his hat. As he was turning away from the window, a horseman riding in from the north on a plodding line-back dun caught his eye. The man wore a wide-brimmed light colored hat, pulled low to battle the sun's rays, and a faded red shirt. Crusted shotgun chaps protected his legs, and a six-gun in a Mexican loop holster rode his right thigh in a cross-draw position. These things Kittery noted as the rider drew past and pulled up before the cantina, dismounting.

Kittery watched the man disappear behind the wall of the cantina, then stood away from the window. Stepping from the stifling room, he went quietly down the stairs, out the door and to the cantina. This was a man he had to see.

Chapter Five
Before a Storm

The twin doors of the cantina creaked as Tappan Kittery passed through them into the smell of smoke and old wood and alcohol. Crossing to the bar, he leaned up against it beside the man who had just ridden in. He raised a finger for Mario Cardona to bring him a drink. The newcomer's head was lowered; to all appearances he was unaware of Kittery beside him.

A smile parted Kittery's lips as he picked up the glass Cardona brought him. He spoke warmly, "Hello, Cotton. It's been some time."

The newcomer, Cotton Baine, stiffened, then relaxed as he turned to raise smiling eyes to Kittery. He was a young man whose reckless features were blessed with a quiet, homespun handsomeness. His hat hung from bonnet strings behind his head and his thinning, sand-colored hair wisped to one side, revealing a smooth forehead that appeared pale against the wind-blown brown of his face. Alert blue eyes belied the grin that took over his face now, and he spoke with the drawl of the Texas child he was.

"A good long time, Tap. How've you been?" He put out his hand, and they shook, holding on for an extended time.

"I'm gettin' along," Kittery replied. "In years, too."

Baine chuckled and peered at Kittery's sideburns, where the big man knew if someone looked hard enough they could locate the few strands of gray. "Yeah, well too much gallivantin' around c'n do that, Tap. Believe me, I know. You may be grayin'—barely, but my hair's thinnin'—heavily."

"Your hat's too tight then, cowboy." Kittery took a sip from his shot glass as he studied this old partner of his with gladness in his eyes. "I was hopin' I'd run into you down here. Thought you mighta moved on by now."

Baine's smile faded as he clucked his tongue. "Well, not yet. But you might not o' found me next week. My job played out."

"Then what're you doin' to live?"

"Odd jobs, same as the old days. Ridin' the grubline a lot." Cotton Baine's gaze wandered, drifting across the rows of bottles behind the bar, as if there were something there to draw his interest. Kittery had the feeling his old riding partner's thoughts had roamed to the time not long ago when they had ridden the Mogollon Rim together, working for the Bar Cross D.

"Arizona wasn't made for cowboys, Cot. Least not this part of it. Take Texas. There's still all sorts of work back there, I hear tell. Those big ranches are boomin'. An' I daresay it'll be even better up in Wyoming and Montana, if they ever manage to whip the Sioux out o' their land."

Baine grunted. "Yeah, well, I like Arizona. I'd like to stay. Freeze your butt off up north."

"Tried huntin' bounties?" Kittery asked, taking a swallow of whisky.

"Animals, yeah. Not the two-legged kind."

"That's why I'm here," Kittery volunteered. "I'm after Ned Crawford."

Baine nodded. "Yeah, you always was an idealist. Ned Crawford, huh? I'd watch my step there, big fella. Word is, there's some tough ones with him."

Kittery shrugged. He didn't scare easily, not after fighting a war as bad as the one he'd suffered through and after hunting badmen off and on, between respectable employment. He was tired of being warned away from the deserter and backshooter, Ned Crawford. The outlaw might not have killed any great number of men, but he had killed children and given women a taste of things worse before putting them under the ground, too. If he had others with him, what did it matter? That was expected of a coward like Crawford, and whoever traveled with the likes of him deserved to die, too. Not that Kittery wanted to kill them. He would rather bring them in alive. But he planned to see, one way or another, that justice was served.

Kittery's reply to Baine's warning was simple. "The more the merrier. Bounty huntin' is good pay, when you can manage to collect one."

"Sure, but . . . maybe the gain ain't worth the risk, when you work alone."

On impulse, the cowboy drew makings from his shirt pocket. Kittery refused a smoke, and Baine, without skipping a beat, fashioned himself a cigarette. "Messicans hooked me on rollin' my own," he said. He lit the smoke, then turned in unison with Kittery at the sound of boot heels inside the batwing doors. Luke Vancouver stood there, his sweat-soaked shirt attesting to the heat.

"Glad you're here, Tap," the lawman said. "Beth's finally done shopping, and she sent me over to invite you and Joe to supper and a horse ride tonight."

Kittery grinned. "I'll take you folks up on that."

"Good. Come up to the place about six. You're welcome, too, Cotton," Vancouver added with a smile.

"No thank yuh, Sheriff. I got an engagement here at the bar."

Kittery grinned at his partner and waved goodbye to Vancouver as he left.

"How do you know the sheriff?" Baine asked.

"I just met 'im this mornin'. He was givin' me some information on Crawford's habits."

"You make friends fast, Tap. Awful fast."

"Well, they're nice folks. An' if you'd ever eaten with Beth Vancouver you wouldn't have turned the offer down."

Baine laughed. "Yeah, well, I recognize an invitation of politeness when I hear one, Tap. Sheriff just asked because I happened to be here."

"Maybe so, but you were a fool to turn down a meal cooked by Beth Vancouver. Even if you don't eat, it'd be worth goin' to look at her an' listen to that voice of hers."

"Oh-h-h! You best be careful if you're thinkin' that way, Tap."

Kittery chuckled. "Oh, no. You're mind's wanderin'. She's nice to look at, but she's a trapped turkey, an' I don't intend on messin' with that."

Baine laughed. He knew his friend at least that well. "Hey, you ever run into that brother of yours you used to talk about all the time?"

Kittery returned his friend's laugh. He knew he *thought* a lot about Jake, even after all these years he'd been gone, but— "Did I really talk about him *that* much?" he asked.

"Seemed like it." This time the corners of the cowboy's eyes crinkled, but he didn't laugh outright or even smile. He knew how much Kittery had loved his brother and how he missed him.

The truth was, Kittery figured Jake had died in the war. That always had been the rumor, and he had never seen anything to prove different. He changed the subject. "You gonna be stayin' in town?"

"I'd hoped to. I need a room that's next to free."

"Well, I got just the thing." Kittery caught Cardona's attention and motioned him over. When the Mexican stopped before them, Kittery brought up his earlier offer.

"The room is still empty, *sí* . . . for the fifty cents."

"Well, I found a room in the hotel, myself. But my friend here could sure use that one."

Cardona turned to glance at Baine. "For Cotton? *Sí,* of course. The room is his."

"Fine." Cotton produced a fifty-cent piece, which he placed in Cardona's empty palm. "Reckon I'll move my stuff in now."

U.S. Marshal Joe Raines put down the book he was reading and stood up from the bed, his boots lying beside it on the floor. He went to the window to draw open its faded curtains, which splashed diffused light across the fly-speckled floor. As he looked, he saw Luke Vancouver pull his wagon away from the dry goods store and start it toward home. He watched Beth put her arm around her husband and squeeze him close, and an ache came up from deep inside him. He put his fingers up to the window, steadying himself as he watched them disappear around the water tower.

Beth Vancouver was quite a woman. Beautiful, charming, well bred and graceful. Everything a man could want in a wife. But she wasn't his Emily.

He thought back over the years, back to the woman he still loved. There could never be another Emily. She had been his soul. His reason for life. There was nothing he wouldn't have done for her.

But Emily had been taken. Taken away in a mahogany box—he had insisted on that, instead of pine. It had cost most of their savings to pay for it, but without Emily he hadn't needed the money anyway. Now Emily was gone, buried far away in Maine, her home. He missed her. He wished her grave were here. But she had wanted to go home. He could never have kept that from her and been able to live with himself.

Moon Tarallo. That was the name that ate at his soul. That was the man who had killed his Emily. Moon Tarallo. He walked back to the bed and picked up his book, reading the front cover. *The Book of Mormon.* He flipped through the pages absent-mindedly and dropped it back to the mattress, sitting down to lower his face into the palms of his hands. *Oh, Emily. Sweetheart, I hope you know I did all I could.*

With sudden rage, Raines erupted from the bed and made a quick circle about the closed space. He had to stop thinking before he went insane. He walked back to the window. Down the thoroughfare he could see the Mexican, Gustavo Ferrar. He and several compadres lingered about the warehouse doors, passing around a dark brown bottle and smoking cigarettes. Now and then Ferrar would rub his head, then glance toward the hotel, an ugly sneer across his face. He would snarl a few words of anger, then spit in the direction of the hotel. On impulse, Raines made a point of committing to memory all the faces down there in case there was trouble later. If anyone ended up hurt, this time he wanted a perfect case. He could never let a man walk away guilty the way Moon Tarallo had.

Over a skewed chest of drawers hung a flyspecked, framed mirror with a big crack all the way down its center. Having never liked the itchiness of a whiskered jaw, he removed his paisley vest and lavender shirt to shave. As he did, he studied his torso. Was he growing old? He was forty, but he didn't look it. His body was lean, and his chest, arms and abdomen muscular. He didn't feel old. Did he look old to others? How would Emily look now, after all these years?

When he finished shaving, he splashed the remainder of the water from his chamber pitcher over his face and matted chest, patting himself down with a musty towel. Out of habit, he unloosed a long-barreled Smith and Wesson American .44 from its holster, cleaned, oiled and returned it, assured the hammer rested on an empty chamber.

He was getting sticky under the arms, and sweat stuck his hair to the back of his neck. He picked up his shirt and shrugged back into it. He had never been a claustrophobic man, but the room was closing in, with heat, stale air . . . and with memories. It overpowered him, and he finally went along the hall and downstairs, still in his stocking feet.

The lady proprietor sat behind her desk, a young olive-skinned girl beside her. "Hello, ma'am. I apologize for my stocking feet."

The woman smiled. "This is my daughter, Emilita. An' what you wear leavin' your room for a breath of air is no concern of mine."

"Thank you, ma'am."

He stepped through the back doorway, moving to one of two huge clay ollas that hung there in the shade. From a dipper, he drank long and deeply of the clear, lukewarm water, letting his dark eyes trail south, out across the wide, barren plain to the maze of canyons and arroyos and the high cliffs and mountains that looked blue and fog-laden and distant in the sun. He planned to be deep within those mountains in a matter of days. And he planned to be alone.

He had told Vancouver and the others there would be a posse. But he had lied. He had let Moon Tarallo slip through his fingers, the most important criminal he'd ever brought to justice—lost because of a technicality in the courtroom. Walked away free. People had seen his failure, and they had cursed him for being so legal-minded and so foolish as to trust the laws of the land. Now he cursed himself.

No, there would be no posse. Josiah Raines knew where the Desperado Den lay, for he had seen it. He hadn't told Vancouver that; he had seen no point in it. But he would be there soon, lurking in the rocks high above it. And by the time he was done killing the guard, and done lighting off a score of dynamite sticks and sending them down, the Desperados Eight would be no more.

And Josiah Raines himself? He would either be dead, or he would be the most famous lawman in the Territory of Arizona. Maybe in the entire West. Either way, Emily Raines would be avenged by him, after a fashion. And there wasn't anything else that mattered in this life.

His mind drifted to the thought of death out there in the Baboquivaris. It was going to be a large enough task riding out of the clutches of the Desperados' country, but riding out of it with all of them dead was a monumental undertaking. It was almost a joke, but he couldn't find it in his heart to laugh.

"Em, I'll have them," he said quietly. "I didn't do right by you when I had the chance, but I will." With that, he turned back inside.

Five minutes later, he cleared the front doorway with a string tie drooping down from beneath his collar. He strode along the weathered porch, past a Mexican asleep in a chair. He stepped into the dusty street and to the cantina, pushing through the doors. Moving up beside Kittery, he ordered a whisky.

He eyed his friend. "Keep an eye on that Ferrar. I don't think you beat him bad enough. He's working up to something."

"What makes you think so?"

"Oh, he's at the warehouse—where he's supposed to be, I guess—but he's got himself a bottle, and he's talking big to his friends. He keeps looking up at the hotel, so it's obvious we're heavy on his mind. Or at least you are. And I don't imagine he's whispering sweet nothings."

"That's a good bet. He ain't the kind to let a thing like that lie."

Raines gave him an odd look. "Would you?"

"No, I reckon not."

Catching Baine's puzzled stare, Kittery explained to him the afternoon's events at the hotel, the eviction and the fight. Then someone's passing through the batwing doors made Raines turn. He lowered his eyebrows and glanced back at Kittery.

"Here they come."

Gustavo Ferrar stood inside, six men flanking him. A fiendish look filled his eyes as they swept the room, merely skimming over the three men at the bar. Motioning his compadres to follow, he swaggered up to the bar and ordered drinks.

The oppression of silence clutched the room. Kittery stared straight ahead. Ferrar did likewise, but the fire between him and Kittery was a living thing.

The silence continued for some time as the Mexicans sipped their drinks. Then, for no apparent reason, Ferrar backed away, turned and stalked out the door. Faces puzzled, the six would-be gunmen looked at each other. As one, they gulped down the remains of their drinks and followed their leader out.

For seconds, the quiet ran on. Cotton Baine ended it with an amused chuckle as he dropped his hand away from his gun. "Now, what was that all about?"

Kittery scowled. "I wish I knew. Reckon it was Ferrar's idea of a show of force."

"I think he lost his nerve, gentlemen." The corners of Joe Raines's mouth lifted in a faint smile, but then his face turned once more sober. "But I bet he'll find it again in a bottle of whisky. Be on guard, Tap. He wants you."

"Thanks, Joe. He's gonna wait for me to be alone." He turned to Baine. "You in town long?"

"A while."

"Let's go for a ride tomorrow—talk about the old days."

"Sounds good," Baine said with a smile. "Maybe you c'n think of some more stories about yer brother you ain't told me yet!" He jabbed his friend teasingly in the ribs.

"The way you talk I musta told 'em all! I'll see you around seven, before it gets hot," said Kittery.

"I'll be here."

"I'm gonna get a little rest," Kittery decided. He nodded at Joe and turned away, calling "so long" over his shoulder.

The marshal rose and followed, announcing, "I'm going with you. For a while, two guns will be better than one."

Once more in the shade of his quarters, Kittery peeled off his dark blue wool shirt. And "peel" was the apt term for it, since sweat had plastered it to him like the skin of a banana he'd once eaten in a shipping yard on the California coast—the *only* banana he had ever had occasion to sample. He laid his shirt on the back of the chair, then flipped the mattress over on the bed to reveal its cleaner side. Giving it a dubious once-over for bed bugs and lice—seam squirrels and pants rats, as they called them in the army—he stretched out upon it, his feet dangling over the end.

At nearly five-thirty, he opened his eyes again, rubbing the sleep from them. He had to force them open several times, for he had slept too long and they wanted to keep slamming shut.

Levering himself up, he sat on the edge of the bed, blinking and rolling his eyes, smearing the sweat from his face with his forearm and feeling it bead up again. He hadn't looked at his watch when he lay down, but he felt like he had slept for three hours. He must have needed it, but he felt worse than before his nap, not better.

But one thing the nap had given him was a powerful hunger. Sleep always did that to him. Although Beth's meal had been filling, it had also been the only one of the day. He was well ready for another. Shaking dust from his shirt and wishing it clean, he shrugged back into it and slapped at his arms, knocking more dust loose. He moved across to the window, where he let his gaze fall to the warehouse.

As he looked, a big, raw-boned freighter was pulling his high-boarded rig away from the loading dock, heading a six-mule team and its load of corn freight for Tucson. They called it corn freight because unlike oxen mules couldn't live off the land. They had to have corn to haul a load, and they brought their own supply with them. Wild roadside corn was sort of a scarce commodity in the Sonoran Desert.

As the wagon pulled away, five of the Mexican workers came into view, along with a muscular colored man who disappeared inside the building. Ferrar was one of the Mexicans, and he glanced toward the hotel, that perpetual sneer on his face, and said something in anger.

Kittery gave the big Mexican a measuring look. Ferrar seemed too cautious to face him in one on one combat again, so perhaps he planned something else—a little bonfire, maybe. The hotel was weathered and would take flame and burn well, killing both Kittery and the marshal in one decisive blow. Or would his vengeance be aimed at the Mexican woman? She was weak enough for him to handle, and, after all, had it not been for her the trouble would have never come about. In any instance, Kittery knew Ferrar would do something. It was a matter of time.

Kittery turned to his dressing table. He splashed tepid water on his face and ran his sleeves over it to dry off. Snatching his hat and clamping it on, he stepped into the hall, clicking the door shut behind.

Down in the street, the furnace blast had begun to wane, and the air was able to circulate, unlike in his prison cell of a room. People had come from their houses and were making their final business transactions of the day.

In the livery stable, Satan blew as Kittery drew his cinch tight. The stallion was restless, stamping his foot in eagerness to be on the move. "All right, boy, let's go."

Kittery swung aboard, nudging the black toward the door, and on the street he put him into a trot. People stood aside as the grand black horse chose his path, and they turned to look after him in admiration, watching his wavy mane and tail lift and fall and glisten with every step, and the sunlight ripple off his curved neck and powerful muscles.

In minutes, Kittery drew out onto the crest of the lane that led to the Vancouver spread, and the horse slowed to a shuffling walk, halting at the porch. Vancouver stepped out to meet them.

"You're just in time, Tappan. Tie your horse to the post there."

Kittery did this, and he and Vancouver went in the house. Joe Raines hadn't been inclined to join them that evening. He had seemed to Kittery to be in a melancholy mood, and Kittery hadn't wanted to press him. After giving Luke the marshal's apologies, Kittery dropped into a chair at Luke's invitation.

Beth stepped into the room, and Kittery stood and squeezed her hand, holding onto it perhaps longer than he should have. "Mighty nice of you to have me up again so soon, Mrs. Vancouver."

"It's our pleasure. And please call me Beth."

"Yes, ma'am—Beth." He guessed he should have told her to call him by his first name but didn't know if that would be proper.

Once more, Beth stepped out of the room but returned momentarily and set the table. Then, as that morning, the two of them carried in platters of food, and the three sat to eat the meal of rice and breaded chicken. Like the meal before, it was excellent fare. Beth Vancouver was either naturally talented or had been schooled in the art of tasty cuisine, and Kittery complimented her on it. Beth laughed lightly and thanked him, and the conversation ran on light and easy and friendly.

In no time supper was over. With one good hour of riding light left, they stepped outside. The Vancouvers' horses stood saddled and waiting in their stalls. Kittery unhitched the black and brought him forward. The three mounted, nudging the horses forward at a quickening pace, south along the road that traced the top of the clay bluff.

The late afternoon sun splashed brilliant gold-orange light like artists' color along the cliffs and canyon walls of the surrounding mountain ranges, creating a beautiful scene of solitude. An almost fluorescent green tipped the half-shaded trees and saguaro cactus. Spanish bayonet menaced the soft sky with its needle-sharp points, and the distant, sun-gilded dome of Baboquivari Peak beckoned across the desert floor. On the horizon to every side, in fact, the mountains stood soft and clear, promising water and shade and trees and grass, and in the calm evening air they seemed very near.

Mourning doves made their soft cooing in the paloverdes, and far away the loggerhead shrike cried, accenting the stillness. A hundred yards distant, a kit fox dashed across their path, beginning the nightly rounds of his territory. He had slept all day, and for him and many other creatures, the darkness was time to wake up. His path led him through fields of Mexican poppies blooming so brightly orange it was as if the hills were afire. The blue of lupine, purple of owl clover and yellow of brittlebush augmented that illusion of flame bursting across the face of the desert, making the word "desert" seem like a ridiculous misnomer for that luxuriant landscape.

"The way of the desert," said Vancouver with a smile. "No lack of beauty after it's rained hard for a couple weeks, then it dies and it's like death took over the land."

"But even then it's pretty," observed Kittery.

Beth agreed. "I think we'll get along just fine, Tappan Kittery."

The riders gazed at the landscape reverently. This was Castor's church, in a way, for that was one thing that hadn't come to their town yet. They had a service in the warehouse on Christmas and Easter, but that was the extent of Castor's attempt at organized religion. If a man wanted to be close to God anywhere near Castor, he went north ten miles or so to the Catholic mission San Xavier del Bac, home of the Papagos, or he came to the hills—God's intended church.

They stopped to gape when the sun bedded down along the western rims. It set with a breath-taking and glorious beauty, a beauty so marvelous it left them silent all the way back to the corrals. Here, man and wife unsaddled their horses and turned them loose with Satan to water. When they had drunk their fill, Kittery took the black and tethered him in the shadows over a patch of spring-greened grass. Walking inside, they left their hats dangling from a rack on the wall.

Their talk began and ran on for two hours. They talked of horses for a time, and Kittery confirmed one of Vancouver's suspicions, that he intended one day to start a horse ranch of his own. It was the only reason he had never had Satan cut, when few people would have put up with the complications of traveling with a stud out in open country.

The Vancouvers spoke of a family—a family Kittery hadn't guessed they had. There were two sons and a daughter, Morgan, Milo and Myranda. They had departed for Prescott six months past to live with Travis Vancouver, Luke's older brother and a marshal in Prescott. Kittery also learned he had misjudged both of their ages, assuming them younger by five years or better. Luke was forty, and Beth was thirty-five—three years older than Kittery.

"The kids are scouting out the country up there, you might say," Luke told

Kittery. "If they like it, that's where we may be moving, within a year or two." Kittery felt an inexplicable twinge of regret at the revelation. The Vancouvers had been kind to him in the short time he had known them, and even if he was living in Tucson it wasn't so far to come and visit. But he didn't get up to the Prescott area very often, and if they moved there was little chance he would see them again.

Beth Vancouver sat and listened to her husband and Tappan Kittery talk. She felt the same odd sensation she had while watching Tappan earlier in the day. He made her feel things she couldn't understand. He was so large and obviously powerful, his voice smooth and full of kindness but deep and resonant as a desert well. She had never seen a man so . . .so *perfect*, at least if one judged perfection by his build and his looks and his voice.

In spite of herself, a smile started to break out on Beth's face, and she had to look down to conceal it, pretending to clear her throat. Perfect? Well, of course Tappan Kittery wasn't perfect. Maybe she was only remembering her childhood fantasies, reaped from tales of Sir Lancelot, King Arthur and Odysseus. Hercules. Samson. Gawain. With little effort, she could visualize Tappan Kittery in those roles. So what was his dark side? There had to be something wrong with a man who seemed so far above criticism. What was *his* Achilles heel?

At last, when the time came to go, Kittery rose, standing there silent in the center of the room and making it look small. "Thanks for the good meal," he said finally, his eyes flickering to Beth, then away. "And the good company."

Together, they moved to the door, and Beth handed the big man his hat, imagining it with an ostrich plume in its band and Kittery with his dark hair curling over his shoulders instead of clipped conservatively. She smiled at herself, and Kittery caught the smile and gazed at her curiously, then dropped his eyes.

"Tappan"—Beth leaned closer—"I want you to know you always have a chair here . . . if you ever need someone to talk to, please come. We will be here, and I speak for both of us when I say you will be welcome in this house forever."

Beth sensed the desperate loneliness of this tall, quiet man, and she knew Luke well enough to feel comfortable saying what she had. Her husband might not have been as tall and daunting in appearance as Tappan Kittery, but he couldn't have been more self-assured, and as far as Beth was concerned no man could be more handsome.

That evening, three hearts had been bonded in lasting friendship. "Luke . . ." The big man put out his hand and shook the sheriff's, then squeezed Beth's again. "Good night, Beth."

They walked outside, and Kittery stepped alone from the porch into the soft, cotton-like dust. He said, "Thanks," knowing he could never thank these two enough for the way they had taken him into their home. Then he walked off into the night.

When Beth heard the soft hoofbeats of the stallion on their lane, she and Luke said "Goodbye," into the darkness and listened to the fading sound.

The night was a thick, inky black as Kittery rode into town. Three lights glowed along the street when he reined the black into it: the cantina's, Greene's and one in the jail. Silence reigned, even the usual barking of dogs absent. The black plodded past the dark, glaring eyes of the other windows, his hooves making clopping sounds on the hard-packed street. A deep, drowsy sleep blanketed Castor, a sleep Kittery was careful not to disturb.

Ignoring the night's rest the hotel promised, he steered Satan past and to the tie rail in front of the cantina. Dismounting and tying the reins, he stepped into the dim-lit building. One glance informed him that no one, not even Cardona, was in the room, but he moved up to the bar.

The Mexican proprietor appeared from a door behind the bar and stepped forward. "Señor Kittery."

"One small mug of beer, Mr. Cardona."

Cardona complied, placing a clay mug before him.

"Tell me, when did my friend Cotton go to bed?"

"He was gone for a ver' long time. He's no 'sleep. He came back jus' before you."

Curious, Kittery picked up his mug and went to tap on the door. It swung open. The cowboy flashed him his friendly smile. "Yo, Tap. Somethin' on yer mind?"

"What're you doin' out cattin' this time o' night? Got a girl somewhere you didn't tell me about?" Kittery asked jokingly.

"No, no," Baine uttered, his eyes flickering away. "Just out exercisin' the hoss."

"Oh. Well, I was just makin' sure you remembered that ride tomorrow."

"You bet. I'll be right here. Drop by when yer ready."

"Okay. See you here tomorrow—early." Kittery backed away so Baine could shut the door.

As he returned to the bar, he was thinking about his friend's reply to his query. *Exercising his horse?* That horse had looked plenty *exercised* when he rode in. He could have used a day of rest, and that was something Baine, as a horseman, knew as well as Kittery. Would Baine lie to him about it? And if so, why? Puzzling over this, he veered away from the bar and took a seat at one of the corner tables.

Cardona gave him an impatient smile from across the room and glanced overtly at the clock on the wall. It said ten minutes past midnight. Catching on, Kittery tipped back the rest of his beer and stepped into the tranquil night.

The first thing he smelled was tobacco smoke. As he heard the inner doors shut behind him and the bar drop into place, he glanced left and right. For an instant, he thought he saw the glow of a cigarette tip at the far end of the hotel's porch, but then only murky shadow remained.

His brow furrowing, he unholstered the Remington and stepped to the black, moving to the opposite side from the hotel. Except for handfuls of stars dashed loose across the sky, only the cantina's light remained. Then it, too, went out. Kittery could hear nothing, could see no movement.

He thought about taking Satan to the stable—he would have to before the night was over. But instinct told him to take the black with him. He hated to use the animal as a shield. But although it seemed pretty selfish, a bullet that would kill a man might only injure the horse. And as happy as he was with the horse, he could replace the dead stallion a lot easier than he could his dead self.

Easing the reins from the rail, he again moved around the horse and started up the street, his body shielded from view of anyone in the vicinity of the hotel. Then, from inside the hotel, the blast of a shot split the night. Kittery moved as swiftly as that sound, giving the black a hard slap on the rump. With the Remington in hand, he charged for the porch.

Chapter Six
Shots in the Night

Inside the dark hotel, Marshal Josiah Raines eased back the hammer of his Smith and Wesson American. The little metallic clicks he normally found a comfort sounded at that moment like a mortar's roar.

For the second time came the muffled cry from below, as of someone trying to scream through a hard-clamped hand. The marshal found absolute silence impossible; the stairs whined at his every step. Nevertheless, he managed to work halfway down the staircase unnoticed. He paused here, leaning an ear toward the direction from which the cries had come.

And then, with sudden harshness, a door slammed back against the downstairs wall, and in its wake came a terrified scream. The pound of a single gunshot followed, and in the shot's fleeting orange flash Raines glimpsed the face of the gunman.

He had no chance to fire, no chance to move, for someone else was down there—probably the Mexican woman or her daughter—and the chance of hitting an innocent was too great. He had to wait.

Tappan Kittery knew by the shot that danger lurked inside the hotel. But the danger that overtook him was unforeseen. The snap of a small caliber pistol firecrackered to his right, and then came a second. The bullets winged past, narrowly missing him. He turned, now hearing footsteps race up behind. The man in front with the pistol moved like a shadow, reaching Kittery in an instant. He swung at Kittery's head with the pistol, grazing him before Kittery sliced upward with the Remington. He felt its hammer catch the man somewhere in the face and send him backward.

Kittery whirled toward the man behind him in time to see his hand slashing downward, and he felt the sharp pain of a blow to his left shoulder. He lashed out hard and fast with the Remington again. It caught the assailant stepping back, and Kittery felt metal strike bone and bone give way. Then there were others close behind him, three or four of them, and he felt something heavy strike him across the back of the skull. His senses seemed to somersault as he wheeled about, swinging with the still uncocked Remington. The blow took a man in the side of the head, and he heard him hit the porch with a soft whimper. He was fighting now with the instincts of a killer. There was no place here for Jake's pugilism.

It was then he realized his left arm was soaked with blood. The first man who struck him must have wielded a knife. Now others swarmed about him. They swung again and again, fists and clubs battering his back and arms. Then they broke away, moving to the front door in a wave. He moved after them, dazed. Three of them crashed inside the hotel as one, and he stumbled right behind. He felt a blow to the side of his head, and then he was falling . . .

Along the street, lamps flickered to life, throwing a sickly yellow glow across the wheel-furrowed dust. But in the alleys that no windows faced, the shadows seemed to deepen.

Miles Tarandon was Luke Vancouver's deputy, and tonight he worked night shift. He lay half-asleep on his cot as the commotion across the street began. Jerking awake, he swung his bare feet to the cold floor, trying to orient himself. He shook his head and reached out to tug on his boots. He could hear the faint scuffle as he stumbled to the gun rack and drew out a Remington-Whitmore twelve-gauge shotgun and jammed in two shells, grabbing a handful more for his pocket. He slipped out the back door and moved up through the alley.

Across the street, no movement. That skirmish was finished, yet he waited. Had it merely been a scuffle between miners? He thought he had heard gunshots, but he'd been asleep and couldn't trust his ears. If he had to bet money, he would have guessed it had something to do with Gustavo Ferrar, for he had heard about the fight at the hotel. After a close scrutiny, he made out the form of a man pressed against the hotel wall in the shadows between buildings.

He strode down the front of the buildings on the west side of the street. When out of view of the hotel's alleyway, he crossed to the other side. Running along the rear walls, he rounded the back of the cantina and reached the opening of the alley next to the hotel. There, the man crouched in the shadows. He could make out no detail, but assumed it to be one of Ferrar's friends—or Ferrar himself.

Miles Tarandon issued the hushed command, "Drop your weapons, greaser, or you'll be whistlin' an ode to Santa Anna through your chest."

Startled by the sudden voice, the man stood up slowly. "I don' have no weapon," he said.

"Then back up towards me. Hands high!"

The Mexican complied, and three feet away the deputy struck out with the butt of the shotgun. He stepped back as the gunman dropped at his feet. Tarandon looked down at the man's chest to check for signs of breathing. He saw none in the darkness and shrugged. Whether the man was dead or alive would make a difference in his report to Vancouver; otherwise, it was inconsequential. He had killed a man before; it was the end of the man's world, not *his*.

Tarandon confirmed the man as one of Ferrar's cronies, then he stepped over him, moving shadow-like along the alley.

At the street, he saw people watching from the far porch. The first one he recognized was James Price. The bearded man was a territorial hangman from Tucson who ran a gun shop out of his home on the south side of Castor. An expert marksman, he carried a rifle now and appeared ready to enter the fray.

Tarandon managed to catch Price's attention, and by hand signals he warned him to watch for gunmen on the opposite side of the hotel. Price nodded and started across the street. Midway, he stopped and jerked his Winchester rifle up chest-high.

James Danning Price was a straightforward man. An educated gentleman, in grammar as well as in war, he had studied four years at Princeton, then served as a sharpshooter during the Civil War. Straight as a gun barrel, with a dark, stern eye, he was a citizen not only respected, but feared.

"Drop your weapons, gentlemen, and raise your hands high. *Solten sus pistolas. Manos arriba*," he repeated his commands in Spanish. "I'd just as soon shoot you as look at you."

Five seconds of dead silence followed his words, but James Price would give no second warnings. The big rifle leveled and boomed once, spouting a foot of flame. The sound of retreating footsteps came from the alley as Price jacked in another shell and fired once more. A startled yelp ensued, and then the sound of a body sliding in the gravel. Once more, silence.

Inside the hotel, Gustav Ferrar's mind raced. He had little doubt he had killed the girl, for he had fired at her chest point-blank. The little devil had bitten his hand! Yet what he had done to her and her mother before he killed her would have gotten him hung by any jury in the territory. Most of Castor's citizens were fond of little Emilita, and they would have gladly put a bullet in his head—even most of his own people.

Now he held the Mexican woman, Maria Greene. He had an arm crooked around her throat, and one thought raced through his drunken mind: he must take her as a hostage and escape out the back. He had friends in the Mexican quarter of town who could provide him with a horse and escort him into the mountains. Even then, perhaps he would be found and gunned down, but that was better than the hangman's noose that waited here. And, moreover, he would have ample time to make the woman sorry she had ever been born, sorry for turning on her own people.

But the Mexican's mind was fogged with drink. When the front door burst open for the second time, he flung the woman aside. A lantern borne by the man in the doorway cast a quivering glow across those close quarters, revealing Ferrar's form as he broke for the back door. Three successive shots shattered the night, and the last slug smashed Ferrar in the thick meat at the back of his thigh, sending him into a crazy lunge. He went down, skidding along the floor to crash against the rear wall. There he lay writhing, clutching his wound.

A moan tore from the Mexican's throat, and he cried out, "I don' fight no more! *Mi madre!*"

Then Marshal Raines stood over him, silhouetted against the lantern's light. The barrel of his forty-four aligned with the exact center of Ferrar's chest, and his stocking foot jammed down with crushing force on the hand that still clutched a battered Remington Beals revolver.

"Push away the pistol with the tips of your fingers—*pronto!*" Raines ordered.

The wounded man obeyed, whimpering.

"Get up." Though not believing he could do that, Raines took no unnecessary chances.

Ferrar looked up in shock, eyes widening and mouth hanging open. "My leg—it's broke!"

Ignoring the plea, Raines stooped and retrieved the discarded revolver.

"Get up," he growled. "Get up now." He drew back the hammer of the American for emphasis.

Ferrar grunted and whined as he made his best effort to push to his feet. He failed and dropped back. Nodding satisfaction, Raines released the hammer of the .44 and holstered it. He had started to turn away when a boot scraping on the floor behind him warned him to turn around.

He whirled, drawing and cocking the American in time to see Ferrar rising up to his knees with a knife held high. The .44 bucked in Raines's hand and spurted flame. Its two hundred grains of lead tore through the Mexican's chest, throwing him backward to the floor. He rolled over, and on reflex he started to come up again. Raines's second bullet took him in the head. Ferrar shuddered on the floor and was still.

Several citizens scuttled into the lobby beside the man with the lantern. James Price and Miles Tarandon shouldered through them, their eyes flashing about the dim-lit room. Raines backed away from Ferrar. He took in a quick breath as he saw Kittery stretched out near the door. With a curse, he rushed over and knelt by his side. The big man's breaths came ragged, but at first glance there didn't appear to be any serious damage. A shoulder wound and red marks that would likely turn later into bruises were the extent of it, unless there was something hidden.

Raines stood and turned to the others. The little girl lay where she had fallen after Ferrar's shot had penetrated her chest. Maria Greene sat in shock. By her face, Raines knew the girl was dead, but to be sure he bent and checked her pulse, then nodded grimly and stood up.

"You"—Raines pointed to a bystander—"go get a doctor, if there is one. Pick someone to go with you and fetch some blankets, too. And tell the doctor we have a knife wound and a woman who's been beat up pretty bad." The man turned and ran out.

Raines glanced at Deputy Miles Tarandon. "Deputy, doesn't this woman have a husband?"

"She does. Thaddeus Greene."

Raines was surprised. He remembered Thaddeus Greene as the obese drunkard who had run into Kittery on the cantina's porch. Dismissing his surprise, he said, "Why don't you clear these people out of here. I feel closed in."

Tarandon dipped his head in reply and cleared the room. When he was done, Thaddeus Greene had appeared and stood alone near the door. He walked across the floor, face white.

"Sorry, Mr. Greene," Raines spoke softly as he walked past him. Greene didn't even glance over at the sound of his voice. He moved close to the body of his daughter, staring down as his hands began to shake. His wife had been facing to the side all this while, but she turned when her husband moved near and whispered her name. Raines saw for the first time that the front of her dress was slashed and torn almost completely off. She held the tattered material in place with her arms. A long, bloody gash ran across the top of her chest, and the blood trickled down over her folded arms.

Greene came out of his trance, and together he and Raines moved Maria to her room. But Greene didn't stay to comfort her. Once they had her sitting on the bed, he stalked from the room, eyes livid. He walked over and stood glaring down at the dead Mexican, where he lay by the rear wall.

"Is that who did this?" Greene's cheeks had become almost sanguineous, and his jaw muscles pulsed.

"That's him," said Raines. "He's taken care of."

Greene whirled, turning murderous eyes on the marshal. "In the old days, I did some trading with Comanches and Apaches. There are ways to take revenge on the dead. If I wanted my revenge, the devil himself couldn't stop me."

With that, the merchant turned and stalked back to the dark room where his wife waited. He threw the door shut behind him.

Raines turned to see that someone had come with blankets and covered Kittery, and also the girl's body. The man still held one blanket, and he looked a question from Raines to Ferrar.

"Don't soil it. It's too late for him to benefit." Raines turned and spoke to Tarandon. "What's keeping the doctor? You do have one?"

"He lives on the ridge, Marshal. The sheriff will likely be with 'im when he gets here."

Raines nodded, then whirled around at the sound of footsteps on the porch. In came a fifty-year-old man whose jet-black hair taunted the effects of time. He wore a mustache and a rounded goatee, and beneath his sheltering arm was a nineteen- or twenty-year-old woman with his same black hair hanging loose well past her shoulders. Her eyes were large and frightened. She wore a russet-colored skirt and a much too-large white shirt with rolled-up sleeves.

"I'm Randall McBride," the man introduced himself. "A runner came from the doctor. I'd like to be of service if I could. This is my daughter, Tania"—he pronounced it *Tawn-ya*— "She's helped the doctor nurse some of his patients in the past."

"That's considerate of you, Mr. McBride," Raines said. "Thank you. Mrs. Greene's been beat up pretty bad, and we could use your help with her. But her husband's in the room with her now, so it'd be best to wait here."

Tania nodded agreement, her dark eyes lighting on the bedroom door. "I hope she's all right," she said, her eyes darting up to her father's face.

Luke Vancouver came striding in with Beth at his side. Seeing Tappan Kittery on the floor, Beth rushed to his side with a concerned cry. She put her hand on his chest to feel for breathing, then turned to Raines. "What's the matter with him, Marshal? Was he shot?"

"No, the blood's from a knife wound in the shoulder. I'm hoping he's just bruised a bit. He's a tough bird, Mrs. Vancouver."

A man of medium height and build with shaggy brown hair and a full, pointed beard had come in behind Luke and Beth. "I'm Doctor Hale," he said to Raines, opening his black satchel. "Who needs the most immediate attention?"

"The one on the floor there has a knife wound in the left shoulder, and he's been hit in the head a few times. But I think he'll be all right. I'm more worried about Mrs. Greene. She's awake, but she's got a gash across her chest that doesn't want to quit bleeding. Looks like it was done with a dull knife. She's been beaten up, too."

Doctor Hale turned to Beth. "Beth, I'm glad you came. I'll see that you're paid if you'll take a look at Mrs. Greene for me." He turned to the black-haired woman. "And, Tania? Can you help me with the man on floor, here? We'll need to stop this bleeding right away and sew him up."

As the doctor and his assistants went to work, Luke Vancouver moved toward Ferrar and glanced over his lifeless body. He walked to the small form beneath the blanket. "Emilita?" He looked at Deputy Tarandon, who nodded. "Ferrar wasn't alone, I gather from the talk in the street. Who else?"

Tarandon's eyes darted to James Price. "Well, Luke, just some more of his friends. Two of 'em are dead, and I give one a nasty headache. They all had a chance to give up."

"Those are the facts, Sheriff," Price nodded in agreement. "I gave the pair an ultimatum, and they chose to run. I was forced to shoot."

Without being asked, Tarandon named the two dead men, raising a finger as he spoke each name. He also gave the name of the man he'd clubbed unconscious. Vancouver nodded and looked away, flexing his jaw muscles. "I hate to see this in my town."

When Doctor Hale had looked Kittery over, he went to Ferrar. After a brief inspection, he looked up at Raines. "That problem was solved conveniently, I see."

Raines grunted. "I guess it was. He drew a knife on my back."

Hale smiled. "Mexicans have an affinity for knives, they say. My experience bears it out."

"What about Tappan—the other patient?" Raines asked.

"Well, he'll need rest after the young lady and I stitch him up. And I would suggest we place him in one of these downstairs rooms. Unless of course someone wants to carry him upstairs, or maybe over to my office." He smiled wryly and looked down at the fallen man, running his eyes over his two hundred fifty-pound frame.

Kittery slept downstairs that night.

When Hale had finished stitching up the big man's shoulder, he went to Maria Greene. There were two beds in her room, side by side. They placed Kittery in the one beside the woman's, for ease of caring for them both.

After a detailed examination, Hale came out into the main lobby, where Vancouver, Raines and the others waited. His expression was a little lighter now. "The baby will live, thanks to God. What the woman is going to need is rest and someone to stay with her. Beth and Tania both said they would stay, so Luke and Randall, it looks like the two of you will have lonely houses tonight. What worries me about Mrs. Greene is—" He tapped the side of his head with a frown. "I suppose you've already guessed what the Mexican did to her, and it was all of that and more."

The others nodded grimly, some of them looking over at the dead Mexican.

"Well, you can't guess it all," Doctor Hale said, "and you really shouldn't want to know. The only reason I tell you at all is so you won't have any regrets about killing Ferrar. That gash on her chest was made by the front sight of his pistol, by the way. Not a knife."

When the town had turned quiet again, and only the occasional coyote chorus trilled in the night, Deputy Miles Tarandon stood at the corner of the hotel's porch and smoked a cigar. He shrugged into a black frock coat that brushed the tops of his calves, putting his hands deep in the pockets to keep them warm.

The time had come when the Castor Vigilantes would be called upon to act on a crime within their own community. They had tried so hard to preserve tranquility here, to make this a sanctuary. But now they had failed. There were white and Mexican, black and Indian bands raiding back and forth across the international border. They stole stock in one country, sold it to crooked buyers in the other, then turned around and repeated the act on the other side. They killed anyone they wished, then hid in the rugged mountain ranges that spanned the border country like gigantic, inexpert stitches. And the vigilantes knew about these men. They were the reason for the organization of the vigilantes in the first place. The lawlessness had far outreached the capacity of the law. But the vigilantes, at least in Tarandon's opinion, hadn't expected to deal with something like this in their own town. This was supposed to be the one place where they could find peace. And now Ferrar and his bunch had violated that peace. Some of them had paid, and the rest would follow.

Going to the sheriff's office, Tarandon picked up a lantern and lit it, keeping the wick down low. He returned across the street to the rear of the hotel. After searching either side of the building, he found what he was after. Crouched close to the ground, he scrutinized the dust, then looked off toward the Mexican quarter of town. That was where he must go.

As plain as day in the thick dust were two sets of tracks, two men who had escaped the bloody outcome of the hotel skirmish. One set had been made by a man stumbling along, leaning against the other as if being held upright. All along the trail were splotches of blood. This proved Tappan Kittery's fight had not been futile.

Tarandon followed the two sets of tracks until they led him to the hovel of Raoul Mendoza and Juan Torres. They went inside, and there was a streak of blood on the blanket that shielded the doorway. Tarandon backed away from the home, blowing out his lantern. He backed to the next house, then turned around it and got out of sight of Mendoza and Torres's place.

As he walked back briskly to the office, he thought of Mendoza and Torres—two more men who must die before this affair had ended. But Luke Vancouver would never know who had killed those two . . .

Chapter Seven
The Vigilantes

Tappan Kittery slept well through most of the night. But early in the dawn his mind whirled awake. A fever burned deep in the muscles of his shoulder, and his head throbbed mercilessly. He lay there too groggy to move, wanting to go to sleep but unable to because of the pain.

After a few minutes of lying still, he was able to look to left and right without feeling like his head was going to explode. When his vision began to clear, and the bursting lights in his eyes died to a dull gray, he made out the form of a chair toward the foot of the bed. He closed his eyes to rest them. When he looked again, the shape of a person asleep in the chair began to register on his brain.

Curious, Kittery eased onto his right side, forcing himself not to think of the pain that jogged through his head. There was a lamp burning across the room, its wick down as far as it could go without extinguishing. By the dull orange gleam through its tarnished chimney, he saw another figure supine on the floor beside the bed. His eyes were working well enough now to make out that one as Beth Vancouver, peaceful in sleep. Kittery smiled and closed his eyes for a moment. What a pleasant surprise to wake up and see someone who cared that he was hurt—especially someone like Beth Vancouver.

He lifted his head to try to make out the other person. He could make out dim features, enough to tell it was a female, but he couldn't even see if her eyes were shut. It almost looked like she was staring right through him, but . . .

"Hello." The word came without warning, but the gentle voice didn't startle Kittery. For a moment, he fancied he had imagined it.

"Hello," he finally answered, feeling foolish for waiting so long. "Where am I?"

"In Maria Greene's room."

He pondered the response a moment, screwing his eyes up to clear his vision a little more. "Maria Greene's room . . . That's in the hotel?"

"Oh, yes. I'm sorry. Maria Greene is the Mexican lady who runs the hotel. I assumed you knew." There was a prolonged silence before she leaned slightly forward in her chair and spoke again. "I'm Tania McBride."

Even in his groggy state, Kittery found himself enjoying the appealing voice from the dark. "Tappan Kittery," he replied.

The woman stood up from her chair and walked to the side of the bed. She held out her hand. Easing his arm from the blankets, he squeezed the hand, warmed by the sensation of its tenderness through the sun-tanned roughness of his own.

"They tell me you're a captain."

Kittery chuckled. "Well, I was, in the war. Who's 'they'?"

"Oh, everyone. You're the talk of the town since yesterday afternoon." A pause. "How does your shoulder feel?"

"Fine, compared to my head. Just a little throb to let me know it's there. What happened last night?"

"Some of the Mexicans who work at the warehouse got drunk and attacked Mrs. Greene and her daughter, Emilita. Well, you met them—it was the man you…fought. Emilita was killed."

"I guess I didn't meet her."

"Oh. You might have seen her at the dry goods store. Blond hair but very dark eyes."

"You mean . . . Ahh . . . Not the little girl that swept the boardwalks!" He remembered the girl smiling at him on his way to the cantina the day before, remembered the hope and the life and the dreams in her young girl eyes.

"Yes. Emilita Greene," said Tania McBride. "Thaddeus Greene was her father."

In spite of his pain, Kittery tried to sit up in shocked surprise. A wave of dizziness reined him back down to the mattress, where he pondered the woman's words. Emilita, Maria Greene's daughter . . . Thaddeus Greene was her father, so . . . In spite of the common last name, he hadn't put Thaddeus and Maria Greene together. Even now it was hard to accept the idea. Maria Greene was a pretty woman, and a lady, from what he'd seen. She and the fat, drunken Greene didn't seem a likely match. And that beautiful little girl! How could she have come from a man like Thaddeus Greene?

Tania McBride finished relating the night's events, and Kittery lay there in a daze. The queasy feeling in the pit of his stomach grew almost unbearable. Had he caused that little girl's death? And the Mexicans. Had there been some who escaped? He hoped he hadn't been taken by a mere three of them. It had seemed like an army.

They were speaking quietly, but Beth Vancouver stirred awake and sat up. "Getting acquainted?"

"Oh, I'm so sorry, Beth! We didn't mean to wake you," said Tania.

"Tania was telling me about last night."

Beth shook her head. "That was a terrible thing. Just awful."

Kittery's jaw hardened. "Guess it was my fault. If I hadn't been so rough on Ferrar, maybe . . ." He lowered his eyes and cursed himself silently.

"Don't be hard on yourself, Tappan." Beth reached out and squeezed his arm. "With that man, something had to happen sooner or later."

Of course Beth would say that, but it didn't make Kittery feel any better about the fact that he was the man who caused it to be sooner instead of later. That innocent little girl! The pain that racked him was a familiar pain. It was a pain that could never go away.

Beth stood up and walked to the bedroom door, opening it a crack to look out. "The sun will be up before too long," she said when she turned back to them. "I'd better go fix something for Luke for breakfast before he goes to work—he's back on the day shift. Is there anything I can get for you, Tappan?"

"No, Beth. Thanks," Kittery said with a smile. "You go along."

"I'll go with you," Tania McBride said, rising from her chair and laying the blanket that had been over her legs across one of its arms.

They made one last check on Maria Greene and found her sleeping soundly. Then they said goodbye to Kittery and stepped out, easing the door shut behind them.

In his dreary little room in the back of the cantina, Cotton Baine put on his faded red shirt and tugged on knee-high brown boots with jingling spurs still attached to them. His hat was already perched aslant over tousled hair, and now he reached for his gun belt and buckled it on like he'd done a couple thousand times. He shrugged into his vest and patted the pocket to make sure his makings were there.

Shaking the sleep from his head, he stepped into the main room, met by a wave of morning aromas—fresh coffee, bacon, and frying tortillas and eggs. In the background was that rested smell a room of entertainment takes on after a few hours of cool darkness when the tobacco smoke has settled into the rafters, the spilled liquor soaked into the floor, and the smell of sweaty men has taken refuge in all the dark corners.

Ignoring his hunger, Baine moved through the open doors, rolling his first smoke of the day—the best of the day, when his mouth had forgotten how good it tasted. The street was quiet. A wagon waited at the warehouse, tarped and tied down, but its team was still sleeping at the livery stable. Three vigilante horses stood near the sheriff's office. He hoped their owners had a good explanation if someone asked them what urgent business had brought them to town so early.

The cool dust of the street cushioned his steps as he crossed and stepped into the sheriff's office. Men crowded the little room. James Price, the hangman; Miles Tarandon, the deputy; Rand McBride, the banker; Thaddeus Greene and four others had found any available space to seat themselves or leaned against the walls. They glanced up cautiously as the door opened, but relaxed when they recognized Baine. Each of them but Thaddeus Greene nodded their acknowledgment.

"Good morning, Baine," James Price said with a succinct nod. Then he turned back to one man of forty or so with a thick head of white hair and a drooping gray mustache. The man was Zeffaniah Perry, who owned the Double F horse ranch, northwest of Castor. "Go on, Perry. You were saying?"

"I just think things haven't gotten out of hand," Perry said, shifting his eyes around the room before settling them on Price. "We started out with the idea of ridding the border country of horse and cattle thieves, but I think somethin' like this, right here in town . . . well, there shouldn't be any question whether we do something or not."

Price nodded, trying to gauge the feeling in the room. By the looks Cotton Baine could see, everyone seemed to agree. "I think we all know why we're here, Perry. I understand your being upset, but listen to me for a moment. The border country you mention is the reason this country needs vigilantes. Luke Vancouver and Miles, here, have control of this town. We aren't here to take over their positions, only to look after problems that are too far-reaching for them to handle."

Thaddeus Greene was silent, but Baine could see the hate burning in his bloodshot eyes as he stood at the back of the room staring at James Price.

The town blacksmith, "Bison" Sabala, studied Price for a moment from under shaggy blond brows. "The marshal killed the one who started that ruckus, if I'm hearing correct. And you killed two more of them, Price. Do we need to go any farther?"

Price shrugged. "Do we? You people tell me. I think it was handled in a pretty reasonable fashion, considering the fact that we had no time to settle on a plan."

"But," Perry countered, "if those Mexicans had been taken care of before, that little girl wouldn't have had to die."

Perry's and Price's eyes met, and Baine saw the disdain in the hangman's face. He took a moment to form his words. "They hadn't done anything before that made them need *taking care of.* Come now, Perry. We can't go about punishing men for something we think they might someday do! When we formed this committee, it was with the firm understanding by all of us—and I know, because I wrote the oath—that we were here to support the sworn officers of the law if they got into something they couldn't handle. We're working *with* the law, not against it. Had we 'taken care of' Ferrar and the others before, as you put it, what would set us apart from the lawless element we are trying to contain?"

Zeff Perry went silent, as well as the rest of the room. Cotton Baine nodded his respect for the tough, bearded hangman and looked around the room to see how feelings ran. For the most part, the people here were pretty clear thinkers. He didn't doubt they would make the most prudent decision.

Baine thought back to the day in February when he had signed the oath that joined him to this committee of vigilance. Word had arrived in town that a little cavalcade of freight wagons had been sacked on the Camino Real, and the drivers all killed. After taking everything they could carry, the raiders burned the wagons to the ground. They also burned the wife of a driver and her infant son, who had hidden in one of the wagons at the order of the woman's husband. The raiders fled into Mexico, where the law couldn't touch them.

Baine had always been a law-abiding man. But he had realized that day that Arizona and New Mexico Territories were both well beyond what any organized law could handle, even if the army became involved—which it didn't seem to want to do. He had never liked the thought of taking the law into his hands, but it was time someone did. He had signed James Price's oath with no guilt, and he had felt no remorse since.

"I'll agree with you on that point," said one wiry, black-haired man who leaned against the back bookshelf and had stood silently smoking a cigar up until now. "We can't read somebody's mind and know what they'll do. If we could, let's face it: this would be a pretty crime-free country. But I'm sick of sendin' men to jail to have them sit and live off the government until we get around to givin' 'em a trial. I'm of the firm opinion if we killed a few the word would get around. Even if Tarandon and the sheriff can put these two Mexicans in jail, it's a long way to justice. We need to make a statement here. This kind of thing won't be tolerated. People aren't stupid. They know we exist. Let's let 'em see we mean business."

The man was Adam Beck. At thirty-six, he had gone from hunting bison to hunting men, after his Mexican wife had been butchered by an Indian trader named Monk St. Clair, who escaped the clutches of the law until Beck gave up on legal justice and took up the trail. It was rumored that Beck could put a hole through a tomato at two hundred yards—which put him on a par with James Price.

The Castor Vigilantes had their own government, their own code, and their own regulations. They held elections, and James Price had taken the title of captain with ease. Miles Tarandon was appointed his first lieutenant. Castor's committee of vigilance was anything but a mob. They followed their own rules strictly, and in an orderly fashion. And when they put a man to death it was only after a unanimous vote, except in extenuating circumstances like the night before. There were men such as Zeff Perry and Thaddeus Greene who would go against that, but the others had always been able to hold them in check.

James Price pondered Adam Beck's statement for a long time before making a reply. "So what do we do, Beck? How do we make a move without someone finding out who we are? We're not out on the prairie here. And those Mexicans aren't likely to leave town."

Thaddeus Greene cursed Price. "You better open your eyes! We can run this country if we work it right. We can make sure our power is unquestioned, whether they know about us or not. Those greasers need killing. It's as simple as that. They killed my daughter last night, and they'll strike again, sure as you're born. I say we prevent it!" He was livid by the time he finished his speech. Baine, on the point of being embarrassed for him, shifted his attention to Price.

"Strike again! Don't be so dramatic, Greene," Price scoffed. "Get a grip on yourself. This was an isolated occurrence, and three men have died over it. I'm not saying at least Ferrar didn't get what was coming to him. He did, and I'm glad he did. I'm even more glad it was done by a marshal, so there won't be any questions. All we have to worry about is the two I had to kill, and I believe that will be cut and dried. But not all Mexicans are demons, Greene. You should be the first to admit that."

"You low-life," said Greene with a snarl. "Leave my wife out of this. We all know those Mexicans at the warehouse, and we know they'll do something again."

"Well, I don't think, *Mister* Greene, that we are in place here to suppose we can read the future, see who will break the law next, and punish them in advance. That would be nice, but I, at least, am not a prophet. And I'm reasonably certain that neither are you."

Price didn't raise his voice. He never needed to. There were nods of approval about the room now, even from Zeff Perry, who stood.

"I agree with you, Price. We can't say who may one day deserve killin'. But it seems pretty reasonable to me to say that them two greasers Tarandon followed from the fight intended to commit murder, whether they got the job done or not. They didn't go over there thinkin' it was a fish fry, that's for sure. Yeah, sure the sheriff and Tarandon can handle them, but like Beck said, why should we all have to pay for it, as honest citizens?"

"It's either pay for it from your purses or from your conscience," Price said blandly.

"Well, my conscience lets me sleep good at night," Perry countered. "We know them two Mexicans that got away were in on the murder of Greene's kid. I say they die."

"I understand the way you feel, Perry. Don't think I don't. But we happen to be in a precarious position here. We could jeopardize our entire future. It's not like we're an impressive force as of yet." He glanced around the room with a wry smile. "But I'll leave it up to all of you. I would suggest a vote."

Randall McBride, the banker, spoke for the first time. "That sounds fair to me, Price. I can't imagine there would be any objection." He looked over the faces, searching for argument.

"Then we had better vote," said Price. "The sheriff will be in here any time. So . . . from what Miles learned, there were just the two men who escaped: they went to the house of Juan Torres and Raoul Mendoza. One of them was injured for certain, and perhaps both—we don't know. If we find that one of them has no marks on him, it may be that he is innocent. It could, in fact, be that both Torres and Mendoza are innocent. For all we know the tracks belonged to friends of theirs. My heartfelt suggestion would be that we only act upon proof."

"We can prove it by their tracks," Adam Beck brought up. He was an accomplished tracker.

Price shrugged. "Perhaps we could. If that's true, so much the better. But as I was saying, I hope you will all agree that we need proof of Torres and Mendoza's involvement before any punishment is rendered."

Greene spat on the floor. "Everybody in this room knows Torres and Mendoza were friends of Ferrar. That makes them guilty."

Price ignored Greene, and the merchant's face grew redder.

"Who is for acting only upon unequivocal proof?" Price asked.

Everyone but Greene raised his hand. The fat man scanned the crowd with glittering eyes full of hate but remained wordless. Price didn't even ask for his vote.

"You're all a bunch of cowards," Greene said, glaring about the room. His hands shook with rage. "If you don't kill every friend of Ferrar's, you aren't worth spit. Everybody here knows how close they stick together. If they haven't done anything yet, guaranteed it won't be long."

Price looked at Greene and folded his arms. When the fat man finished talking, Price said, "I'm your captain, Greene. Everyone here voted on that. I expect you to do what I tell you until I'm no longer your captain. Is that understood?"

"Then you're no longer my captain!" Thaddeus Greene spat the words out like poison darts. He took one last sweeping look at the faces in the room, then turned and stalked out, slamming the door.

After a long moment of silence, Price shrugged. "He was being a nuisance anyway."

"He was that," Rand McBride agreed. "But if he's planning on leaving the committee, I hope his liquor doesn't loosen his tongue too much. He could damage us seriously."

Price nodded. "He'll keep quiet. We'll remind him constantly—if he breaks the oath, he dies. Every one of us will be held to that. It wasn't put into the document as a joke.

"Now," he went on, "I'm going to take it upon myself to complete our business, if that looks feasible. The men we're looking for may find justice in the next hour or so. In case that should happen, I would like all of you on the street, creating as much of a disturbance as you can to keep Vancouver from sniffing me out. Until then, keep a low profile around town. Even after it's over, try to lay low, at least for a week or two, until I can gauge how much Vancouver has wind of. He may not suspect any large movement of vigilantes yet, but he suspects something. Mark my words. So sift out of here. And I would prefer that you not gather in groups of more than three at a time without good reason. Also . . . remember your oath."

When it was understood the meeting was over, everyone stood to leave.

But Price spoke Baine's name and asked him to wait. Then the bearded hangman turned to catch Adam Beck by the arm. "Just a moment, Beck. I'll need your assistance. I'm going to see to our little problem this morning—early. I'd like you to go up to my shop and keep an eye out for our man at the warehouse—or men, whichever it may be. Keep the closed sign on the door so you won't be bothered. And use my telescope to identify them, if you need. When you know who it is, come for me at the cantina."

After Beck left, Price turned to Baine. Price didn't ever smile much, but in his eyes was about the closest thing he ever showed that resembled gladness. Cotton Baine didn't know why, but he knew James Price had taken a liking to him from the beginning. He counted himself among the lucky few—if lucky was the word for it.

"Let's take a stroll and talk," Price suggested. The two of them nodded goodbye to Miles Tarandon and stepped out onto the porch, standing under the sheriff's office awning with new sunlight spilling over them.

"How is your friend, the captain?" Price asked.

"Don't know yet. I haven't looked in on him. I didn't even get a chance to look at him last night, with the marshal shovin' people out of the hotel the way he was."

"Well, I want you to work on him, Cotton. Stay close to him. I've heard a bit about him, up in Tucson. And I've seen a bit more here. He's a man of action, Cotton. The kind of man I could use. Don't let up, all right? This means a lot to me."

"I'll do my best," said Baine with a shrug. "But I'll tell you now, Tappan's always been his own man. I don't know how he'll feel about vigilantes. We'll have to see."

Price dipped his head and seemed to catch himself almost smiling. "Your best is good enough for me, Cotton. I have faith he'll be with us in no time at all."

Baine wasn't so confident, but he grinned anyway. "Thanks for the trust. I'll try not to let you down."

"It's time to make your move."

The manhunter, Adam Beck, leaned over the scar-topped table in the cantina, his eyes flickering from Baine to James Price. "The only man I saw who looked to've been in a fight was Juan Torres," Beck went on. "I didn't see Mendoza come in at all."

"Really?" Price glanced over at Baine, then back to Beck. "Are you certain it was Torres, then?"

"Oh, yeah," Beck said, raising his brows. "Absolutely, unless the captain got both of 'em. The side of Torres's face looks like he was kicked by a bull."

Price looked down and turned his hand over on the table, seeming to study his palm for an extended time as he nodded his head. At last, he clenched his fist. "I would imagine the captain took a piece out of both of them, then. Probably a bigger piece out of Mendoza than Torres—which is why Mendoza didn't come in. That means we'll have work to do later."

Standing up, Price dropped a silver coin on the ash-littered table and snuffed out his cigar in a clay dish. Looking neither left nor right, he strolled from the room and listened to the doors clacking shut behind him.

Leaning against an awning post, he glanced over at the warehouse, putting the fingers of his right hand deep into his trouser pocket. It didn't take long to pick

Juan Torres out of the crowd of Mexicans. A large purple bruise spread across the left side of his face, from beneath the hair of his sideburn almost to the corner of his mouth. The outside corner of his eye was also swelled shut.

Juan Torres most definitely was not beautified by his encounter with Captain Tappan Kittery. But it was something he wouldn't have long to rue.

Chapter Eight
Swift Retribution

James Price arrived at his gunshop with his heart trying to rattle him loose inside. He couldn't shake the feeling that he was back in the war, that he was about to embark on some mission that might lead to the loss of his life. He had been a sharpshooter, and a good one. He had killed men—he didn't know how many. Too many. But in time it no longer mattered. That was part of why he accepted the position as territorial hangman. He was used to taking lives, and he wanted to prevent anyone who didn't already know the feeling from having to carry that load. Killing men changed a person, no matter what they tried to claim. Any optimism he had found in life had died with the first young Confederate soldier to fall beneath his Sharps rifle.

Price sat at his kitchen table, a little pine affair he'd crafted himself. He let his eyes travel the length of his gun cases, admiring every cold piece of mated steel and wood displayed there. Some of them worked, others were in various stages of repair. But he knew them all intimately. They were his babies—his life's joy. Yes, guns, not ropes. Hanging was his way to make a living, to legally rid the earth of some of its vermin. Guns were his release. His love.

Standing up, he walked to the stove. A fire was dying deep in its belly. He picked up the coffeepot with callused fingers and poured a cup to the brim, spilling none. It was two day-old coffee, and taking on the flavor of the porcelain-coated pot. That was good—he needed some iron in his guts.

Passing the table, he set down his cup and continued on to a backroom, where his eyes fell upon a bull-barreled rifle in the corner. Ballard's Number Seven "Creedmoor A-1" Long Range rifle—one of the most expensive rifles on God's green earth, and one of the deadliest known to man. He hefted the sixteen-pound bull-thrower and studied it grimly, then reached up with his left hand and withdrew two .45-100 caliber shells from a little wooden box.

He had ordered this rifle earlier in the year, and it had arrived by freight two months back. It was fitted with the octagonal bull barrel that wasn't standard, and with a vernier tang peep sight and spirit level wind gauge front sight that were. With checkered pistol grip and the most exquisite cut of English walnut, a forend tip of carved horn, and hand-engraved scroll work designs on the action, it was the fanciest gun he had ever picked to take a man's life.

So far he had put fifty bullets through the barrel, ten of them to hone it down to fly-killing accuracy, thirty-five through a whisky keg six hundred yards away—five through a water barrel at fourteen hundred. The factory claimed accuracy up to thirteen, but for anything as important as a man over five or six hundred yards away he would have needed to mount it with a telescopic sight. At six hundred yards he was already pushing the limit's of mortal eyesight.

The handloaded shells he sifted in his fingers, searching each one for flaws he knew didn't exist. He let the shells slide down into his vest pocket before returning to sit at the table, this time with his back to the gun racks. Now he could see out his window, but without getting closer or stepping out onto the little balcony he couldn't see down into the street. He could only see the rooftops of the warehouse and the hotel. His shop sat another fifty feet higher than the town.

Looking down at the big rifle that lay before him, Price sipped his coffee and grimaced at the heat of it against his tongue. *When will this cursed country be clean?* He looked at his heavy-veined hands. *When can I stop killing? And then what will I find to earn my keep?* He swallowed deeply of his cup and let the dark liquid burn all the way down his throat. Curse his conscience, but another man was about to die.

In Mario Cardona's cantina, Adam Beck closed his eyes and tried to doze. Cotton Baine had gone to his room, leaving the manhunter alone with his thoughts.

His eyes snapped open. He blinked to clear them and looked over at the whisky bottle Baine had left. It was four fifths full, and it had to stay that way. But it looked good. It called to him. Dang thing wasn't going to draw him in. But then, would one more shot kill him?

The sudden boom of a rifle slammed its harsh reverberations against the walls of the cantina, its flat finality and the silence that ensued relaying the end of the story. Beck jumped up and hurried from the room, tagged by Cotton Baine, who had charged from his sleeping quarters and almost run into him. Making a show of scanning the street, the two of them let their attention fall on the warehouse, where a group of men was gathering on the loading dock, looking down. Several of them were vigilantes.

Baine and Beck rushed across the street, and what they found there stopped them cold.

Juan Torres had been hit, all right. But he was not dead. The distance from where James Price would have had to fire to the warehouse dock had to be better than five hundred yards. Price was an expert rifleman, his weapons unparalleled for performance. No one was better at judging bullet drop, range, wind-drift and other factors that added up to a perfect shot. But there was always some factor of chance in a shot that far, even with a telescopic sight, which Price didn't like.

To Torres's poor chance he was alive. But he lay there screaming in terror and pain, clutching at his thighs while two of his friends tried to hold him down. His thrashing finally ebbed. He had passed out from the pain or the loss of blood, which spattered the porch and boot-worn steps.

Adam Beck studied the downed Mexican and swallowed hard, grimacing. To a man who had lived with death, it was easy to make quick judgments, and his prognosis for Torres was not favorable. The Mexican had a chance at living, because Dr. Adcock Hale was a fine surgeon. As long as the shock and blood loss and chance of infection were treated, the Mexican might survive. But he would not walk again. The slug had shattered the bone of his upper leg and lodged in his right knee. His chances of sporting two stubs before his medical treatment was over were good.

Tappan Kittery sat up in his bed. He looked over at the other bed and saw Maria Greene, her eyes wide. She, too, had heard whatever sound had brought him awake.

The door burst open, and Tania McBride stood there holding a tray.

"What happened?" Kittery asked.

"I'm not sure. I heard a shot, but I was already inside. I didn't go look."

"Let's go see," suggested Kittery.

Before Tania could raise her hand or open her mouth to protest his rising, Kittery had rolled out of his blankets and sat on the edge of the bed, clad only in his pants. A wave of dizziness swept over him, and his head began to pound as the blood shifted. Ignoring the feeling of sickness, he reached for his boots and tugged them on. He glanced around for his shirt, then looked up questioningly at the girl.

"Oh, I'm so sorry! Your shirt. We meant to get you another, but—we didn't think you'd be up so soon."

Frowning, Kittery stepped past Tania and out the door. He was embarrassed by his bare torso, but more concerned about the shot.

On the street, Kittery and the girl saw the crowd gathered down around the warehouse. A minute or more had passed since the shot, and Luke Vancouver had reached the scene. His voice rang out. "It's all over, folks. Go home or at least get off the street while we figure out who did this. If there's a sniper around I'd hate to see someone else get in his sights."

With the sheriff's tactful persuasion, the crowd dispersed to the perceived safety of the other buildings and continued to watch from there. Cotton Baine was among the crowd, and he left and came toward Kittery and Tania. When he drew close, Kittery drove a concerned glance at him. "What happened?"

Baine waved toward the warehouse. "The dangedest thing! Somebody shot one of the warehouse workers."

"But why?" asked Tania.

"Nobody knows. Don't even know who did it. I guess he didn't think his reasons were good enough to stand in the street and do it."

Kittery glanced past Baine to see a black-haired man with a mustache and goatee, dressed in a business suit, walking toward them. When he stopped before them, Tania introduced him as her father, Randall McBride, and Kittery shook his hand.

McBride looked him up and down. "There were people worried about you last night. I'm happy to see you're doing better. I see my daughter's got you out on the street half naked."

Kittery felt himself blush, and his eyes flickered down as he remembered being shirtless. "Yeah, I guess they stole my clothes," he said, chuckling but wanting to disappear. Hoping to draw attention away from himself, Kittery motioned down the street. "This Mexican, is he one of Ferrar's bunch, or what?"

Baine looked thoughtful. "Well, now that you bring it up, I think he is."

"Pretty coincidental," Kittery said, glancing from Baine to McBride. "Kind of hard to believe it was just coincidence. Who else was involved last night besides me?"

"Your friend the marshal, for one," said Baine.

Kittery nodded. "Yeah, Miss McBride told me that much."

"And there was James Price and Miles Tarandon," Baine added. "Price owns a gunsmithin' shop here in town, and Tarandon's a deppity. Price is also a hangman."

Looking at the few people left on the loading dock of the warehouse, one of whom was crouched beside the Mexican, with a black bag at his side, Kittery shook his head. "Makes a man wonder how a fellow can shoot somebody in broad daylight and nobody sees him do it."

That time Baine glanced away and cleared his throat. "Yeah . . . well, sometimes that's just how it goes. By the way, you runnin' around naked reminds me I got somethin' for you." He flashed his fun-loving grin. "Be right back."

When the puncher had gone, Rand McBride smiled and excused himself, too. "I'd better be getting back to work. Tania, I'd rather not have you on the street for a while."

"I'll be with the captain," Tania said with a nod. "Don't worry about me."

McBride's eyes turned on Kittery, studying his face for a moment and taking in his eyes with the critical glance of a caring father before he returned his glance to her. "Well, you have things you could be doing at home, too, don't you?"

"Maria Greene's in the room, too, Mr. McBride," said Kittery, a smile tugging at his lip corners. He was amused, yet glad to know Tania had a father who cared about her well being.

McBride cleared his throat. "Well, just the same. Tania, you won't be too long, I hope. I'm sure Mr. Kittery could use his sleep."

Tania looked at Kittery, and he laughed. "I've *been* sleeping. Now I'm awake."

McBride's face broke into a smile. "Well, all right. Don't mind me, Mr. Kittery. I guess I'm just used to having her all to myself. I'll expect to see you at home later then," he said to Tania. "And, Kittery" —he returned his eyes to the big man— "it was good to make your acquaintance. I'm sure we'll speak again."

With a slight bow at the waist, the banker stepped off the porch and strode across the street to the bank, going inside and leaving Kittery and the girl standing alone. "Your breakfast is getting cold," she said.

"I reckon it is. Join me for a while?"

Tania McBride smiled, casting a nervous glance toward the warehouse, where some of the men were lifting Torres onto a blanket. "I'd like that," she said.

Taking her elbow, Kittery turned her away from the bloody scene at street's end, and they stepped once more into the hotel and into Maria Greene's room. The Mexican woman had fallen back to sleep.

Tania McBride watched Kittery pick up the tray of food and set half of it aside for Maria Greene. Then he sat down on the bed and picked up a fork. Tania took a seat in the armchair across from him and watched him eat the meal of beans, tortillas and corn bread with honey.

She folded her hands in her lap and waited for the captain to finish. She tried to avert her eyes from him, to keep her attention elsewhere in the room. But somehow she couldn't stop looking at him. For one thing, it made her happy to see him eating with such gusto the meal she had made him. And then . . . she enjoyed watching him. He was . . . well, quite a picture.

She noted the big, brown hands, how they contrasted the whiteness of his muscular chest and arms. His shoulders were the widest she had ever seen, and they were dusted with a trace of the same dark, curly hair that matted his chest and belly, giving him an earthy appeal. The years he must have spent behind a plow or with an axe or a sledge hammer in his hands showed in his build. Yet she had seen other men who worked hard and who didn't look this way. Nature had had a lot to do with the way Captain Tappan Kittery turned out, much the way of a prize bull or quarter horse stallion. It embarrassed Tania to think about it, but she couldn't draw her eyes or her mind away from Tappan Kittery.

He smiled as he finished the last morsel.

"That was good. You made it?"

Tania nodded and opened her mouth to speak, but a brisk knock at the door stopped her. Cotton Baine pushed inside without an invitation. He passed a light blue cotton shirt to Kittery, then took a step back and folded his arms in front of him. "From Greene's store," he said. "Sheriff's wife asked me to buy it for you. She took yours to mend and wash—it was obvious you hadn't done it for a while," he said, grinning.

Kittery laughed, reddening a little. "True enough." He held the shirt up with his good arm to look it over. "Thanks, friend. It'll do just fine. But don't be offended if I wait till I'm alone to try puttin' it on. I don't want you to see me in pain."

"Bet it hurts, huh? Well, if you need anything else, just send for me over at the cantina. A bottle might do you good, for that matter."

"And it might do me bad. No bottle, Cot. But I do need somethin' else." His words turned Baine back around. "I'd like to know more about this shootin'."

Baine's glance shuffled between Kittery and Tania. "What about it?"

"Is that Mexican still alive, for one thing?"

"Yeah, to his poor chance. But maybe not for long." Baine told about the damage done by the bullet.

"Too bad. Nobody deserves that."

"You're right there," Baine said with a sigh. "I gotta agree. He deserved to pay a price, but not like that."

"Pay a price?" Kittery stared at Baine. "Pay a price for what? Being Ferrar's friend?"

"Well . . . not exactly. By the look of his face, he was with the bunch that attacked you last night. Looks like you drew a little blood of your own. Maybe his killer knew he was one that got away."

"Sounds like somebody's done a little investigatin'."

"S'pose so. Anything else you wanna know?"

"Guess not."

Tania sat and listened to the two men talk, not saying anything. She felt a twinge of guilt, and she wondered if Cotton Baine did, too. They both knew about the vigilantes, and Baine was one of them. It was hard to keep them a secret, but a person couldn't tell how someone thought about something like the vigilantes. They couldn't be too careful.

Tania's heart was torn when it came to the vigilantes, especially over her father's being involved. But because of the nature of her mother's death, she couldn't keep him away from the organization. And in truth she didn't know if she wanted to. Murderers and thieves had been allowed to run rampant in that country for far too long.

When Kittery spoke again, it was of Castor, questions of curiosity about its people, its events, about what held it all together. Tania warmed to that, telling the man things about the town that others might have considered secrets. But she didn't feel guilty. She was drawn to this big man. He made her comfortable. He made her feel warm. She might have been intimidated by his size, but she had seen him unconscious. She had taken a hand in undressing him for the doctor to dress his wound. It was hard for her to be frightened by a man she had nursed, and whose words came across so honest and gentle—and caring.

When she made ready to leave, Tappan stood up with her. "I don't mean to seem sudden, Miss. But would you go to dinner with me today?"

"But—what about your shoulder?" she asked.

"I don't eat with my shoulder."

Tania laughed, feeling foolish. "Then yes. I'll join you. Would one o'clock be all right?"

"Sounds good," he said.

After Tania left, Kittery sat alone on the edge of the bed, lost in thought. Tania McBride. The name fit her, he guessed, except the McBride, perhaps. Her hair was black, and with a name like that he would have expected a blonde or a redhead.

Why did that woman seem so permanent in his life when he had just met her? That wasn't his way, to get close to someone so quickly. But she was easy to get close to. She had to have a beau somewhere, though. She was too nice not to.

He lay down on top of his blankets. For a time, he thought of Tania and listened to the sounds of the settlement—the barking of dogs; horse hooves clip-clopping along the street; children playing; the laugh of a woman; and the crow of roosters answering each other across the morning rooftops. The last thing he remembered was the sound of wagon wheels rattling up the street.

When Maria Greene woke again, she stared about her in confusion. "Emilita," she whispered. "Emilita? Oh, baby . . ." A tear appeared in the corner of her eye and rolled down onto her cheek. "Baby, baby, I thought you were dead . . ."

Her gaze fell upon the man who lay asleep on top of the bed across from hers. She shuddered and jammed the back of her hand across her mouth, biting down. The tears seeped out as she looked at the big man. A lump formed in her throat. "Emilita," she sighed. "Not my baby. Not my baby."

She lay staring at Kittery for five minutes, thinking about her daughter. It was true. None of it was a dream. The stranger in her room proved that. Emilita was dead! Tears welled up in her eyes, and she sobbed until her emotions felt dead.

A sobbing intake of breath wracked her body, and she sat up. Her eyes swung to the gun belt and revolver that hung from Kittery's bedpost. She stared at them, her eyes wide. What she would have given to have that gun last night! She could have stopped those horrible things from happening to her and Emilita. But nothing stopped them. No one stopped them. And none of them would have happened if not for this big man, this Captain Kittery.

She climbed out of the bed, her hands shaking, moving like a shadow across the space that separated her bed from his. Standing silently, she reached out and stroked the butt of the pistol hanging there. Her gaze shifted from it to Kittery. She couldn't stop thinking if it hadn't been for what he did, her baby would still be alive!

Taking a firm hold on the pistol grip, she drew it from its holster and pointed it at Kittery's head. *You killed my baby,* she thought. *You should have to die.* But then she slammed her eyes shut and shuddered. She lowered the weapon, placing a hand over her mouth. How could she think such a thing? Kittery had been defending her! She was the one who had asked them to remove the men from her hotel. If anyone was at fault, she was.

Glancing into the mirror on her chest of drawers, she was sickened by the sight of herself. In effect, she had killed her own daughter. It wouldn't do to try and blame it on this man who had acted only in trying to help her. If she killed anyone, she should kill herself.

She looked down at the pistol and imagined bringing it to her own chest. But that was as far as she went. She reached up with one hand and seized the white cotton of her nightgown, looking at the bulge where her baby grew inside her. If she killed herself, she also killed this baby. She could never do that.

She reholstered the weapon and then simply stood there, both hands on her abdomen. "I will not let you die, too," she promised.

A soft moan broke from Kittery. Maria Greene moved closer to see his face. It was twisted in pain. Perhaps he was in the middle of some horrible dream. Wistfully, she looked down at him and sat beside him on the bed. She stared at his face and at his long, powerful frame. His chest and shoulders were so big! His hands would make two of hers. He was an admirable man, to say the least. She tried not to think about him except as a kind stranger, but that was so hard. It had been so long since she had had a man to care for her, and never a man with looks and a voice like this one.

"Why couldn't Thaddeus be like you?" she whispered. With a trembling hand, she reached out and brushed the ruffled hair from his forehead. The feel of it sent a tingle through her arm, making her want to touch his cheek, his shoulder, his hair-covered chest. Making her want to touch those hands that could have held her and kept her so safe and warm.

Another tear touched her cheek, and she stepped away from him. In the same silence that she had stood over him, she turned, and with a heart that beat fiercely against her chest, she took off her nightgown while watching him, pulled a gray wool dress on over her head, put on her shoes and left the room.

The sun splashed fire on the Baboquivaris.

The little man called Shorty Randall sat his plodding white horse in his own swirling dust, unnerved by the tawny rims that loomed around him like gargantuan skeletal vestiges. The same ragged castles of stone climbed ahead of him in broken crowns of grandeur. Rivulets of sweat picked their muddy way through the dust on the little man's cheeks and soaked the shirt underneath his tan duster. Sweat pooled in the toes of his boots and turned sour against his salt-streaked saddle.

In the pale sky above, three vultures made their lazy circles, watching the bleached land with the keen eyes of those who depend on their senses to survive. Their unending patience kept them always within sight of the little horseman who traveled alone. Yet their deathly silence made them seem like specters, angels of death.

Shorty rode into the canyon of the Desperados and along the precarious trail midway between the canyon's rim and its rock-strewn floor. The echoes of his horse's footsteps against either wall of the canyon were like death's own song.

In the Den, on his boulder perch, Paddles-On-The-River, the old Papago guard, plucked a shard of bone from between his teeth and cocked his head to one side. The faint, resonant steps of Shorty's gelding carried to his keen ears through the stifling heat of that windless day. Paddlon came to his feet in one smooth motion and strode to the mouth of the grotto, staring north. After that, he didn't move for some time, until the Desperados returned their attention to their meals, knowing if Paddlon was unconcerned then the one who approached must be friend, not foe.

It was Desperado Eight, Slicker Sam Malone, who finally spoke. "That must be Shorty a-comin'."

The rider drew up presently, and Paddlon walked back to sit on his rock. After a minute or so, Shorty Randall stepped into view, his eyes seeking out forms in the dark cave. Baraga watched him for a moment, then spoke in the gruff voice of one accustomed to people following his every word. "When you can see, Shorty, walk over here and sit down."

Shorty blinked his eyes once more then came near and sat on an upended oak log. "What's brought you here, Shorty?" asked Baraga. "You barely left. It had better be important for you to come here."

Shorty glanced about. "You know about the U.S. marshal."

"You told us."

"Well, word's out that he's gatherin' a posse and makin' plans to come after you pretty soon."

A mirthless grin parted Silverbeard Sloan's lips. "Hell, let 'im come, boy. An' he better have one big posse. What makes 'em think they'll fare any better than the rest?"

Shorty shot a glance at Sloan, and his mind unwillingly pulled up a memory of the pile of sluffing bones that lay at the bottom of the canyon, downstream of the spring. The bones down there were of different ages, all still wearing vestiges of clothing, some of them army blue. His eyes flicked back to Baraga. "I knew you wouldn't want me out here, but I wanted to be sure you had the word. It'll be soon, I think."

Once more, Sloan chimed in. "He'll never find us, boy. Much less hurt us." This time when Randall didn't even glance at him a heavy scowl slashed his features.

"They say this marshal's the best," Shorty said to Baraga. "He don't give up—ever."

"No lawdog is the best!" barked Sloan, erupting to his feet.

"Simmer down and shut your mouth . . . *boy*." Baraga's imperturbable voice was more ominous than if he had snapped the words. His use of the word "boy" was an overt jab at Sloan for using it on Shorty.

Major Morgan Dixon broke the tension, freeing Sloan from having to make any regrettable retort to save face. "Sit to dinner, Shorty. This one's on me." The Major's own meal was finished, and his tin plate made a dull thunk as it landed in the sand, the fork clattering across it.

With no objections from anyone, Shorty dropped to his haunches and retrieved Dixon's plate. He had to brush off some grains of sand before helping himself to a plate of rice and venison. When he had finished eating, he laid the tin aside.

Baraga eyed him coolly, wiping his mouth with thumb and forefinger. "Shorty, you might as well stay the night here. Then tomorrow, ride back early. I want you to watch this marshal. If he makes a move, you ride out here fast and warn us—understand? What did Mouse say about the money deal, anyway?"

Shorty shrugged. "He got mad. But he didn't say much, except tellin' me to get out."

Baraga nodded and turned to Paddlon. "Injun, are you done eating?"

Paddlon nodded.

"Then what are you doing still here? Get back to your watch."

"Yeah, go on, Injun," Slicker Sam Malone threw in. His smile died when Baraga gave him a warning scowl.

The old Indian didn't look up. He kept his eyes to the floor, and without a word he rose and stepped outside. He scaled up to his ledge above the cavern, where no movement escaped his vision.

Sitting down on the wind-swept rock, the old Indian stared out across the canyon, eyes glittering like coals. With a sneer, he whipped out a long-bladed knife and began to design eight stick figures in the dirt. When he had finished, he stared at them a moment, then slashed the knife blade through them all—once, twice, three times. With a brisk nod, he sheathed the knife and returned his gaze to the trail below.

There was going to come a time when he would no longer listen to Savage Diablo Baraga and his hated henchmen. He would leave here one day and return to his people. It had been so long, and he missed hearing his own tongue. He was not an Injun! He was one of the *Parientes,* the kinsmen. One day he would go, and he would take Baraga's power with him.

Inside the cavern, Silverbeard, like Paddlon, brooded with his own silent hatred. The words of Baraga still clung to his brain like cobwebs. No one told Silverbeard Sloan to shut his mouth! At least no one who would remember it for very long. How could Baraga treat Silverbeard Sloan with such disdain? Hadn't he seen him draw? Hadn't he seen him shoot? No one here could beat him. He was indispensable. He should be respected. Even Baraga should fear him. Well, one day there would be a showdown. It was a matter of time. Then he would be the leader here, not Baraga. Who would want to follow a one-armed man anyway, once they saw that Sloan was the better man? He would gun Baraga down, then perhaps Bishop, too, unless he was smart enough to follow him.

Noble Sloan had suffered his disrespect as a child. He wouldn't take any more. His stepfather had called him "boy" too. Always "boy." He had never considered him a man, no matter how good he became at fist fighting, or whatever he chose to undertake. Now, all these years later, things were no different. But now he had the power to change things, and he would! One day he would lead the Desperados . . . or they would die.

As Silverbeard throbbed with hate and Paddlon invoked his medicine, the other Desperados sprawled about the cavern floor. The vultures circled the sky, and the sun sizzled like molten brass across broken granite rims.

Tappan Kittery's strength had redoubled by the time he awoke again. He had to attribute that to the nourishing meal, and he was more than ready for another one now. His stomach ached to make up for lost time.

But the black stallion was his first priority. Where had Satan ended up? Surely everyone in town would know who the animal belonged to by now, but what if he had run *out* of town?

Anxious to find out, he sat up on the bed and then stood up too fast. He sat back down to let a wave of dizziness pass over him, and when it was gone he pulled on his boots. Then he picked up the new light blue shirt. It was agony sliding his wounded arm through the sleeve, but he felt much better having it on. He didn't like the thought of being seen without a shirt. This was the Victorian age, and that wasn't proper, even for a man.

While looking with a jaundiced eye into the mirror, he brushed his hair with an ivory-handled brush he found lying on the nightstand. Then he turned and buckled the Remington around his waist and went into the lobby.

He had no idea what to say to the Mexican woman who was the first sight to greet him. He was taken by surprise to see her sitting there, like nothing had ever happened in those dark quarters. Should he speak at all? Would she acknowledge him? Why was it words never seemed to come at times like these? He tried to smile—a sad, apologetic smile—then turned and hurried out the door.

His first thought was that perhaps the old hostler, Jarob Hawkins, had caught the horse and put it away for him. Or maybe the sheriff or Cotton Baine or Joe Raines. He didn't have time to go and check, though, for the first person he saw was Luke Vancouver. The sheriff was in conversation with, of all people, Jarob Hawkins. When he glanced Kittery's way, he excused himself from Hawkins and met Kittery mid-street.

Seeming to read the big man's thoughts, he said, "I thought you might want to know a fellow brought your horse in to me this morning, and he's over at the stable—the horse, that is. The man's name is Adam Beck. If you make his acquaintance before I have a chance to introduce you, you might make sure you thank him for his trouble. How's your shoulder?"

"Fine. Doc says it's a clean cut. The knife barely sliced along the outside layer of muscle. It was the hit to my head that put me out."

"Well, I'm glad it wasn't bad. By the way, I talked to Rand McBride—he said the two of you had met. He asked me to see if you would come to supper at his place tomorrow night with me and Beth. He was concerned about Tania spending time with you, but I vouched for you."

Kittery laughed. "Puttin' yourself out on a limb, ain't you?"

Vancouver returned the laugh. "I trust my judgment. And by the way, how about coming up to our place at seven o'clock tonight?"

"Well, that's one request I couldn't very well turn down," Kittery said with a grin. "I hang around this town long I'm likely to get fat."

"We'll do our best," replied Vancouver with a grin. "See you tonight." He moved off, making his rounds of the town.

With the stallion safe, Kittery had no more need to worry about him. Turning, he eased back into the coolness of the dim-lit hotel. Without a glance Maria's way, he climbed the stairs to his room.

At first, he thought he had stepped into the wrong room. But a second check proved it was his. What he saw, however, took him back. Vanished was the look of squalor; where the day before the room had resembled a pigsty, now it was spotless, or as near so as it could ever be again. New sheets adorned the bed, and clean blankets, along with a big, soft pillow with a fresh cover. The floor was swept and scrubbed, the writings on the wall nearly obliterated. The curtains had been washed and put in place, the window was clear and on the washstand sat a new porcelain basin and a matching pitcher filled with fresh water. It was strange to see a room take on such changes in so small a time. Maria Greene must have done this the evening before, while he dined with the Vancouvers.

In spite of the stifling heat of the room, this time when he lay down it was in comfort and on the edge of style—a complete contrast to the day before. With the unexpected new feeling of luxury, he forgot to check for pants rats and seam squirrels.

Chapter Nine
Tania McBride

A rectangle of light splashed across Tappan Kittery's bedspread, and he blinked his eyes against the glare. He lay there soaked in his own sweat, staring at the ceiling for a few moments while he gathered his bearings. The room air was stale with heat, and again he contemplated the foresight of building with wood in a land known for such constant vicissitudes of temperature.

Rising, he pulled off his shirt and pants, using the washcloth on his nightstand to give himself what had to pass for a bath. He dressed again, then clamped on his hat and left the room before the close air could start him sweating again.

He felt the swift, mighty push of heat as he opened the front door. But there was a touch of a breeze, too, and it offset the onslaught of sun. Somewhere behind the jail a slamming door attracted his casual attention. By the height of the sun, he placed the time around one o'clock, when he and the girl had agreed to meet for dinner. Tania had assured him the café owner, Bartholomew Kingsley, was as American as they came and didn't partake in the Mexican custom of siesta. Kittery sat down on one of the chairs on the porch to wait. In spite of Kingsley's attitude toward siesta, the street was quiet. The only living thing in sight was a tan hound asleep beneath the steps of the dry goods store.

Then he heard the soft clop of hooves across the street and looked up to see Tania passing between the bank and dry goods. She sat sidesaddle on a gray Arabian.

"Right on time, Miss McBride." He smiled as he gave her a hand from the saddle.

"Oh, no I'm not. It's one-thirty."

"Well, I just woke up from a nap. So you're timing's perfect."

Tania laughed as Kittery took the gray's reins, leading it as they walked together to the café. Kittery took three loops of the reins around the hitching rail. He never tied a horse by the reins; a couple of straps of leather wouldn't stop a horse from getting free if he wanted to. It was insult enough trying to find a fugitive horse without having to put out another dollar for a new set of reins besides.

Kittery stopped and opened the door for Tania, and she looked up at him and met his eyes as she passed. He was so tall! Her father wasn't a short man, but Tappan Kittery made him seem small. And yet he didn't look awkward or bulky like many big men she had seen. His shoulders were very wide, and his waist and hips narrow. He moved with every bit of grace possessed by a longhorn bull, lean and in its prime.

Inside, he took her elbow and led her through the empty room to a table against the left wall. He seated her with her back to the room, then sat down with his own back to the wall.

Bartholomew Kingsley not only owned the café, but he was also the head cook and spent most of his time there. He lived in an apartment built onto the back of it. Kingsley was a chef of the highest order, which was odd for a town of little importance like Castor. Tania had heard bits and pieces over the years since she and her parents and the Vancouvers had taken part in founding the tiny community. There were different stories, but the one that seemed most accepted had Kingsley learning his culinary skills in a specialized cooking school in Boston. He had gone to the head of his class. From there, he went on to prepare meals in the finest restaurants of New York City, New Orleans, and Saint Louis and as far west as San Francisco. Needless to say, his skills were varied, and when he could manage to have the necessary ingredients brought in to so remote a location as Castor he could prepare a meal for the most demanding customer, the most exotic taste. And what had dragged him here at last? The gentler people said he had come because of the mild winter climate. Rumormongers said he had killed a man in the Bay area and come here to evade the authorities. Whatever the reason, Bart Kingsley was the pride of the town. His prices, even for the finest cuisine, were more than fair.

Bart Kingsley walked into the room wearing a too-small apron over a tall, clumsy frame. He wore spectacles and a well-trimmed brown mustache, and the sparkle of mirth shone in his eyes like candle glow.

"Howdy. So you're the Captain," Kingsley said by way of greeting. "I'm Bart Kingsley." He grinned mischievously and held out a big hand, looking all the time to Tania like he was about to make a joke.

Kittery raised his hand to shake, and right away he knew he would like this Kingsley. "They call me the Captain, but I prefer Kittery—or Tappan, if you like."

"Well, I'll be whacked with a wooden spoon if I don't cater to my customers' will, Tappan. Good to meet you."

Tania was embarrassed by Kittery's brash decision to order her the day's special meal. He seemed to think he was rich. Tania had already given him her leave to make the choices, and she regretted it a little now. She couldn't imagine a traveling man like Tappan Kittery had much money to spare. That day's standard menu was roast beef and potatoes; the big man passed on it to order the other choice—filleted, breaded rainbow trout, French bread with garlic cheese, a tossed salad and a slice of dried apple pie.

Bart Kingsley brought the meal out with well-deserved pride and set it before them, then stood back grinning and folded his hands in front of him. "Wale," he said, putting on an exaggerated southern drawl. "Enjoy."

The meal was delicious. Tania wished she had the means to come here more often. Of course she couldn't complain about any lack of money, being the town banker's daughter. But her father had never been frivolous.

As she ate, Tania noticed that the big man was very dainty in the size of bites he took. She slowed her own eating accordingly and savored every morsel—and every moment.

They talked of their lives, of things they had done, and of others they hoped to do some day. Kittery spoke of the ranch he wanted, where he would breed champion horses with the Friesian bloodlines of the black stallion. His house would be of two-foot-thick adobe, if he chose to live in the southwest, or of big logs, if the north country was to be his home. Either way, his front porch would face west, toward the sunset. A wide awning and tall trees would shade it, and there would be a swing there where he could sit and watch the sun simmer down into the horizon. He wanted a large family, and a home where they could feel the same love and acceptance his own mother had shown him and his siblings.

Tania had always been shy, but she once more found herself telling Tappan Kittery things she hadn't intended. She agreed with his feelings on family, although her brothers and one sister had been dead for many years, killed by the Apaches who had taken her mother. In spite of this, or perhaps because of it, she dreamed of many sons and daughters. She wanted a home in the hills to raise them, with plenty of space to grow and to make their mistakes. She dreamed of keeping a fire in the hearth on a cold winter evening. There supper would stay warm for when her man returned from the range, or from a hunting trip.

She didn't tell Kittery this, but she often dreamed of the pleased look on her husband's face when he came in after dark, dirty, tired and hungry, to find the aroma of chicken and dumplings or hot apple pie wafting in from the kitchen. He would kiss her, and she would take his coat, and rub his back, and fill his stomach, and love him.

As the words flowed from them, Tania thought how odd it was that they had just met. It seemed they had grown up together—or at least *grown* together. She was dazzled with wonder. Perhaps it was more than chance that had brought them together here. The only man she had ever loved was Alan Peck, and when he had gone away he was no more than a boy. Now he was a soldier, far away in the land called Dakota. He was so proud to write her of his adventures, and of the admiration he had for his commander, George Armstrong Custer. She always replied to his letters, but more and more he seemed like a distant memory—a boy she used to love. She wasn't sure she even knew Alan anymore. But she was sure of one thing: she had felt love for Alan, but he had never stirred her the way Tappan Kittery had from the first time she saw him.

Tappan Kittery dabbed at the corner of his mouth with his napkin and watched Tania as she spoke to him about Castor and its little "secrets," secrets everybody in Castor knew. He took in her teeth that seemed a dazzling white against the dark of her skin. He had seen those teeth plenty of times, for she was quick with a smile—always ready to laugh.

A preference toward dark-haired women was ingrained in Kittery, and it came no darker than Tania's. Each time he had seen her, the hair had been in braids above her head, or fixed there in some other fashion. He wondered how it would look flowing loose.

There may have been men in cities, where there was an abundance of manufactured beauties to look at, who would think of Tania as plain. She didn't have the look of a classical beauty, but he had never been drawn to the look portrayed in classical art, anyway. The plain fact was Tania was beautiful to him, and her beauty grew the more he talked to her and studied every feature of her face.

He had been drawn from the first to her frame and features. She had a bearing that was uncommon to the rugged frontier, particularly to a girl who couldn't be far over twenty. Her body, although not slender, seemed firm, judging by the little that a Victorian outfit revealed. She appeared to be strong, and he admired that in a woman. A weak one wouldn't survive for long in Arizona.

The way her hair was raised above her head accentuated a long, lean, but not slender neck, the kind of neck a man might caress without any worry of hurting it. Her mouth was full, and her skin was soft and clear, and so deep-colored her blood could not be entirely Anglo-Saxon. Her father's hair was black like hers, but his skin was light. That led Kittery to surmise that Tania's mother was of Mexican or Indian heritage, but he didn't have the audacity to ask.

But above everything else, when he looked at Tania he was taken in by her eyes. No one, not even the highbrow boys of Harvard or Princeton, could deny their beauty. Large and dark, they were full of mystery and framed by long, curling lashes. Kittery caught an unnerving sparkle in them now and then when she looked up from eating to catch him watching her. It turned his insides to clay, made him stare at her like a lost pup gazing at the moon. He had seen women, many of them. He had been with Mexican women whose beauty he didn't think could be surpassed. But those eyes of Tania's held something different, something he couldn't remember having ever seen. It was difficult to guess what thoughts might lie behind them, about him—or about *them*, together.

They finished the last of their meal and pushed the plates away. When Bart Kingsley stopped at the table, Kittery smiled up at him. "So how'd we do?" the chef asked. "Will you be back?"

Kittery chuckled. "You must know I will. That was one of the best meals I've had in a long time." That was in big part due to the company, but he wouldn't embarrass the woman by voicing that fact. Anyway, he had a feeling she knew.

They stepped out onto the porch into the dwindling shade of the awning. Some people moved along the street now, but most still guarded themselves against the day's heat. The guttural tones of a Bantam cock's crow broke the stillness, soon answered by another, hidden off in the catclaw on the ridge. At least somebody still had enough energy to throw insults.

Kittery and Tania had spoken during the meal of their planned get-together the next evening. "I'll come down to meet you at seven o'clock tomorrow night," Tania said. "You'll need an escort."

Kittery laughed. "Thanks, Miss McBride. You'll be most welcome." He smiled at her, touching her hand.

When the girl had gone, Kittery took a deep breath and stepped into the dim light of Maria Greene's hotel. He was hoping the Mexican woman would be absent, but she sat behind her desk, her eyes staring blankly. His first thought was to turn and climb the stairs in haste. But Maria Greene became aware of his presence, and her dark eyes flashed upon him. He had no choice but to acknowledge her now. He had no choice, but all he wanted to do was shrink under her dark gaze.

Chapter Ten
A Father's Challenge

Tappan Kittery stood transfixed by Maria Greene's gaze. It almost became a contest to see who would look away first. Finally, he swallowed and moved up to the desk.

He had no idea what he was going to say or do, or why he had walked up to her. Without any conscious decision, he reached into his pants pocket and came up with his dollar rent for the day, clicking it down on the desktop before her. But he didn't stop there. He continued putting down coins until he had laid out four more.

"I wanted to thank you for cleanin' my room, ma'am."

Maria looked back and forth between his eyes, then shifted her attention to the coins. "You'll be staying four more days, then?" she asked, sliding the coins across the desk and into her palm as nonchalantly as if there had never been any trouble.

"Prob'ly more than that. But I'm only payin' for yesterday, today and tomorrow. *Two* dollars a day. You're prices aren't too high."

With an uncomfortable nod, he turned away and climbed back up to his room. His feet seemed to drag on the carpeted stairs.

At four-thirty the following afternoon, Tappan Kittery closed his well-worn Bible and moved to the washstand. It never got cooler in the room, that was for certain; he was soaked with sweat from head to foot. He splashed water from the basin over his face and neck and torso; the tepid fluid was mild relief.

He kneaded his shoulder, and its progress pleased him. Between the grace of God and some fine care, he should be back to doing anything he wanted in no more than a couple of weeks. But the truth was, all he wanted to do was see Tania. A day of sitting around reading and sleeping—and thinking about the young woman—had driven him near crazy. It had managed to drive Ned Crawford right out of his thoughts.

Fishing among his gear, he came up with an ivory-handled razor in a leather case. He worked up some lather in a shaving cup and took off his several days' growth of beard, feeling ashamed of himself for not doing it sooner, before his meal with Tania.

Later, he stood in thoughtful silence on the front porch, watching a golden eagle soar on untiring wings on the air currents a hundred feet above where the

McBride house must sit on the ridge. The usual gathering of Mexicans was absent from the porch of the cantina, perhaps afraid of another sniper bullet. In their place was a big colored man Kittery had seen several times loading crates, a job the colored man tackled alone, even though he was handicapped by a missing hand that had been replaced with a silver-colored steel hook. Kittery stepped over to the cantina and halted in front of the colored man where he leaned against the doorframe, watching him.

"Where are your friends today?" asked Kittery.

The black man glanced to left and right with a surprised look on his face. Then he looked Kittery up and down, his surprise changing to an amused twinkle in his eyes.

"Friends? Who you callin' friends, them Messicans? Maybe you ain't noticed, but I'm a Nigger."

"Colored man," Kittery corrected with a faint smile.

This brought a grin to the face of the stocky Negro, who looked Kittery over again. "Ain't caught yore name," he said.

"Tappan Kittery's the name, sir. And yours?"

"Solomon Hart. I'd shake your hand, but . . . I'm a— I'm colored."

Kittery chuckled. "Colored, huh? You don't say." With that, he held out his hand, and Solomon Hart took it haltingly and smiled.

Kittery had turned to go, but Hart's smooth, purring voice raised a warning. "Mistuh Kit'ry, I'd keep a cat's eye out, I was you. Ferrar's still got him a few friends around, though it don't seem possible."

"I appreciate the advice, Mr. Hart. I'll have to consider us friends, you and me." Kittery smiled, and with a nod he turned to walk across the street toward the livery stable, leaving Hart with a wide grin.

"'Sir', he says. 'Sir'. Now, I like that." Solomon Hart stuck his hook and his right thumb into the armholes of his loose-fitting vest and walked away chuckling.

The black stallion slowed from a trot to a walk as he reached the rear of the livery stable, coming back into town. Kittery was guessing at the exact time, but he had cut the stud's exercise short, figuring it must be time to get ready for supper. He didn't want to miss a moment of Tania's company.

The heat of the day had drawn down now, and newness was in the colors of the landscape. Even the blue of the hotel and the green of the dry goods store seemed deeper, and a sweeping, brash shade of those colors emboldened the sky and the plant life. The dust brought to life by Satan's black hooves took on a bold golden cast in the slanting evening sun.

The street was packed this Saturday evening. Ranch hands, miners and farmers were in town early to celebrate the week's end. They would spend the night drinking and gambling and visiting with the señoritas who lingered in the shadows on a lively evening such as this. They had come to do anything to wash away the bleak memories of a week's weary labor.

After taking care of the stallion at the livery, Kittery left him with a bait of grain and crossed to the hotel. On one of the chairs on the porch there lay a stray copy of *The Arizonian*, a leading Tucson newspaper. Picking it up, Kittery took the seat it had been on and began to skim the pages while he awaited seven o'clock, when Tania McBride would come.

The wrinkled pages of the paper were saturated with stories of Indian outbreaks in the north country, and some that also told of the Apaches. But mostly, since Cochise had died and the rest of the Apaches remained in a restless state of dormancy, the stories were of the Sioux and Cheyenne in the Territories of southeastern Montana and northeastern Wyoming. The articles talked of gunfights throughout the booming mining camps and cattle towns, political treachery in the east, corruption and thieving on the west coast. After several pages, they all looked the same—documentaries of a young nation running all out toward a goal of self-destruction. Yet by some miracle each new day found the United States of America not only alive, but prospering. Maybe this country was not that different from any other. No better, no worse. Just a place where people fought daily to survive.

Kittery reached the end of the paper with a sigh and folded it in two. It drained him to read of the violence that beset this land. And Tappan Kittery? He was part of it.

He looked up in time to see Tania McBride crossing the crowded thoroughfare on foot. Whistles and calls of delight rose from the onlooking revelers, but the noises died when Kittery met her in the center of the street and took her elbow. "A bit spooky out here for a girl to be on foot, isn't it?"

Tania laughed lightly, glancing around at the miners and cowboys who crowded the street. "They wouldn't hurt a woman, Captain. Some of them may have big mouths, but let one try to hurt me, and see what becomes of him. They can't scare me with their harmless jokes."

Kittery laughed his agreement. She was right. The West wasn't a place where a woman generally had to concern herself with safety—not in a town crowd, anyway. Few enough women inhabited the barren parts of the country that men tended to treat them like fragile wards. It was more the dark shadows between her place and the town that worried him. He told her that as he led her between the bank and Greene's Dry Goods.

She studied him a moment as they started up the winding ridge road, but he didn't meet her glance. After a moment, she looked forward again and spoke quietly. "Thank you for worrying."

The McBride place was another of the kind of prim residences he had observed from the road below. Typical of the greater number of Castor's homes, it was constructed of whitewashed adobe brick, and the white exterior gave it a measure of coolness during summer days when the temperature could soar to a hundred and ten or higher.

A rare white picket fence surrounded the house, and a flagstone path meandered among wildflowers to the door. Red and yellow roses grew along the sides of the house, and though the sun had burned the edges of their petals, Kittery admired them aloud.

"Thank you. Mother planted them not long before she died," Tania replied, glancing wistfully at the bright blossoms. "They've been in my care since."

She forced a smile and opened the oaken door to step into the house in front of him. It was a large, open room, with one doorway leading through a hallway to the right. The floor was of polished oak, with two braided rugs and what appeared to be a woven one of Indian manufacture. A huge, carved oak table, stained darker than the floor, sat in the middle of the room, surrounded by ornate chairs with whicker backs and seats. The far end of the room was set aside for all the activity that would take place in a kitchen—a porcelain stove, cupboards with plenty of counter space, and a bread board. Lastly, he observed a pine pie safe with images of pineapples punched in the pieces of tin that were inset into its doors to allow for air circulation. A pair of dueling pistols and a Kentucky long rifle were arranged in a manly display on one wall, and pictures adorned the opposite, old photographs of people with sober, shadowed faces.

Kittery said hello to McBride as he walked in from the long hallway. When the older man took a seat on a sofa against the wall where the guns were displayed, Kittery seated himself in a cowhide chair that sat at right angles to it, ignoring the other couch. They fell to conversing over politics and current events while Tania toiled over the stove in the far corner of the room.

Beth and Luke Vancouver arrived, and Tania welcomed them. She served supper, and they sat to eat at the expansive table that squatted beneath a cluster of candles. The candles were unlit, and late afternoon shadows filled the room. After they had finished, they lit four coal oil lamps. The Vancouvers took one couch, McBride and his daughter the other, and Tappan took the same chair he had been in earlier. Its white hide with a liberal smattering of black spots suited his simple tastes.

As he had expected, Kittery was the outsider when it came to the conversation. Everyone there knew everything about everyone else, so it was he they chose to pick on. He was able to squirm out of speaking of his family, however, when one of the first topics they broached was music.

"I remember around the end of harvest time the whole mountain would get together at our place to dance," Kittery said. "Nobody in my family could hold a tune. Except maybe my older brother, Jake; he could do just about anything well. And maybe my sister Eliza. But most of us could play." To himself, Kittery had to admit the part about his singing wasn't altogether true. Many people had begged him at those gatherings to sing them a song or two and claimed he had the mellowest, most beautiful singing voice of them all. But saying what he had would keep them from asking him to sing tonight, and that was his main intention. Unfortunately at the same time, saying what he had said got him notice he hadn't counted on.

"Oh, you played an instrument? What did you play?" asked Beth.

"I played the fiddle."

Rand McBride perked up and glanced over at Tania as he scooted forward on the couch. He was poised as if to spring up, yet he managed to keep his seat. "This is my night. Captain, I have an old violin that belonged to my wife. It's been still since she died, but I'd admire hearing you play a piece or two on it."

"Well, I reckon I could. But it's been a long time since I played. I'm not sure I'd remember how."

McBride stood up, now able to look down on Kittery for the first time. "Captain, I don't believe a man ever forgets a talent like that."

"It wasn't exactly a talent," Kittery began to explain. But the others were adamant to hear some music, and at last he gave in with a sigh. He knew he wouldn't be able to face them again if they didn't hear him make a fool of himself on the fiddle tonight. Or the violin, as McBride referred to it—he could play it either way. "All right, you win," he said. "But don't expect much."

Doing a poor job at hiding his excitement, McBride hurried to retrieve the instrument. It was a beautiful piece, cherry red in color with time-yellowed ivory keys. The strings, however, made a horrible wrenching sound of agony when Kittery first drew the bow across them. After re-tuning the instrument, he tried it again, and the mellow tone sent shivers up his spine. It had been so long since he had done this. Two years, maybe more. And since his childhood every instrument had been a borrowed one. But he remembered none nicer than this.

After playing a few half-hearted notes, he lifted his head and looked around at the others. "I've forgotten the songs I used to play."

"Play 'Barbara Allen'," suggested Vancouver.

Kittery looked over at him and gave him a wry smile. "Thanks."

But "Barbara Allen" was one of his own favorites, and when his hands began to move, and the bow glided across the strings, he forgot how uncomfortable all this attention made him and lost himself in the piece.

Tania McBride watched in a silent reverie. She felt the back of her neck begin to tingle, and her eyes dimmed with tears. She remembered how her mother had played this song. She found herself looking around the room at the others, marveling at the rapture in their faces as they gazed from the musician's broad hands to the sun-dark skin of his face.

And then her eyes were drawn back to Tappan—to Captain Kittery. His eyes were closed, and his mouth moved ever so slightly with the deep emotion he poured into the song. How strange to see a man of such physical power, a man who looked so natural atop the black stallion, a rifle in the boot and a pistol and knife on his hips, as he fell into a piece of music with the rapture this man did! What kind of deep-souled man was this Captain Kittery? What other secrets did he hold?

He played the war hymns next, both from the North and from the South: "Bonnie Blue Flag," "The Battle Hymn," "Dixie," "When Johnny Comes Marching Home Again," and "Tenting Tonight on the Old Campground," her favorite crying song. Tania remembered these songs from her childhood, from the trail out from West Virginia, fleeing the war with her mother and father, the Vancouvers, Bison Sabala, Albert and Sarah Hagar, Adcock and Priscilla Hale and all the others who were either dead now or had drifted off to parts unknown.

The Captain played "Lorena," "Aura Lee," and "Kathleen Mavoureen," and Beth Vancouver added her lullingly beautiful voice to "Shenandoah." When he laid the violin aside, to remarks of admiration and regret from the Vancouvers, Tania looked at her father. He was staring at the old tintype of Argentina McBride, her mother. His eyes shone with tears.

Luke Vancouver glanced over at Beth, then at Rand McBride, before slapping his knee. "I guess Beth and I should be going. It'll be a long day tomorrow, with the hanging going on."

He and Beth thanked Tania for the work she put into the delicious meal, and they turned to Kittery as one. It was Beth who spoke. "Tappan . . . I wish we could properly thank you for your music. No words could describe how perfect your talent made this evening."

Kittery felt himself blush. "I only wish you had joined in sooner. You could have sung 'Shenandoah' without the fiddle, with a voice like yours."

"Thank you," was her reply. "Thank you so much."

Kittery shook hands with Vancouver, who held on for a moment. "You're no ordinary man, Tappan. It's good you've come to Castor. It was always our dream to fill this little piece of Heaven with extraordinary people."

Kittery chuckled. "Sure. Like Thaddeus Greene."

The sheriff laughed. "There's always an apple trying to spoil the barrel."

They all walked outside, and Luke helped his wife into the waiting buckboard. He climbed on himself, then, and tipped his hat to Kittery and the McBrides. "Good night."

The rig pulled away, melting into the darkness of the lane as Kittery watched. "I reckon I'll be goin', too," he said.

"You're hat's inside," McBride reminded him.

"Couldn't forget that," Kittery said.

He paused, half-expecting McBride to go get it for him. But neither the man nor his daughter made any move until Tania spoke. "I'll be right back, you two. I need to see to the horses before I turn in."

Kittery called his offer of assistance after her, but she refused him and disappeared into the shadows, leaving Kittery and McBride alone.

"I'll get my hat," the big man said.

The two of them walked back inside, and Kittery heard the door shut behind them as he reached for his hat. He turned to see McBride studying him.

"Have a seat, Captain. Please."

Kittery hesitated, then walked over and sat on one of the couches. Rand McBride sat on the one opposite and scooted to its edge. He folded his hands, looking at Kittery. "I'd like to tell you about my daughter, Captain. She's a whimsical g— a whimsical young woman. I know she's not a girl anymore, no. And it scares the devil out of me. But I'll be frank, Captain. I'll fight the man who tries to take her away from me only to lead her into a life fraught with danger. Tania is my life. Her mother is gone, rest her angelic soul. And Tania is all I have left. I'll not see her hurt."

Kittery waited while McBride said his piece. He had no children of his own, but he knew what it must be like to be a father. He had practically raised his youngest sister and had seen her die a tragic death. Because of that, he couldn't be angry with McBride. But the man was jumping the gun.

When the older man finished speaking, Tappan smiled. "Mr. McBride, your daughter's a sweet young lady—a breath of fresh air in a country so full of violence. And I find her beautiful. I won't lie about that. But I barely know her. I doubt I mean one thing to her. Just somethin' new from an outside world she prob'ly don't see much of."

McBride shrugged and resituated his hands. "I said my bit, Captain. Now I'll say this: you're a fine specimen of a man—there's no denying that. You've an honest eye and a handshake that speaks of great restraint, judging by the obvious strength of your hands. Captain, I haven't seen Tania so addled since . . . I don't know when, to tell you the truth. She has a beau by the name of Alan Peck who went off to fight with Custer, but she hardly talks of him anymore. She thinks you're something special, if you couldn't tell. You had best beware of that, whether or not you have any feelings for her."

Kittery chuckled and dropped his head, scratching the back of his neck uncomfortably. Was McBride's intent was to keep him away from his daughter or to drive them together? "Miss McBride's a fine young woman. Please don't get me wrong. I find it hard to believe she would be . . . *addled* over me, as you put it."

"No, you don't," McBride said flatly. "You must be thirty years old,

Captain—or more. You've been around, and you must know how women perceive you, unless I miss my guess."

Kittery was embarrassed, but he met the older man's eyes squarely. "I'm thirty-two, Mr. McBride—"

"Call me Rand."

Kittery paused, surprised again. "All right. Then I'm Tappan. Anyway, I don't see any point arguin' over how women might look at me. I haven't been around many I'd like to talk about—leastways not single ones. I'm honored if your daughter likes me . . . Rand. I'm honored, because I think a lot of her, too. But I'm a busy man. After tonight, I may never see her again. Would knowin' that comfort you?"

Taken aback, McBride stared at him. "On the contrary, Tappan. I'd like to see more of you. Much more. What I'd like is to see you find a respectable job. A man couldn't ask any more."

Kittery sat there suddenly realizing what Rand McBride was all about, and why he had wanted to talk to him. In some kind of snap judgment, McBride had come to the conclusion that he wouldn't mind having him vie for a position as his son-in-law. At that moment, the front door opened, and Tania walked back in. Kittery took that as his signal and stood up, holding his hat in hand and gauging Randall McBride. "I'll think strongly on what you said, Rand. I promise you that."

He held out his hand, and McBride eagerly took it. The older man held on for a long moment, searching his eyes as if for some sign. At last, he smiled and gave Kittery's hand one last, sound pump and dropped his own away.

Tania stepped off the porch with Kittery, and when the door closed behind them they stood alone in a cloak of black, lit only by the single window at their backs. Bats flitted against the starlit sky, and not far off a poor-will cried.

Kittery could see a strand of hair hanging in front of the girl's face, and with a slightly shaky hand he reached out and brushed it away, resting his hand against her cheek. Before he could pull his hand away, she leaned her face into it and raised her hand to press his palm and fingers more firmly against her face.

"Miss McBride . . ."

"I heard you call my father Rand," she stopped him. "Couldn't I become Tania?" She let her hand fall back to her side with a look of reluctance.

"I'd be honored," said Kittery. "I hope to see you again, Tania. Very soon."

She smiled. "I hope so too." She reached out and took his hands, making one small step toward him before stopping herself. She raised her chin, and he thought he should kiss her. Or perhaps he should hold her. She was so close now, of her own will, that he could smell the soft rose perfume of her hair.

But he simply squeezed her hands and let go of them. "G'night, Tania." With a deep breath, he stepped past her and heard her voice behind him.

"Good night, Tappan Kittery."

He swallowed hard and trudged back down the ridge road.

He could hear the merrymaking below long before he reached the street, and he saw the yellow lights smeared up and down the thoroughfare, eerily illuminating the tied horses and the occasional rover who couldn't seem to make up his mind where he wanted to be. Upon reaching the street, he went to the cantina, where he knew Cotton Baine would be.

The dim glow of a cigarette from the front wall drew his attention, and a sultry female voice stopped him. "*Buenos noches, el Capitán.* You would like to buy a drink for me?"

"I don't know you."

The woman laughed as she stood away from the wall shadows and came into the light of the cantina window. She was pretty, with a long sculpted neck and high cheeks. But there was too much rouge on them. "No, you don' know me. But everyone knows el Capitán."

"Sorry, but I have a señorita," he said, and with that he stepped into the smoke-filled barroom.

He spotted Cotton Baine at the bar and crowded in to lean beside him. "How 'bout that ride tomorrow?"

"Name the time, Tap. I'll be ready," the cowboy replied above the din.

Kittery decided on late afternoon, and after several more minutes of idle conversation the cowpuncher excused himself and headed to bed early for the night. Kittery was baffled by his friend's departure. Baine had never been one to leave a party before everyone else, and this was some party.

Kittery turned without interest to the smoke-hazy room. Card games were in progress at every table, with stakes for the wealthy or for the poor. But Kittery had never cared for the odds of gambling—maybe because he never won. It seemed an exercise in boredom and frustration to him, and he was glad to remain at the bar an idle onlooker.

Deputy Miles Tarandon leaned against the far wall. He was dressed in his best tonight, with a long black broadcloth coat and tan *calzoneras*—Mexican-style flare-legged pants, these with black stripes and big bronze conchos down both sides. When his eyes fell on Kittery he lifted a hand and touched his hat in cool acknowledgment. Kittery nodded a reply, then let his gaze swing away at the sound of a loud-voiced gambler. He recognized the gambler as Jed Reilly, the man whose pistol he had lifted on his way to Castor. Reilly sat at one of the card tables, and judging by his curses he was not faring well.

A wiry man with dark hair and a huge curling mustache that did little to hide a naturally smiling face pushed away from the bar and walked straight to Reilly's table. Jed looked up when the man stopped beside him and placed a hand on his shoulder. "Well, howdy, Joe." His words were slurred.

"Sorry, boys," said the newcomer to the others, "but I'm gonna have to take little brother home before he gambles away everything our sainted mama left us."

One of the other men laughed and slapped Jed Reilly on the shoulder. "Take 'im, Joe. *We* have! Took 'im fer everything he owns. Now he needs his mama."

The younger Reilly swore at the man but stood up in one ungainly lunge when Joe tugged on his arm. He reached down and picked up the one silver coin he had left in front of him, throwing his cards at the man who had spoken. As they walked away, the man looked at the cards and guffawed, holding them out for the others to see: a deuce, two fours, a five and a King.

Kittery chuckled, and then his eyes fell on the man who had been seated opposite Jed Reilly. He was a huge, barrel-chested man wearing a red undershirt. His hair was black as soot and smoothed straight back, and several days' growth of gray-speckled beard littered his jaw. The biggest pile of coin on the table was his.

Bison Sabala, the town blacksmith, was one of the other players. He was a heavily muscled man every bit as impressive as the stranger, with shaggy, sand-colored hair and a choppy beard. He, too, wore a filthy longjohn shirt, its sleeves rolled to the elbow. He now scooted to Jed Reilly's deserted spot to give himself more room.

The game continued for several minutes before the black-haired stranger won another hand. Without warning, Bison Sabala towered to his feet, swaying from too much to drink. His momentum scooted the table forward until it touched the belly of the black-haired man. The scuffling of boots marked the retreat of the other players.

For a moment, the blacksmith stood there staring at the black-haired man. Then he spoke. "You've played cards some, stranger. Your tricks don't seem to stop."

Blackie eyed him. "I've played some." His voice was a soft, deep purr, contrasting the gruff one of the man who faced him.

Sabala nodded. "You've been a might lucky tonight."

"That's right, mister." Blackie's eyes glinted like obsidian under a noontime sun. "*Lucky.*"

"It'd be best if you rode out," Sabala said.

"Best for who?"

Sabala glowered at him. "This town don't need your brand of luck."

"What're you sayin', exactly? That I cheated?" Blackie tipped his head forward, and now his eyes were glittering beads set back into his skull, deep shadows cast over them by the lamp above.

"Maybe you did, maybe you didn't. But it makes no difference. We want you out."

"Looks to me like *we* means you. Ain't heard nobody else complain."

"They're cowards."

Blackie's eyes never left the smith. "Well, sorry. I'm just beginnin'."

Sabala came around the table in a lunge and latched his powerful hands onto Blackie's shirtfront to rip him from the chair, knocking it and the table aside. They stood face to face, eye to eye, two men of equal size and strength. Kittery tensed. This place was going to erupt.

Chapter Eleven
Bison and the Bear

For no apparent reason, Bison Sabala backed away in careful steps from the big, black-haired stranger. Kittery couldn't tell why until they came into just the right angle of lamplight.

A bone-handled butcher knife held by the stranger was digging its tip into the blacksmith's ribcage, almost tearing his shirt. Sabala backed away until he came to a stop against another table. He relinquished his handholds on Blackie's shirt, and his hands lifted to the sides.

The exaggerated blast of a gunshot cut through the tension, ringing away to cast itself against the hills. The heavy silence that ensued seemed louder than the shot itself, as everyone turned to locate Miles Tarandon clutching a smoking Smith and Wesson .45, his arm rigid at his side.

"Let's break up the party, boys. Cardona don't like his place shot up, so the next bullet won't go in the floor." The lawman spoke calmly, but an icy danger lay in his eyes and in the tone of his voice.

Bison Sabala, grateful for the intervention, pushed the table out of his way and backed up several more feet from the stranger.

"What's the problem?" Tarandon asked.

"He was cheatin'," Sabala growled. "Cheatin' us blind."

Tarandon shifted his eyes from player to player, and the others shrugged. "What do you say, stranger?"

Blackie sheathed his knife. "I was playin' cards, just like everyone else."

"So the question seems to be, who gets the pot. Is that it?" Tarandon looked from one to the other. Sabala nodded, while Blackie only looked on. "Well, all right," Tarandon went on. "Simple enough. Since you two seem to get on each other's nerves so good, I have an idea. Bison, you say this man cheated. In my book, that's fightin' grounds, unless you c'n prove it. And you're the only one at the table who thinks he was cheatin', or at least the only one with the guts to say so. So here's the deal. You both look like you c'n use your fists. I wanna see you both after the hangin' tomorrow, back at the arena. Then we'll see who takes this money home. And until then, I keep it."

After Kittery watched the two rivals depart, and the noises of the room began to return to normal, he heard a voice at his side. He turned to see a man with a thick but well trimmed beard and short, wavy black hair underneath a cavalry-style hat looking up at him. At a glance, the man's face had strong lines, and his body was well balanced and wiry. He held out a sun-browned hand.

"Hello, Captain Kittery."

Shaking the hand, Kittery cocked his head to one side. "H've we met?"

"No, but there can't be another man in town that matches your description. The name's Adam Beck."

Kittery looked the man over anew. "So you're Beck. Sheriff tells me you brought my horse in. I wanted to thank you for that. What'll you have to drink?"

"No need to thank me," Beck said with a smile. "Nothin' you wouldn't 've done for me."

"You're pretty sure of that. And I judge you to be a *cerveza* drinker." Without awaiting a response, Kittery held up a finger to Cardona and repeated the Spanish word for beer.

As Kittery stood there and shared conversation with Adam Beck, he was pleased to learn everything they had in common. Beck was from over the Carolina line, in Virginia. He, like Kittery, had had strong family ties that drew him to the Confederacy. But he couldn't justify being a part of weakening his country, so he had sided with the Yanks. His family had forsaken him after that, and he drifted west for a stint in the regular army. He had seen much of the same country Kittery had, from the war-torn east to California, from Montana to Arizona, and half a dozen states and territories in-between. Beck had been a buffalo runner in northern Texas, harvesting the great herds along the Cimarron, while Kittery cowboyed for a time in Arizona, along the Mogollon Rim with Cotton Baine. Both Beck and Kittery seemed unable to fit in one place, yet each of them claimed a fondness for this little desert burg of Castor.

The hour grew late by the time most of the crowd had drifted out of the cantina, and Kittery said so long to Beck and trailed out with the last of the customers. In his room in the hotel, he drifted off to sleep.

The following morning, the sun's pale lemon glow burned a void in the fading purple of the eastern sky. Marshal Josiah Raines rose before that sun and was on the gray street watching as its light struck the western rims. The air smelled of cool dust and wood smoke.

Going to the livery stable, Raines walked past a sleeping Jarob Hawkins and brought his dark bay out of the corral in back, saddling it up. In the quiet of early morning, he rode up the ridge road, which was still clothed in blue shadow, and made his way up past the last house and the end of the road to the rocks emboldened by early sunlight.

He drew his pipe from his coat pocket, lit and puffed on it while he sat on the rocks beside his horse and gazed over the vast land. The distant rims of mountains and some not so distant were made hazy by the sun behind them. The massive Santa Ritas loomed close across the valley, half hiding the Empires and the Whetstones beyond. In the great distance the Dragoons were a thin jagged line, a blue not much darker than that of the sky. To the northeast hulked the Rincons, the Tanque Verdes, the Santa Catalinas. From up here a man could see across the continent, it seemed.

And behind him, but over the rise, awaited the Baboquivaris, home of Baraga. Home of his lurking doom . . .

He wanted to go there, to meet his destiny or to claim his fame. He wanted to go there, yet he couldn't tear himself away from Castor—not yet. It seemed there were things left to do there. Well, besides Jason the Coachman's hanging that day. That was a given. That was responsibility. But there was something else. A lonely longing to spend one more day, one more night among friends before he went to join his Emily, wherever she might be.

A movement on the ridge road drew his attention. It was Sheriff Luke Vancouver riding toward town, toward a new day of upholding laws that did little to protect anyone, only attempted to right wrongs after it was too late. Vancouver was an idealist, and he still believed in those laws. But if he lived long enough the day would come that upholding the law was just a job.

Raines waited until the sheriff was out of sight, then knocked the tobacco crumbs from his dead pipe and climbed onto the bay. He rode down to the Vancouver place and rapped on the door. Beth opened it and smiled with surprise.

"Good morning, Marshal. I'm sorry, but you must have passed Luke in town. He's only been gone a minute or two."

Raines nodded. "Well, that's all right, ma'am. I came to talk to you."

Beth looked puzzled. "To me? Oh, well . . . can you come in for coffee?"

"No, ma'am—this won't take long. I just wanted to say something to you—about that man of yours. He's a good one. I hope you realize how good he is to you. One in a million. And, ma'am . . . that's what I want to talk to you about. He's too good to be in this line of work. And because I like and respect the two of you, I tell you this—that move you're planning? Try your best to get him to make it. Get him away from this job and out on a farm somewhere. There are too many men out there who will be glad to put him in a grave. I don't wish to upset you, but that's how it is, even in a little settlement like Castor. I don't want to have to hunt down the man who kills him."

Beth stood there in shocked silence, her hand on the doorframe to support her. Raines gave her a sad look, not wanting to do this to her or make her feel badly of him. But he felt a responsibility to voice his opinion. He owed them that for their kindness.

"Remember what I told you the other day? About lawmen leading a war that never ends? That's true—and it's a war that never pays off. A man slows down as he gets older. Sometimes he gets careless. Then the enemy closes in. The law sometimes seems a wasted war, because there comes a time when a lawman no longer cares whether he lives or dies, after seeing so much death. He finds out he's not doing a whole lot of good anyway. I'm at that point, but it's too late for me to back out. This is my life. I have things to take care of before I'm done, for justice's sake. I'm one of those lawmen who knows he'll die violently. But Luke doesn't have to be. He has too much to live for. So get him out. Don't let him waste his best years."

Beth's fingers played at her throat as she stood there close-mouthed and listened to him. When he stopped, she raised her eyebrows and began to shake her head. "Marshal, I . . . I don't know what to say. I—"

Raines raised his hands to cut her off. "Just heed my words, ma'am. That's all I ask. If we don't meet again, I'm honored to have known you."

Without another word, Raines turned and climbed onto the bay and rode back down the ridge road. Beth Vancouver walked into the house and poured herself a cup of coffee. Today was Jason the Coachman's hanging. Luke would take a hand in killing another man. She sipped her coffee and burned her mouth, sat there and cried.

Tappan Kittery stood on the hotel porch and enjoyed the early morning smells of Castor: wood smoke, breakfast cooking, desert plant life enlivened by the young sun. A train of freight wagons had just pulled in. Twenty-four big mules towed three high-sided wagons with wheels six feet high. Tully, Ochoa and Company, the prosperous freighting firm of Tucson with which he had terminated employment to take the trail of Ned Crawford, owned the wagons. Yelling a hello to freighter Tule Simpton, as his wagon rumbled past, Kittery went down the porch and to the café.

While he ate his breakfast of steak and eggs, he engaged in conversation with Adam Beck and James Price, who had come in after he did. Somehow, during the conversation, Price's occupation came up.

"How'd you come by the job?" Kittery asked.

Price glanced up at him, a fork of beef poised in mid-air. After a thoughtful moment, he finished taking his bite, chewed and swallowed. He leaned back in his chair, a half-amused smile on his face.

"Well, Captain, I studied at Princeton to be a hangman."

Kittery glanced over at Beck to catch his reaction, then shrugged and took a sip of coffee. It was Price's business, he guessed. He wouldn't press it and be toyed with. But the hangman grew serious after a moment.

"I don't mean to be frivolous about the subject, Captain. It's not something I talk about very often. To tell you the truth, I did study at Princeton, but to be an English professor. I don't know what it was about the East, though. The people, the smells—I couldn't say. But something got to me, and I headed west. I was a sharpshooter in the war, so I decided to try my hand at buffalo, like Beck, here. But before I could do that I ended up in a shooting scrape over some settler's wife being brutalized by two scamps in Abilene, Kansas. As it turned out, one of them had a reward on his head for five hundred dollars. That reward changed my life.

"I decided to hunt men, which I understand is what you're up to now. As you probably have learned, bounty hunters aren't often looked upon highly, but I brought an outlaw into Tucson once against whom Governor Safford had a personal grudge. He invited me to dinner to show his gratitude, and we became friends. When the job of hangman came open he contacted me, and the rest is history, you might say."

"So. Four years of college, studying English. Looks like we have a real genius here in our midst," Kittery said with a smile.

Price scoffed. "Genius? Sure. That's why I kill men for a living."

That comment put Kittery to silence. He looked down at his plate, mulling over the words. At last, when no one else seemed inclined to speak, he broke the silence. "Someone has to bring justice to this land."

He was looking down and thus missed the look of satisfaction that passed between Beck and the hangman.

Joe Raines lay on his bed reading a book Kittery had given him his first day in town. It was called *The Book of Mormon*, and some traveling missionary had given it to Kittery's folks, back in Carolina. They had taken such a liking to the book, which was supposed to be a true account of how Indians—called Lamanites in the book—came to live in America, that they had named his two older brothers, Abinidi and Jacob, for characters in the book. Raines knew some about the Mormons. Everyone did. Some called them an evil scourge, but if this book was theirs he didn't see how that could be the case.

He had been reading it since his return from the Vancouver spread, and as he closed it he pondered on the profound wisdom he found between the covers. Judging by its first hundred pages, it was a book every man should read. It put a lot about life into perspective, whether it was true or not. If it wasn't true it had been written by a man who was vastly wise. Maybe a committee of them.

He would have liked to finish the book, but there were other affairs to tend to. Mainly, he was to oversee the hanging of the "Coachman" at noon, and it was surely past eleven o'clock now.

Going to the street, he saw it was nearly empty. Almost everyone had already headed back to the gallows in order to get as close as possible for the neck-stretching. He stepped across to the sheriff's office and walked inside.

Vancouver looked up from coffee. "Hey, Joe. You ready?"

"Is a man ever ready for something like this?"

Vancouver shrugged. "He's a killer. Anyway, it's fifteen minutes to noon, so while we wait . . ." He stood and unlocked the padlock that secured the chain running through the trigger guards of the shotguns and rifles in his rack. "It'll help prevent trouble if we appear anxious for it."

Raines agreed, and he withdrew a Winchester carbine, while Vancouver picked a shotgun. When they had loaded them, Vancouver snapped his shut and looked at Raines. "Let's go."

They went to the condemned man's cell. Jason Hall, alias Jason the Coachman, sat in brooding silence, staring into a corner of the cell. The sheriff opened the cell door, and the Coachman seemed to shake himself out of a trance. He looked over at Vancouver and smiled, showing two rows of perfect, sparkling teeth. "So who's goin' ta do the dirty deed t'day, Sheriff?" he asked in what sounded like a Scottish brogue. "Come on, now—donut be leavin' me a-hangin'." He laughed loudly and good-naturedly, and when the sheriff only smiled in reply he turned somber. "I'm sorry, my man. I wouldna be doin' this to ye if I cud go back in time."

Sheriff Vancouver met his eyes and nodded. "I know, Hall. Don't worry about me. You have other things to think about."

"Such as soilin' me pants 'n' a-buggin' out me eyes?" He laughed again. "I'm sorry. Let us go now."

Vancouver just nodded again and led the bearded man out to where Miles Tarandon and James Price had come to wait by the desk, heavily armed.

Hall's eyes lit up when he saw Price. "Aye, so it's you'll be doin' the deed t'day! I've seen your work b'fore, sir. Ye stretch a fine neck, ye do. An' there ain't noon as be finer than mine."

Price gave a half-hearted smile to the Coachman's hearty laugh, and the five headed out behind the jail to the gallows.

Surrounding the gallows, a crowd numbering easily three hundred people sat on the arena boards or leaned against the nearest structure they could find. Many had brought chairs, but they were of little use when others crowded in front of them for a better look. A few families had brought picnic lunches and didn't seem to notice or mind the dust that drifted down over their fare. Dogs and little children weaved through the crowd.

The lawmen herded the miner through the throng, nudging aside those in the path. They climbed the faded wooden steps with grim deliberateness. When the murmuring died down, Luke Vancouver stood on the edge of the platform and raised his voice for the ears of those on the fringes of the gathering.

"Jason Thorpe Hall has been sentenced to be hung by the neck until dead for first degree murder on this, the twenty-first day of March, eighteen hundred and seventy-six."

He turned and walked to Hall, accompanied by the padre from the mission San Xavier del Bac, nine miles south of Tucson. The Coachman, pale and sweating, stared at his feet. In the time it had taken to walk out of the jail his entire demeanor had changed. "Any last words, Hall?" asked the sheriff.

The condemned man shook his head, one drop of sweat dripping off his chin. His face was ghostly pale. He looked up and worked up a scared little laugh. "When they asked me to hang around, I'm afraid I didna know exactly what they meant. 'Twas a sad day fer me, the day I pulled the trigger."

The priest prayed for him in Latin and made the cross on his chest. Then Hall looked up and gave a quick nod as Price slid the black hood over his head. The noose came next. Price settled the knot beneath Hall's left ear.

The hangman stepped back and placed the palm of his hand over the top of the big wooden lever. His hand twitched with tension. He saw the hood begin to shake, at first almost imperceptible, then violently, and a loud sob came from beneath it. Raines nodded at Price, trying to ignore the obvious terror of the man beneath the hood. Price's hand shot forward in one crisp motion. The hinges cracked. The trapdoor swung down. The Coachman's boots plummeted, then jerked to a halt mid-air and twitched three times. The scaffolding creaked eerily with the final gyrations of his body.

They were calling him Blackie, and some referred to him as the Bear. Bloody Walt Doolin didn't care what they called him. He'd been called worse. If they pressed him for a name, he was Howard Gordon. Until then, they could call him any nickname in the book.

In height, Doolin just topped six feet. But it was the mass packed into that six feet that made him impressive. He had a barrel chest, legs like oak trees, and a slight bulge to his stomach that, contrary to appearances, was as hard as the granite of his sunburned face. He smoothed his thinning, jet-black hair straight back, and a touch of dark whiskers peppered his jaw. The .36 caliber Spiller and Burr hanging from his hip was outdated, its brass frame worn and tarnished. But Doolin preferred a rifle, anyway, and the pistol was seldom used. Consequently, its working parts had not been shot loose, a common ailment of brass-frame revolvers.

At fifty-five years, Bloody Walt Doolin was the oldest of the Desperados Eight, and beyond doubt the most brutal. For use at close range, he used the butcher knife he had pulled on the blacksmith the night before. And once he had an opponent down, he had been known to carve his initials into their flesh, whether they were dying or dead. It came natural to him. His own father, in a drunken fit, had done the same to him when he was ten years old. It was true that Doolin wore red longjohn shirts, as the papers said. But he had to laugh when he read about his redying them in the blood of his victims. He didn't even like the smell of blood on him, or the sticky way it felt.

Bloody Walt Doolin knew he shouldn't be here preparing for this fight. Baraga would be furious over it, and Baraga didn't often become furious. Bloody Walt Doolin had no business placing himself in the public eye this way. That was the only reason he hadn't sunk his blade clear through the smitty's belly the night before. But curse that drunken blacksmith. He hadn't cheated, at least not much, and he'd be hung if they were going to take his money away from him. Besides, Sabala needed a good licking. He was far too proud of the strength his trade had given him.

Doolin stood in the shadows of a dark shed he had found and smoked his cigar. He stretched his arms and his back, rolled his shoulders and pounded his fist into the palm of his other hand. If Bison Sabala wanted a whipping, he was about to get one. And then after that he hoped the marshal made his move soon. Bloody Walt Doolin didn't like being cooped up like some animal. Let the marshal make his play or be hanged.

Tappan Kittery leaned against the hotel's doorway, chewing on a match and enjoying this first day of moderate heat he had experienced since coming to Castor. A lazy breeze drifted down from the north. This was more the way the weather in March should have been all along, and it was good to see it back for a change. He watched for some sign of the big fight between Bison Sabala and the stranger, a battle most had come to refer to as the fight between Bison and the Bear.

Many of the visitors who had come for the weekend hanging would remain in town until the following day. Some would stock up on supplies while others crowded the cantina and the occasional residence which had prepared its last minute plank bar and whisky keg business, taking advantage of the crowd. This was the day when every merchant in town would not only make his money, but earn every cent of it.

Every man, woman and child awaited the big match in the arena.

Kittery ambled out and sat down on the edge of the porch. His boots rested in the thick white dust as he watched the passersby. A piebald hound trotted up the center of the street, dodging the boot of a miner that swung at him as he came too close. Fifty yards behind him came four boys of eight or nine, hollering at the tops of their lungs. "They're fightin'! They're fightin'!"

As if by an enormous tornado the entire street was cleared in minutes until everyone milled around the boarded arena. Kittery shouldered in until he was leaning against the outside edge of the arena itself.

Deputy Miles Tarandon surveyed the scene as Bison and the Bear squared off. "You two have your fight. But make it a fair one."

As if egged on by the words, the two men peeled off their shirts. Bare to the waist, they looked like a couple of gorillas, their faces, chests and backs covered with curly, matted hair. Beneath the hair on Blackie's chest, several long, pasty-looking scars seemed to form some sort of design, as if they'd been carved into his flesh for that purpose.

Like the night before, the black-haired man's features bore no expression.

He watched Sabala through dark eyes slanted slightly upward at the outside. His shoulders stretched down to join long, powerful arms that gave way to hands swinging like meat hooks along his thighs. At every step, his legs rippled beneath his drab brown plaid trousers.

Sabala looked equally impressive, his hands clenched in fists, his sculpted arms bent at the elbow, his lip curled up beneath his beard in ferocious anticipation. The two faced each other across the inch-thick dust like railroad locomotives.

Then, without warning, Blackie's right arm shot out. His fist skimmed an ear as the other man dodged with surprising speed. Blackie, attempting to turn and face Bison Sabala again, took a blow that smashed like a nine-pound hammer against his ear. It sent him to his knees as dust rose all around.

He came up with incredible swiftness, swinging with both fists, blows that clipped the air. For a moment, while his foggy mind cleared, he fought like that, swinging but hitting nothing. His eyes seemed to clear just as a big ham-like fist drove forward. With a quick block by his left hand, he smashed a blow to Bison's nose, sprinkling blood over the blacksmith's face and chest.

The blow knocked Bison Sabala several steps backward, but he caught his balance and raged back in. He sent a punch that caught the Bear under the ribcage, ripping the wind from his throat. He aimed another at Blackie's forehead, but the man bent forward, and it scraped the top of his skull.

The black-haired Bear lurched close and threw his massive arms about Sabala. Pinning his opponent's arms to his sides, he squeezed with all his might. Sabala's face began to turn bright red, and he tried in vain to free himself from this power that threatened to knock him out. Seconds passed, and Bison looked faint. He swung his head from side to side in the manner of the beast he took his nickname from. His eyelids filled with blood and seemed about to close his eyes over. But in a last effort, he brought a booted foot up and with crushing force ground it down on his opponent's instep. The Bear, emitting a roar, tried to retain his hold on the other man. But with one final surge Sabala tore free and spun away, sucking air in gasps.

Blackie came in with a frustrated snarl, favoring his injured foot. He tried once more for his death hold on the blacksmith. Sabala was wary of this move, however, and he backed off, moving too near the arena's edge. Warnings drifted up from his friends in the crowd, but his concentration was so intense he appeared not to hear.

When Bison hit the boards, they cracked as he lost his balance and went to a knee. Blackie piled on top of him, crushing him to the earth. Grappling, they rolled over and over. They broke free of each other and came up in a swirl of dust. It stuck to their sweat- and blood-covered bodies as mud, lending a more barbaric air to the raging combat.

Everyone watched in awe. These were two machines, nearly perfectly matched, and for either to win he would have to be blessed by a tremendous stroke of luck.

Bear swung a wicked foot forward, knocking Sabala's tired feet from under him. The blond one went down but eased his fall with an outstretched hand, then grabbed for the other man's anchoring leg as the Bear sent another kick at him. The blow took Sabala at the base of the neck, and he fell back into the dust, rolling over and coming back up on his knees to face the black-haired man. Blackie stepped toward him as he staggered to his feet.

Both men heaved for air as they circled and spat blood. The crimson fluid streaked the dust on their chests and faces.

Sabala roared out in anger and issued an oath. He sank his heavy boot upward into Blackie's midsection, jolting him backward. Then he stumbled forward to where the Bear stood coughing. He planted steel fists into his ribs and chest, and the Bear gave them all back. Before the Bear could step out of range, he hit him with a roundhouse right to the jaw, and then two more left jabs.

The stranger shook his head and closed his eyes against the pain, trying to hold his arms up. He took two weak steps toward his opponent, but his knees buckled, and with a ragged sigh he toppled backward, falling like a dead oak tree. In the dust, he lay still as the air ran from exhausted lungs.

In the late afternoon, Tappan Kittery and Cotton Baine saddled their horses and rode toward the Santa Ritas. They made the road in silence for a time as they climbed up out of cactus country into grassland dotted with yucca and mesquite trees hanging with dried up pods. Kittery cast furtive glances at his friend and tried to read his thoughts. Finally, he broke the silence, judging Baine with his eyes.

"There's been somethin' on my mind, somethin' that needs sayin', though you may not like it."

Baine stopped trying to make a show of enjoying the landscape and scooted his rump around in the saddle to face Kittery more fully. It seemed he figured on being in that position for a while.

"Okay, Tap. Go ahead and say yer piece. I knew it was comin'."

"You had somethin' to do with Juan Torres bein' shot. I already know that, so there's no point in lyin' about it. I would hope you wouldn't lie to me, anyway. I'm not tellin' anybody. But I wanna know what the story is. I'll lay my cards on the table. I've heard about some kind of vigilante movement down here. It was nothin' official, but you know how news travels the backroads."

Silence ran its length while the younger man stared down at his bobbing saddle horn, rubbing its smooth surface with a thumb. "Tap," he started, "you been around. Yer a man who knows the world. And you gotta know it's a hard place for decent people. Sometimes a man's gotta do things he really don't want to if people 're gonna keep on livin', and livin' in peace. Yeah, sure. I was with them that shot Torres. It don't sound like there's any use dodgin' the ball. You always had a knack for knowin' everything."

"Why'd you do it, Cot? You crossed a line I never thought you would."

"Oh, use yer head, Tap! Look around you. You think Vancouver c'n handle what's goin' on around here? You think anybody can, if they stick to badges? It ain't happenin'. I hate an outlaw, Tap. I hate 'em with a passion. But it's gonna take outlaws to get rid of outlaws this time, so I'm ready to be an outlaw. I can't stand seein' people hurt no more, not when I'm not doin' a thing to stop it. Torres needed killin', at least for an example. He was one of the ones that attacked you, and they meant to kill. He had a part in the murder of that little girl, and I couldn't let 'im get away with it. The only way to be certain he paid was to do it ourselves. And he paid. For once, and for dead certain, he paid, Tappan. I ain't sorry for it."

Kittery studied his friend for a long time before returning his eyes to the road, wordless. He had heard Torres had died later the day he was shot. That, at least, was merciful. Baine looked at him until he knew he wouldn't speak, and then he went on. "It's a young land, Tap, with wild oats to sow. It's chock full of good folks, but there's no forgettin' the bad. When the law can't handle 'em, other folks have to."

"So the vigilance committee ain't a rumor," Kittery said at last. "I didn't think so. I only hoped. I've read about 'em, in California and in Montana, back in the sixties. They're no good. And you're in it?" Kittery met Baine's eyes.

"I reckon so. Up to my eyebrows." Baine returned his eyes to the road and straightened around in the saddle.

"Who else?" Kittery pressed.

Baine looked over and smiled uncomfortably. "Now, Tap. You're one of my best friends, but you don't really think I'd say, do you? There's an oath a man signs when he becomes a vigilante. I swore to keep my silence—or die. I signed my name to it. I won't go back on it."

Kittery didn't respond. He pondered this new revelation and all it entailed. Was Vancouver with the vigilantes? It wasn't likely, but he wouldn't have figured on Baine, either. Still, the sheriff was too proud of his badge, even if it wasn't big enough for the job. No lawman alive had a badge that big, at least not for Arizona.

Who else would be in it? It was difficult to guess. James Price was a good bet, in spite of his derisive comment about killing men for a living. And Adam Beck. Kittery liked him, but he was a perfect candidate.

Baine turned in his saddle again. "They asked me to approach you, Tap. To see if you'd join us. They want you bad. You oughtta join, now that you know what's happenin'. Yer the kind of man we're hurtin' for right now. With yer fightin' experience and natural leadership, you'd do us a world of good. And it's not like it'd be that much of a change from what you do fer a livin' already. Tappan, we're tryin' to stop the Baragas and all the rest of his breed from overrunnin' the entire territory. We need all the help we can get. It's too much of a task for the law right now. It's outta control."

While Baine talked, Kittery fought against his instincts. He had often thought the only way to cure the country's ails would be a vigilante force, but he was aware of the unruly nature of mobs. That was what vigilantes too often became—a mob. Not everything Jake had ever tried to teach him had sunk in, but he remembered well his admonition about groups of violent men. *When angry men get together to solve a problem outside the law, instead of many minds thinking together, no mind thinks.* All of fifteen or sixteen years later, Kittery still remembered Jake's words.

"No," he replied. "I'm sorry, Cotton. Maybe you think I'm wrong—maybe huntin' bounties *is* the same as bein' a vigilante. But there's some element of law to it, at least when I'm the one at the reins. I try to bring my men in alive. I've only ever had to kill one. And they get a real judge and jury, not a bunch of drunks with flour sacks coverin' their faces, hidin' out in the night.

"I'm not passin' judgment like a vigilante. I walk inside the law, and you don't. The worst thing about vigilantes is the same as the best thing about 'em: their numbers. There's enough of 'em to do a job, but too many to control."

"Come on, Tap! I've seen you deal with people. They'd listen to you."

"Not all of them would. And all it takes is one or two not listenin' to get somebody innocent killed. Somebody has a grudge they wanna take care of. Somebody's got a hair-trigger temper. You can't control that many men."

"The army does it," Baine countered.

"Yeah, the army . . ." Kittery stopped and shook his head. Baine was right. The army did generally pull it off. But still, it wasn't right. The army was authorized.

"Well, Cot, no matter what you say, I won't join you. I guess I should ride out now, 'cause I won't turn in a friend for doin' what he believes is right. But I can't join you. We may as well head back to town."

Puffs of cottony cloud ranged the blue expanse overhead as the two reined their mounts about and followed the clear-cut wagon ruts back into town. They ran into Joe Raines under the awning of the cantina. The three exchanged words, and then Baine excused himself and headed to his room.

Raines went back inside with Kittery, and they found a table in the shadows. Raines had already had a drink or two, but he took another and pulled out his black pipe to tamp it full of tobacco. "You're a good friend, Tap," he said suddenly.

Kittery put down his glass and leveled his gaze at his friend. "So are you, Joe," he said after a thoughtful moment. He wondered what had brought the comment on, but he didn't say anything about it.

"When are you plannin' on goin' after the Desperados?" he asked. "Waitin' for somethin' special?"

Raines was staring out the swinging doors, and after a moment he realized he had been spoken to. "Oh. Uh, just the hanging. And it's done. So I say we start getting our gear ready and think about two days from now. I planned to tell Vancouver."

Kittery nodded. He was surprised Raines and Vancouver hadn't already spoken of it. It seemed a vital matter, if things were to be organized properly.

"I'm readin' the book you gave me," Raines said of a sudden. "*The Book of Mormon*. You say it's supposed to be a true account?"

"That's what this Pratt fellow who brought it to my folks said. Those Lamanites it talks of, those are Indians. You'll read about 'em breakin' up into groups. Those are different tribes. It all makes a lot of sense, if you study it out."

"Well, I think so. If it isn't true, it should be. There's a lot of wisdom in it, once a man opens his mind to it. Tap? Does your life make you happy?"

"What sort of question is that? I guess it does."

"Good. What about that banker's daughter? She sure seems like a nice girl."

"What's got into you, Joe?" Kittery asked. "You're talkin' a little odd t'day."

Raines laughed and took a swig of whisky. "Too much to drink, I guess." He stood up unsteadily and smiled down at Kittery. "You remember me saying a girl was going to come along one day and you'd be tied down like you couldn't believe? Maybe that banker's daughter's the one. You'd better make sure. Every man needs a good woman once in his life."

Kittery laughed. "Maybe so, Joe. Maybe so." He stood up, and when Raines held out his hand he shook it, searching his eyes. His friend held on for a long time before turning away from the table. Then he turned back, when only a few feet away. "You're a good friend, Tap. The best friend a man could ask for."

And with that he was gone.

In the wee hours of the morning, Joe Raines woke up with a headache and swung his bare feet to the hard cold floor. He moved sleepily to the washbasin, where he splashed water over his whiskered face. When he had grown sufficiently alert, and the pounding in his head had receded, he washed himself from head to foot with his wash rag. He shaved, except for his curving mustaches, jaw-brushing sideburns and a chin tuft beneath his lower lip. It had been years since he had worn one of those, but it didn't look bad. Somewhat sophisticated, in a way. He dressed in his black suit and tie, the dust smoothed away.

There was always in his mind that image of the heroic capture of the Desperados Eight. Yet he was a pragmatic man. He knew the odds of his returning victorious from the Desperado Den, and that was the reason for his suit. This was the Joe Raines he wanted them to remember, whether he came back or not. Everything he had done in his life had been with style. If he had now to go to his death because he chose to hunt the Desperados alone, that would be in style, too. If his finest suit got a few ragged bullet holes in it, well he'd never notice. The name of Josiah Pierre Raines would live on in Castor, some recalling a reckless fool, others a gallant hero. He had made some good friends in Tucson and in the quiet little town of Castor. They would remember. Tappan Kittery would never forget and would be proud to say he had been the friend of such a man. He grinned wryly to himself. The man who rode after the Desperados Eight in his best Sunday suit, with two pistols, a rifle and a saddlebag full of dynamite. So let it be. They would all remember the raw courage of this lawman, riding out as he had ridden in . . . alone.

With everything in order, weapons cleaned and loaded and supplies packed, Raines slung the saddlebags to his shoulder, took his bedroll and Winchester in hand and descended the stairs.

Outside, the street lay in its cloak of black, and the air was brisk and filled with the fragrance of morning vegetation, which a cool breeze carried down from the ridge. Stars hung bright and winked as if to say they would remain with him and keep him company until the end. Far up the ridge, a rooster sent its weak message downslope, a far-away, unreal sound in the stillness.

Raines stepped into the livery stable, reaching behind the door for a lantern. When he touched a match to its wick and moved to the bay's stall, he was surprised to see Jarob Hawkins there. Hawkins looked at him with a knowing glint in his eye.

"Kind of early for you isn't it, old man?"

"I got all my sleep yestiday. I plan to stay awake the rest of my life."

Joe Raines smiled. "Yeah. Me, too."

The old hostler watched in silence as the marshal saddled up, not speaking until the work was finished. "You'll be headin' one of two directions, marshal. Which is it?"

Raines eyed him with a faint smile. "You seem like a right perceptive fellow. You tell me."

The old man chuckled, and the lantern light cast canyons and dusty washes across his weathered face. "Straight west into hell. I wish I war goin' with you. If I had two legs, you couldn't do nothin' to stop me."

Raines lowered the lantern, missing the sadness in the hostler's eyes. Hawkins held out his hand, and Raines shook it with a firm, friendly warmth. At last, he smiled grimly. "Tell them all how I was dressed, would you, Hawkins?"

And then he rode away.

Looking over his shoulder, Raines could see the shapes of the buildings, and he thought he saw a light glowing in the hotel and had the feeling it was Kittery's. But then the trail took him out of sight of Castor, and he rode along through the dark.

The ride to the Baboquivaris had just begun.

Fifteen miles into the desert, with the sun climbing an invisible ladder into the sky, Joe Raines was startled by the sound of three successive, evenly spaced rifle shots from somewhere in the bleak stillness behind him.

An instant feeling of elation flooded through him so that for a moment he felt like he was riding on air. Tappan Kittery must have found out somehow that he was gone, and he was coming after him. Perhaps with a posse. Jarob Hawkins had probably sounded the alarm, the old coot!

But yet . . . that wasn't what he wanted. He wanted to do this thing alone. Dang that Tappan! He should mind his own business. Yet he couldn't help smiling. Tappan was the best of friends.

He shook his head, pulling the horse off to the side of the trail and staring back toward Castor. In all that vastness of rock and brush and sand, nothing moved but what was moved by the wind. There was no dust to indicate coming riders. For a moment, he watched the faded red blossoms of an ocotillo rustle on the end of its wire-like arms. The seedpods of a mesquite rustled together when a breeze talked among them. Raines waited, growing uneasy after a while. Finally, he urged the horse on. If they wanted to catch up to him, they could. He was riding slow himself to prevent raising any dust.

A man shouldn't think. It makes him worry. And worry makes the stomach hurt. The more Joe Raines checked his backtrail, and the more he failed to see any cloud of dust, the more worry crept into the edges of his mind and his stomach began to ache. If Kittery and a posse were after him, or Kittery alone, they would be riding fast. Why wasn't there sign of it?

Joe Raines stopped again at trailside. Truth was, those shots were too distant, and the desert too vast. They could have come from almost anywhere. It could have been a hunter firing at game, yet they seemed so evenly spaced, so deliberately slow. Why would a hunter have fired that way?

Ahead were the Baboquivaris, the bald dome of Baboquivari Peak thrusting out of the range like the fist of Atlas. It called him like a beacon—a beacon of death. If Kittery and anyone else from town were coming at all, he wished they would come and be done with it. The suspense was hard on his heart.

Not much farther on, he drew his mount into the brush again and waited, worry gnawing at him like a bored animal. He stayed there for half an hour this time, and still he heard and saw nothing. So he continued.

Joe Raines had come far since the first three rifle shots. He leaned forward as he rode and stroked the bay's neck, and the horse nickered. He pricked his ears forward, eyes searching the gray rocks they had started through. Something was wrong here. Deadly wrong.

And then the shots came again, only this time they couldn't have been from more than four hundred yards behind him. He whirled in the saddle, expecting to see Kittery riding up on him. But again, there was nothing. And then, from out of a wash, came a dun horse. One horse, and on top of him was a man wearing a faded red shirt. He didn't recognize the man, but . . .

Joe Raines jerked around at the sound of metal clinking against stone and the jingle of a spur. His head swung at the ominous sound of rushing feet and rustling trousers, the click of pistols being cocked and a shell being jacked into a Winchester. In that sudden moment, seven men stood all around him, eyes glittering.

One of the men was missing an arm.

Part Two
The Bloody Season

Chapter Twelve
The Fury of the Guns

With the men swarming around him, Joe Raines froze for only a second. It was a second he couldn't spare. Someone seized his reins at the bit. His horse sidled and tried to jerk away, but the man held tight. The man had to be Slicker Sam Malone. He appeared to be in his mid-thirties, a gaunt-faced man with a crooked nose and a hanging scrap of light brown hair shading his lips. He was the least feared—and least respected—of the entire band, known as the last to join a battle or to raise a gun.

Involuntarily, Raines's hand swept to his holster, a brown hand had reached up from beside him and jerked the Smith and Wesson from its holster. He looked around as a huge man who could only be Rico Wells pulled the saddlebags from behind his saddle, then shucked the rifle from its boot. Raines searched the faces of the others, each in turn. Silent eyes stared back. It was almost eerie, the quiet of the entire encounter. No one had spoken a word.

A dark-skinned man with a walrus mustache, undoubtedly Crow Denton, stood spraddle-legged in the rocks, his rifle leveled at Raines's midsection. Morgan Dixon was stationed several feet to his right, brandishing an Ethan Allen shotgun. And Silverbeard Sloan held his two Colts, their barrels staring the marshal down like a snake's eyes. To the left of the horse was the man with one arm, and he clutched a silver-plated short-barrel Colt, besides one in another holster, butt to the front, one in his waistband and another hanging from a homemade gun rig around his chest. Only the man Raines pegged as Colt Bishop held no weapon. His pistol rested in its holster, and he watched the outlaws around him as much as he watched Raines.

There was a clopping of hooves, and the man on the dun came trotting up. He climbed down from his horse and lashed it to a mesquite tree. His face was covered with purple bruises, and one eye was swollen nearly shut. Raines didn't mean to smile, but he did. That brought the first words from any of the outlaws.

"What're you laughin' about?" asked the big man with the battered face.

"You must be Bloody Walt, the man they've been calling the Bear. I heard about you in town. Looks like Bison Sabala walloped you good."

"You'll make a good piece of carvin' wood," growled Doolin, his eyes flickering toward Baraga.

Raines ignored the comment and let his eyes roam the faces of the others. He was struck again by their utter silence. "Hello, boys." A cool unfriendliness swam his shadowed eyes.

Baraga's face remained expressionless. "I've heard a lot about you, Marshal. Wells said perhaps you had come to take Baraga alone, and I said, 'If so, there's a fool.' A fool, Marshal, and now I see we were both right. You came alone, and you are a fool. Who brings you rescue now?"

Raines brushed off the pointless question, studying the man's stub of arm. "So you're Baraga."

"So I am."

"Funny. Somehow I thought I'd be impressed. Now all I see is you, after all these years."

Baraga's lip corners lifted, and then he laughed, his eyes crinkling up at their corners. "Ah, Marshal. You have a wit. But the almighty Federals took away from me anything that should impress a man."

"One arm broke you?"

"The loss of one made me what I am," countered Baraga, his voice going quiet, his face still.

"United States marshal!" cut in Silverbeard Sloan with a sneer. He spat in disgust. "You fell into our hands like a baby."

Raines stared back at him, his eyes bearing down with despite. "Shut your mouth, Sloan. Or give me a chance and I'll be happy to do it for you."

Baraga's eye corners crinkled again as Sloan's mouth clamped shut. Sloan's eyes went wide and crazy, and his hands tightened around the grips of his Colts. "Watch yourself, Beard," warned Baraga. "I'd hate to see you betray me." When he could see Sloan wasn't going to do anything but glare, Baraga looked back at Raines, demanding, "Where's your posse? As I was saying earlier, Wells placed a bet that you might come alone, but that didn't make sense. I figured you'd have ten or fifteen men with you."

"Yeah, so we could feed our pet buzzards again," interjected Slicker Sam Malone.

"They're riding down your neck right now," lied Raines, completely ignoring Malone. "Maybe you'd better let me go."

Baraga let another chuckle escape him. "Ah, Marshal. That just means we'll have to kill you quicker."

Raines lifted his chin. "There was never a posse. This is my job. I came alone."

"Then you *are* a fool," said Baraga, shaking his head in disbelief. "Well, I respect a nerve like that. I can't say much for your brains, though."

Raines ignored the jab. "There are a few things I'd like to know."

"Ask."

"What became of Blue Bell Smith?"

"Wells killed him after he talked to you. No, it didn't take us long to receive that information."

"So Randall—he works for you then?"

"Shorty? Yes, he's a good hand—for a worthless cuss," said Baraga.

A lengthy moment of silence ensued, when the only whisper was that of the wind in the rocks and the swishing of the bay's tail. Somewhere back in his mind, Raines heard Baraga speak. "Marshal, it has been a pleasure, but . . ."

The marshal's eyes followed the lift of the Colt's muzzle, and to the sides he saw several others follow suit.

Strangely, his mind turned then to Tappan Kittery's *Book of Mormon*, resting against the spare revolver in his saddlebag. He remembered names of men in the book who had died for their cause: Abinidi, Gideon, Teancum. He had never had the opportunity to read the final few chapters and never would, for he would now join the ranks of their martyred dead.

The wrinkles flanking the edges of his mouth deepened in an ironic smile. The smile deceived the outlaws, for Raines's mind was galloping ahead. The other outlaws seemed to be waiting for Baraga's opening shot, but the outlaw leader's thumb rested on the uncocked hammer of his Colt.

Ducking low in the saddle, Raines jammed the spurs along the bay's ribs. It hurtled into a leap straight toward Sloan. Sam Malone, still clutching the reins, was jerked from his feet and thrown to the ground. A flying rear hoof took him in the ribs as the bay lunged past.

Lead whipped about Raines. The horse's shoulder slammed into Sloan, sending him in a whirling spin against the rock wall. He lost his grip on the pistols and went down hard, skidding on the rocks.

The marshal whipped his horse with the rein ends. He spurred him cruelly, leaning down over his neck. He dared not look back. He made thirty yards. Behind him curses filled the air.

The shattering crash of Dixon's shotgun out-thundered the other weapons in the confusion. A string of buckshot smashed into the lawman's back and shoulders. The bay ran on, but with one more barrel Joe Raines let go and fell to the dusty rocks. He lit heavily, and more lead sank into his flesh. In a fever, he rose to his knees, facing the outlaws. He clawed at his empty holster.

Two slugs ripped into his upper chest, knocking him onto his back. Blood rose in his throat. He tried to yell in defiance, but choked and coughed instead. He shuddered under the impact of three more bullets, and clutched the empty holster, his knuckles turning white with the pressure. His hands twitched. His muscles contracted and relaxed.

Silverbeard Sloan was up now and triggering his revolver over and over again into the lifeless body. The outlaw's shirt was tattered from falling against the jagged rocks. Crimson ran into the silver of his beard from a bad gash on his forehead.

Sloan ceased firing when his guns were empty, and silence dropped onto the bloody scene like a blanket. The only sound was the ringing memory of gunfire. Then Walt Doolin stalked forward, a gleam in his black eyes. He drew his butcher knife, testing its razor edge with a fingertip. Then he knelt, the knife in his right fist, and gripped the lawman's shirtfront, tearing it open . . .

Tappan Kittery glanced up at the glaring sun. By now he knew he couldn't be far behind his friend. He had left perhaps as much as an hour behind him but had pushed the stallion at a brutal pace to close the distance.

It hadn't surprised Kittery in the still of that morning to find Raines gone. The way he had acted the last time they had talked had been more than just the effects of alcohol. Kittery knew that instinctively. He was also convinced, from thinking back over the course of their conversations since he had come to Castor, that Raines had never intended on taking a posse with him into the Baboquivaris. Too many things he had said suggested this plan to go alone. Unfortunately, Kittery hadn't put it all together sooner.

It was Kittery's sixth sense that awakened him that morning. If not for that, he would be even farther behind the lawman. Nothing tangible had brought him to. The world was still asleep. But he had a feeling, and he went to the lawman's room, and he was gone. Within fifteen minutes, Kittery, too, had taken the trail, after ascertaining from Jarob Hawkins which way the marshal had gone.

Just shy of nine o'clock, he crossed a sandy wash and saw where the marshal had stepped down to stretch his legs. He also found a second set of tracks, these from a larger horse than the marshal's. Some of them ground out those of the bay. There was someone besides himself trailing Raines!

Wishing he had brought Vancouver or Beck or Baine or Price with him, but knowing there had been no time, Kittery rode on, trying to concentrate on the trail ahead. Later, he drew the stallion in short, certain he had heard the faint whisper of a gunshot. Yes! There it was again, and again—shots evenly spaced. Whatever they meant, they came from somewhere far ahead, and there was little he could do now. So he kept the black to the fast, long-legged walk he was accustomed to. He had to keep telling himself that a gallop would not only endanger himself and the horse, but possibly Raines, too. If no one knew he was on his way, Raines might live longer.

Less than half an hour later, he heard a brief, furious barrage of gunfire. He swore helplessly. He *had* to reach the marshal now or it would be too late. Perhaps it already was . . .

Within ten minutes he heard hoofbeats drumming toward him across the desert. Drawing his Remington, he wheeled the black and sent him crashing into the brush.

Chapter Thirteen
A Graveside Promise

He seemed to have just taken to the brush and barely had time to wheel the stud around against the thorn-choked branches when the horse came bearing down on them. One glance told him it was Raines's horse, and his heart leaped. The animal carried an empty saddle.

Drumming boot heels against Satan's sides, Kittery made him lunge headlong into the trail. They were just in time for him to reach out and grab what was left of a trailing rein as the frightened animal squealed and tried to spin away. Kittery lost his hold, but they caught up to the bay and cut it off, forcing it into the heavy brush. The loose horse had no choice but to stop.

Snorting and rolling its eyes, the bay threw its head, spinning to avoid Kittery's touch. It took a minute, but the horse calmed to his soothing voice and Satan's presence. Kittery reached out and grasped the broken rein, his eyes drawn to the saddle. It was the first place a man looked when he found a loose horse in the days of the Indian wars—first looking for a saddle, then for blood on the saddle. The blood was there, not on the saddle, but on the horse's rump. A smear of it there showed the marshal must have fallen backwards leaving the saddle.

Kittery's heart raced with wild fear. He forced himself to calm, but it was only on the outside. The black sensed his desperation and started stamping his feet. Until Kittery threw his rope around the bay's neck and started along Raines's trail, Satan fought at the bit, wanting to move on.

They had ridden three miles and were approaching a brushy outcrop of rocks when Kittery saw something that made the dread hammer up his spine. Two vultures swung above the rocks in tight, determined circles. The stallion snorted and started to eye the brush around him, and the bay pulled back, fighting the rope around his neck. Kittery forced them on.

Around the ragged lip of rocks he saw it—saw *him*. His friend, Joe Raines. The black hop-walked to a halt, eyes and nostrils wide.

There before them he lay in the grotesque posture of death, the blood dark on his shirt. His arms were flung wide, his shirt and vest torn open to the waist. Numbly, forgetting to look around for signs that the culprits were gone, Kittery climbed from the stallion and lashed him and the bay to a nearby ironwood tree. He walked to the body with no feeling in his feet, no consciousness of the sun raging down on his shoulders.

Beside his friend he knelt, his eyes drawn to the open shirt. There on Raines's chest, among a tattoo of ragged bullet holes, were the oozing initials W and D. Walt Doolin . . .

His lip curled and quivering, Kittery shoved away from the ground, stumbling to a rock, where he sat down. He lifted wild eyes from the fly-harried corpse to the brassy sky, where the two vultures circled. His hands closed into fists of rage, clamped until they should have hurt, if he could feel them. His jaws clenched until his teeth seemed ready to crack. He lunged to his feet and walked back to his lifeless friend to convince him this was real.

And then, at last, he dropped to his knees and stared at the once-proud face of his friend, the lawman.

*Law*man! Joe Raines had lived to uphold the law, yet what had it gotten him besides death? And what would it get him now—justice? No! Baraga's fiends had been killing and raping, robbing and stealing stock for years, and the law was no closer to apprehending them than at the start. Joe Raines would see no justice, not from the law.

Kittery clawed a handful of blood-soaked mud from beside his friend and looked down at it. It appeared rich like Carolina loam against the white of his palm. He flung it against the rocks and stumbled to his feet.

His first thought was to pursue the Desperados, to make them pay. But even in his present state of mind he knew that was useless. He would wind up dead, too. Baraga would go free, as he always had. As he always had until . . .

The Castor Vigilantes . . .

The alteration of Kittery's convictions came with such suddenness that for a moment the change went unnoticed even to him. In the few minutes since discovering his friend's body, he went from a law-abiding man who loathed mob violence to one who almost craved the thought of riding amid an angry, violent horde bent on revenge—or justice, whatever they chose to call it. Bitter hatred torched inside him, an urge to see every last Desperado writhing in his own puddle of blood. The vigilantes could see it done. He would make certain they did. His brother Jake's words of wisdom on mob mentality fled in those moments. Kittery was going to be a vigilante.

And if the vigilantes refused to hunt down Baraga, he would find a way to do it alone.

As he rode toward Castor, leading the death-toting bay, Kittery's thoughts galloped the trail of vengeance before him. Yes, he would become a vigilante. Perhaps he was a hypocrite, considering his words to Cotton Baine the day before. Yet his would be no simple crusade to rid the country of crime. Kittery's bent was to see eight men dead, and any who associated with those eight. He *would* see it, too; he vowed this to himself. In the eyes of the law he would become a murderer, acting as some avenging angel issued a heavenly mandate to invoke justice upon eight men. Yet as Cotton Baine had said, if he didn't do it, who would? He could dredge up no compunction for his decision. He had only to erase Jake's warning from his mind until it was all over.

A quarter mile away from Castor, he dismounted to pull his blanket away from the marshal's body. He wanted his friend displayed so the people of the town would gain a hatred like his own for Baraga's Desperados. They all must see that bullet-riddled, once-proud lawman slung over his saddle in plain view.

Siesta was long over, so citizens crowded the street when Kittery reined around the warehouse. No one could miss the body draped over the saddle of the bay. By now everyone must know the marshal had taken the Desperado trail alone, and they wouldn't have to ask about the stiff, blood-caked body.

Luke Vancouver and James Price stood in shocked silence as Kittery lowered the lifeless form to the dust. Adam Beck leaned against the café wall, watching regretfully. His eyes shifted to James Price with a knowing look, and the hangman glanced over and nodded.

Kittery stood wordless over the body, glaring around him through blood-shot eyes. Vancouver stared at the body for a long time, and then looked at the people all around. His eyes landed last on Kittery. "Sorry, Tappan. Sorry."

Later that day, with the body at rest in the cool shadows of the jail, Kittery sat alone in the cantina with a half-emptied bottle of whisky before him. Miles Tarandon and James Price peeked in over the top of the batwing doors, and on seeing Kittery they stepped through and walked straight to him, taking chairs. Price looked at Kittery. "Mind if we sit?"

Kittery looked at him and rubbed his eyes. "You're sittin'." He poured himself another glass and nudged the bottle across to them. Both of them ignored it.

He studied the two of them, taking a deep swallow of the whisky that had ceased to burn his throat. He was drinking mechanically. He had always hated to do that, but this day called for it.

"I want to speak with you about something private, Captain," Price said.

"Yes."

"Pardon me?" Price looked confused.

"I'm no fool, Price. You're here to ask me to help you . . . shall we say, clean up the countryside."

Price cleared his throat and glanced at his lieutenant. "Well, Captain," he said, looking back. "You certainly are no fool. You think pretty clearly, for having half a bottle of whisky under your belt."

Kittery looked at the bottle, a little surprised at the level of the liquid. He swore and slid the bottle farther away. "Must've been part empty when they brought it."

Price nodded wisely. "I'm sure it was. Now, Captain—"

"I told you yes, Price," Kittery said. "I'll take up with you boys now. Tonight."

"I thought you would. But tonight we won't need you. As it happens"—he glanced toward Cardona, at the bar— "we have unfinished business with one of the Mexicans from the other night, one Raoul Mendoza. But we'll see to him."

Kittery thought of Juan Torres, his legs shattered on the warehouse loading dock. Then his mind shifted to the beautiful little girl sweeping the porch of the dry goods store, and he clenched his jaw. "An eye for an eye."

Price shrugged. "Like you, I'm a captain. Of this vigilante movement. For now, that's all you need to know. We'll be in touch. Be patient."

Kittery remained seated long after the two had gone. Well past dark, when the odor of cool dust was battling the stale smell of smoke at the batwing doors, he stepped onto the street. Tania McBride sat there on a chair, and she stood and moved toward him, searching his blood-shot eyes. Her own were filled with tears.

"I waited for you, Tappan. I'm so sorry about your friend."

Kittery looked down at his clothes self-consciously. They were rumpled and bloodstained from picking up the marshal's body. He looked back up, and their eyes met. He answered with a thick tongue. "Tania . . . I . . . I'm glad you came. I need someone like you."

She threw her arms around him, and he held her and wondered how he could have survived the night without her comfort. He forced back tears as he thought of this sweetest, most sincere girl he had ever known, outside of his own sisters. When he looked past her in the shadows, he could see Randall McBride leaning against an awning support of the hotel, puffing on a slender cigar.

They laid Joe Raines to rest in the morning. Sparrows flitted about the tops of the cottonwood and sycamore trees, and the sun lay soft and gentle on the leaves and grass and wildflowers. Huddled next to the Camino Real north of Castor, the drab little cemetery contained thirty-three graves. Some were marked by heavy wooden or stone markers, others indicated by simple wooden crosses. Fresh mounds showed the resting-places of Emilita Greene, Gustavo Ferrar, Juan Torres and Jason the Coachman, along with the two Mexicans who had died under James Price's rifle fire. By that time tomorrow there would be two more mounds—one for Joe Raines, and one for Raoul Mendoza, who had died by a vigilante knife sometime in the night.

The priest from the mission San Xavier del Bac had come, and he read several verses of scripture over the coffin. Then they lowered it into place, drawing the ropes back out of the hole. One by one, people paid their respects and drifted away. One of them was Francis Goodwin, U.S. marshal. He didn't stay long. Another was one of Joe Raines's equals, at least in official standing: Deputy Marshal Wiley Standefer. When he threw his handful of dirt onto the coffin, the lawman wept.

Beth Vancouver walked over and held out her arms, and Kittery squeezed her close. When she stepped away, Luke shook his hand and clasped him on the arm with a nod.

Joe Raines had no family, at least none anyone knew to contact. Perhaps it was partly for that reason, and partly because they were friends and it was his traumatic misfortune to be the one to stumble on the mutilated body, but most of the people who came to the burial paid their respects to Tappan Kittery. It made him feel self-conscious, but it warmed his heart, too. When Wiley Standefer stopped, Kittery looked down at him and nodded.

"Hello, Marshal. You came a long way."

The marshal gave him an odd look, clearing his throat. "Twenty miles, Kittery. Hardly far for my best friend."

Kittery nodded again and cupped the lawman's shoulder. "I know. He was mine, too. We'd have both come farther."

"Some day they'll allow us the funds to go after Baraga in earnest," Standefer said. "He won't get away with this forever."

"No. You got that right."

Standefer searched his eyes, then at last held out his hand to shake. "Take care, Kittery. And remember our friend died for the badge."

Kittery grunted, his throat too tight to speak his disgust.

Finally, the last swirl of dust from departing mourners died in the road. Kittery stood alone with Tania and Dabney Trull, the gravedigger. The half-breed Mexican watched Kittery from a distance of twenty feet, and at last he stepped close.

"You c'n throw down the first shovelful, if you care to."

Kittery studied him. "Did they pay you yet?"

"Sure. Why?"

"Go buy yourself a drink. If you don't mind, I'd like to cover the grave myself."

Dabney Trull considered a moment, then shrugged. "Sure. I guess that'd be all right. You could leave the shovel at the jailhouse." He turned and looked at Tania. "Miss? You—you came in a wagon that's already gone. You want a ride back to town?" He nodded toward his wagon, parked nearby.

"Go ahead, Tania," Kittery said. "I'll do this alone."

The young woman nodded and smiled, touching his sleeve. It was a hurt smile, and he was afraid he had made her feel unwanted, but he needed to be alone with his friend. Trull gave her a hand into the wagon, then climbed aboard himself and whipped the ribbons. A gentle waft of air swirled dust across the silent road as the wagon rocked out of sight, Tania still gazing back at Kittery. At last, he stood alone, dappled in shade from the cottonwood leaves. His heart throbbed against his chest.

He remained like that for some time, staring at the top of the coffin. Finally, he removed his string tie, vest, and boiled shirt. He grasped the shovel, plunging its rounded blade deep into the mound of soil. With every shovelful of dirt he pitched, a Desperado died in his mind.

It took half an hour before he could step away from the finished task. His hair dripped with sweat now, and his blood-pumped muscles glistened with every movement. On his chest, the curled hairs were beaded, too.

Catching a movement from the corner of his eye, he saw Tania McBride riding toward him from town on her gray mare. He had told her he wanted to be alone, but he was glad she didn't listen. He turned back to the grave while she was still a hundred yards away.

His lips moved for a couple seconds before he found words. Then he spoke with grim deliberateness. "They'll pay, Joe. My hand—and my guns—will make them pay. Or I'll die along the way."

Chapter Fourteen
Shorty and the Skunk

Forty miles to the southeast of the Desperado Den, in the foothills of the Pajarito Mountains, fifteen men pushed their sweating mounts across the brush-choked ground. Crow Denton's sleek, long-legged chestnut kept easy pace with the blood bay ridden by Colt Bishop, several lengths ahead of the nearest contender. The pounding hooves sounded like thunder. The Mexican border was ten minutes behind them, and so were two dead vaqueros. There was no telling where the rest of the Mexicans were, or if they had followed them.

Major Morgan Dixon glanced behind him, then looked at his comrade, Baraga. The man's face was set and looked as hard as the Sonoran desert. It was no wonder they called the man Savage Diablo. The jaw-tracing beard, shaped like a spade at his chin, wasn't the only thing that made him look like an artist's depiction of the devil, but it didn't hurt. Adding to it was the sharp nose with its flared nostrils, the bitter mouth, high, wide forehead and eyes the color of iced-over creek foam. He wore a trout-colored hat, the front of its brim swept back almost as high as the crown. Incongruous with the rest of his outfit, his leather britches, Mexican spurs, and canvas duck coat, was the gray floral shirt he wore. Morgan Dixon had to laugh when he looked at it. It was the only thing about his comrade that looked soft.

Denton and Bishop were getting too far ahead on their blasted racehorses, so Baraga roared out a halt. The horses heaved to a stop, sweating and grunting, amid swirling dust and hoof-flung gravel.

They made their camp there at the mouth of Peñasco Canyon. Within minutes, the smoke of three campfires curled into the dusky evening sky, a sky woven recklessly together by swooping nighthawks making their little *kee-keeh* calls overhead. Fresh Mexican beef, impaled on sharpened mesquite branches, began to pop and sizzle over the flames. Morgan Dixon kicked his bedroll out and plucked a cigar from his pocket, scowling over at the skewered beef as he sank to the blankets. Whatever had happened to the fine cuisine of Charleston? Not to mention the fine, delicate ladies with their flowing, colorful gowns? Where was the life of the Old South? Too far distant from this devil's paradise called Arizona.

Dixon puffed on the cigar as darkness settled in. He studied the fire-lit faces around him, knowing most of these men came from the southern states yet that most of them were scum. Ironically, his best friend was Samuel Colt Bishop, the loner, and Bishop, who came from Texas, had not even fought in the war. But Bishop had the intelligence, the finesse, the style that Baraga and Dixon had. Few, if any, of these other men could claim that.

Thinking of Bishop, Dixon had to smile. His friend had brought heavy tension to the camp for a couple of days. Down in one Mexican town Silverbeard Sloan had knocked down a crippled photographer and taken all the photographs he carried in his display box. Bishop had brushed the skinny gunfighter away and given them back. It had been so obvious Sloan wanted to kill Bishop for that, but he had backed down. And everyone knew why Sloan had taken the photographs. He meant to use them for his target practice. They were pictures of families, one with three little girls with the darkest, most solemn eyes a man ever saw.

Dixon's mind churned back over all the events of the last seven days. Visions of trail dust and gunsmoke haunted his reverie. It began in the mining camps and ranches of the extreme southwestern Santa Rita Mountains. From there, the raid covered a little over twenty miles of Arizona Territory, from the Grosvenor Hills, down the valley of the Santa Cruz, and into the Patagonia Mountains.

Except for the ranchers who were their allies, they drove every head of stock before them until they had formed two large herds, horses in one and cattle in the other. Anyone attempting to stand in their path died, which only amounted to three men, since most men knew of Baraga's hell-raisers and wouldn't dare confront them even at a distance. The Desperado band had suffered one casualty. A hellion named Ned Simmons had died when one of the silver miners near Patagonia had picked him off with a big-caliber rifle at over five hundred yards. Simmons was a newcomer on his proving run, and they had left him behind for the buzzards.

They sold the herds in Sonora, to ranchers who were regular customers. Their biggest business was receiving outlaws' stolen stock and reselling it. These were the ones Baraga would not steal from, and would even protect, if it came to that. They had become his main livelihood.

On the way back to the United States territory, they had gathered herds of Mexican stock, planning on selling these to regular customers in Arizona. But when the vaqueros hit them at the border crossing, Baraga decided to desert the herds. They had come away from Mexico with over eight thousand dollars. The ribby Mexican herds weren't worth anyone's dying.

When supper was over, Dixon watched the others cluster like moths around Baraga's campfire, where the one-armed man had spread a blanket. Here, Slicker Sam Malone dumped out the sacks of silver and gold, and the impressive piles flickered in the firelight.

The take was not divided evenly. To each of the men who rode with the Desperados was paid one hundred dollars in gold or silver, as agreed upon before the raid. Thus, in seven days, they made the same wage as a cowboy trailing a herd of cattle up the long trail for three months or more. Once Shorty and Mouse's take was subtracted, that left close to nine hundred dollars for each of the Desperados, and among them Baraga split the take as equals. That was the luxury of being taken into Baraga's fold. He treated his band as well as he did himself. It was something Dixon had always admired about him.

Once the money had been separated and divvied out, the men who didn't belong to the eight drifted off to make their own camps. Dixon didn't blame them.

The Desperados were a close-knit group, and not the most amiable of company for men they felt did not meet their standards. He had to chuckle to himself at that word. Standards? Could a gang of thugs as disreputable as the Desperados Eight claim to have standards? Now, there was a question for the ages.

When morning came, and preparations were made for the last leg of the trip back to the Den, only one of the temporaries continued on with Baraga's group. Major Morgan Dixon rode to the rear of Tawn Wespin, and he studied him. He was a slender wisp of a man with a VanDyke beard and curled mustaches, whose gray stripe down the middle of his shock of shaggy dark hair had earned him the moniker, the Skunk. Wespin was a man Dixon wouldn't have thought twice about killing, given half an excuse. The Desperados, as a whole, weren't inclined to ravage women and butcher children. Wespin was known for both.

Hunting as they went, the band picked their way north, killing a javelina and two mule deer in the Baboquivari foothills. It was late in the afternoon when they rode almost reverently past the base of the magnificent peak, its dome painted breath-taking gold in the waning sun. There was no wind anywhere, yet there was that great, eerie, horrible sound like a gigantic, invisible flood coming down the canyon toward them. It was that mysterious wind that sometimes swept over the bald dome of the peak. That ghostly wind that seemed so powerfully to bespeak death.

They climbed out of cool purple shadows back into the sunlight as they made the last stretch into the Den. They were greeted by the shrill neigh of Shorty Randall's horse and the rest of their herd, grazing along the side of the canyon among the oak trees. After letting all but two of their horses loose to graze, they went inside to see Shorty seated at a tiny fire.

Dixon was first to speak. "Hello, Shorty. What in the world is the matter now?"

"Well, it might not be much. But then again, it could be a lot." He glanced over at Doolin. "Mr. Doolin? You remember hearin' of a man named Captain Kittery in Castor?"

From the corner of his eye, Dixon saw Baraga's head snap up, but Doolin responded first. "Sounds familiar."

"Well, Mouse tells me he was a good friend of the marshal you killed. He tells me he's joined the vigilantes, an' he's talkin' real big. Says he looks like he could turn out to be one of the worst of the whole bunch, since he hates y'all so bad. Especially you, Doolin. You carved on the marshal, and this feller says he wants you all for himself."

Bloody Walt smiled wolfishly. "Then he can have me."

Baraga shot him an angry glance, then turned back to Shorty. "Tell me that name again."

"Kittery," repeated Shorty. "Captain Kittery. Tappan, or somethin' like that."

"Crappin' Kittery," Slicker Sam Malone said, looking around with a stupid grin on his face for approval of his joke. No one pretended to notice him, and his mirth vanished.

Dixon was studying his comrade Baraga while Shorty spoke. What he saw was a face made of rock turn to iron, and when Shorty spoke Kittery's Christian name, Baraga's eyes seemed to glow with hate. He nodded and reached over unconsciously to rub his stub of arm. "What's this man like?" he asked, his voice quiet like it became only when he was most furious.

Shorty shrugged. "I dunno. He's big—real big." He glanced over at Rico Wells, Big Samson. "Near big as Samson, almost."

Everyone watched Baraga, and no one spoke. If it was possible for any of these men to know fear, there was something about Baraga that could bring it out in them. Dixon scanned the faces and analyzed the sodden way his own heart was beating. This was the Baraga no one, friend or foe, cared to see. It foretold a coming explosion in a man of vehement passion that seldom reached the surface.

"Well, Shorty, I appreciate the word." Baraga turned away and fished into his saddlebags. When he turned back he held two small canvas sacks, which he jingled in his palm. "Something to keep you and Mouse fat and happy. And if it doesn't make him happy, you know what to do. Dissenters won't be tolerated here."

Shorty took the sacks, and Baraga then turned to the narrow-eyed Tawn Wespin. "You've a reputation as a drygulcher."

Wespin's eyes narrowed, flickered away, then returned to Baraga as he seemed to come to grips with the title as a possible compliment, coming from a man like Baraga. "Reckon I don't much like the choice of words, but mebbe it's true."

"Well, then. There's an extra hundred dollars out of my pocket if you kill this Kittery fellow." He paused, his eyes flickering as if he felt the surprise of his confederates. "If he was that good a friend of the marshal's, he may be over-ambitious."

Dixon stared at his comrade. It wasn't Baraga's way to make explanations. Why did he feel like he had to now? Whatever the reason, it was obvious to everyone there that Baraga had heard of this Kittery before. He was an enemy of old, and as such he would die. All of Baraga's enemies did, eventually.

Baraga looked over at Shorty Randall. "Shorty, come outside and talk a bit."

Bewildered, Shorty followed him out into the sunlight, and they walked a ways down the canyon. At last, Baraga turned on the little man. He held out his hand to show Shorty his palm, and in it there shone five double eagles. "That's a hundred dollars, Shorty. A hundred dollars that's yours for almost nothing. Just one thing. Who is the Mouse?"

Shorty looked at the money and swallowed hard. "I got my obligations to him, too, Mr. Baraga. I really wouldn't want—"

"Who is he?" Baraga said, his voice going quieter.

Shorty swallowed again and looked back toward the cave. Then he let out a long sigh.

After Shorty and Tawn Wespin had gone, Baraga sat by himself down the trail from the Den. He wasn't in the mood to sit in the cave with the others. He wasn't in the mood to smell their odors, to listen to Malone's stupid jokes. He had to sit here and think. His first instinct was to think of Tappan Kittery, but he didn't dare. He didn't want to lose his edge today; to let himself fill with that kind of furor would make him an unwary man. So instead, he thought about the war . . .

He remembered the explosion, the mind-shattering blast that put him down on the grassy, blood-soaked hill called Cemetery Ridge. He remembered lying there on top of another man, a man who died beneath him. And he remembered looking at his arm and knowing it could be saved by Colonel Arlo Bell, the company surgeon. He only had to reach him. So with grim determination he had begun to crawl.

But reach him he didn't. The Federals found Baraga first. Off they carted him to their filthy prison hospital, their own "den" of death and destruction. He thought they would try to save his arm. After all, he was a colonel, albeit on the opposite side. It was the gentleman's thing to do. You didn't cut off the arm of an officer. And yet they did. With him kicking and screaming, they readied their saw and begun to cut away, not bothering with morphine or chloroform or even a shot of whisky. He felt the horrible bite of the knife slicing through his flesh. He could hear the saw teeth as they bit into his bone, smell his own bone being severed, and he heard the separation as the last piece of skin gave way. And somehow, by some savage, hateful miracle, he stayed awake through it all, hating the Federals. Hating the world.

He remembered the smoke and the smell of the burning flesh as they put a hot iron to his new stub. Then for so long he remembered nothing. Nothing but hate.

He considered himself at one time to have been an amiable man. He had many friends who looked up to him, many who respected him. But when the burning hate grew in him to consume everything else, he lost his friends. It grew in him to encompass not only the Union, but all of the human race. Then he only had comrades as full of hate as he was. There was no room for friends.

He remembered many days stuffed inside a little iron box while the sun or snow raged outside. The box was too small for him to straighten his legs. It was a miracle he had lived. They did this because he was a "reb," and worse, a rebel among the rebs. Once, upon pulling him out of the box, where he landed in a pile in the bright sunlight and couldn't move his legs to stand, one of the guards had urinated on him.

Lack of nourishment and proper care rushed the process of Baraga's growing, almost demonic, hatred along until the war ended. Then he bought himself an Army Colt pistol, cut the barrel down a couple of inches, and by long hours of tedious practice each day taught himself to handle a handgun better with his left hand than he ever had with his right. He robbed an express office in south Texas and killed the guard there, who was wearing a Union insignia on his hat. And so began the spree of plunder and murder that had led him here, the leader of a band of killers . . .

Seven miles outside of Castor, Tawn Wespin drew in his mount next to Shorty Randall's. "This is far as I go, little man. How'd you like to make somethin' extry?"

Shorty drew himself up in the saddle, scowling. Little man! That was something Baraga and his boys might get away with, but any man called the Skunk had best tread carefully. He hid his disgruntlement when he replied. "Might. 'Pends what you got."

"You know this country any?"

"Every inch of it."

Wespin grunted. "Awright. Tinaja Peak. Esperanza Wash. Know it?"

Shorty sighed. "Said I did. Tinaja's due west of town." He pointed to the northwest. "That's it there."

This time Wespin smiled. "Awright, so you *do* know. Ever hear o' Gila Knob?"

"Uh-huh. Halfway between Esperanza Wash and Demetrie Wash."

The Skunk shook his head with admiration. "'Tween the two of us, we got it made. Awright. Here's the deal. I ride out there now. I don't dare go into town—too many notices hangin' around these days. And that there's our ticket. You say this big captain feller hunts outlaws for bounty, right? Well, you go talk to 'im. Tell 'im yer s'posed to meet me out there at . . . say five o'clock in the evenin'. Make up some story why. Then you go out there with 'im, an' I shoot 'im from on top of the knob. Simple enough?"

"Sounds pretty easy," Shorty agreed. But his mind was already calculating other possibilities—possibilities which hinged on the Skunk's answer to his next question. "So what's in it for me?"

"Well, I figure since I'm doin' the actual killin'—thirty of the hundred for you, seventy for me. That seems perty fair."

Shorty nodded. "Uh-huh. Perty fair. I'll see you tomorrow."

Wespin smiled, baring chipped yellow teeth as he raised a hand and reined his horse toward Tinaja Peak. Shorty touched heels to his little white horse and continued on toward Castor, thinking about his parting words to Wespin. *Yeah, I'll see you,* he thought. *I'll see you dead.* He recalled Baraga's parting admonition to him that morning. If Wespin attempted to kill Kittery and failed, Shorty was to take care of it himself. Now, what was thirty dollars when he could have the entire hundred? Shorty chuckled to himself as he rode along. Any man who looked like a skunk deserved to die anyway.

Glancing up the trail, Randall made out the dust-enshrouded forms of two horsemen in the distance. Wiping the sweat-mud from his cheeks, he straightened in his saddle and rode on.

"There he is," Beck said, pointing west. Far ahead, they could see a squat white horse carrying its rider toward them. "Looks like he's comin' on."

Kittery nodded. "One too many trips to Baraga country will kill this one," he said.

When they reached Shorty Randall, the little man seemed to sense their mood. His worried eyes flickered between the two of them, and his "howdy" was meek. Kittery pulled out a revolver and stuck it in his face.

"You're comin' with us, little man. Or make a move and die right here. It doesn't matter much to me, so which is it?"

Shorty looked from one to the other, finding no mercy. Swallowing hard, he raised his hands over his head.

Half an hour later, they drew up in front of the sheriff's office. Luke Vancouver was there to meet them, leaning against his hitch rail and smoking a cigarette as he watched them climb down with their trophy.

Kittery jerked his thumb toward Shorty. "Mind if we keep this jackrabbit in your jail while he answers some questions for us?"

Vancouver shrugged, trying to disguise the faint humor in his eyes. "Well, Tap, I'll tell you. If you don't have a solid charge, the answer has to be no."

Kittery was surprised, but he shouldn't have been. Vancouver would always be by the book. "Well, then—conspiracy to commit murder. That'll hold 'im, won't it?"

Vancouver nodded satisfaction. "That'll hold him. Take him to a cell, Tap. The keys are in my desk."

During the next few hours, either Kittery, Price or Tarandon stayed with

Shorty, sometimes the three of them at once. And in Vancouver's absence there was little about the questioning that was gentle. But Shorty had gumption Kittery hadn't expected. His stubborn silence and sarcastic replies, even in the face of pain, frustrated them at every turn. When at last he did tell them something, it was something they hadn't asked about: the whereabouts of a Desperado hanger-on named Tawn Wespin, alias the Skunk. Shorty was supposed to meet him the following day out at Gila Knob, in the evening. And Tawn Wespin had a five hundred-dollar reward for murder on his head. They left Shorty with blood still trickling out both sides of his mouth and one eye swollen nearly shut.

Kittery, Price and Tarandon sat in the sheriff's office after their last hour of grueling questions. Dusk was gathering outside, with golden light lying broken along the tops of the buildings on the opposite side of the street. The dust of the street lay in blue and violet washes and shifted lazily whenever human or animal passed by. A hog wandered along the edge of the boardwalks, snuffling in the dust now and then for some scrap.

While they were talking about Shorty and Tawn Wespin, Cotton Baine came marching in with a tray covered with a dishtowel. They all turned their heads to look at him, going silent.

"You boys believe in feedin' your prisoners?" he asked.

"Feeding them?" Price asked. "Who are you, an angel of mercy?"

Baine laughed and shrugged. "Well, I was eatin' across the street, and it hit me you prob'ly hadn't taken time to feed that bag o' bones. You wanna keep 'im alive, or what?"

"That's a loaded question," said Kittery.

"Go ahead and feed 'im," Tarandon cut in. "I wasn't planning on it."

"Better check that tray," Kittery said with a laugh. "There's prob'ly a gun in it."

Baine laughed. "Nope. Only a couple of knives and a hacksaw."

The next day around noon, Adam Beck raised a hand, signaling for Kittery and Cotton Baine to halt. "That's what they call Gila Knob, that biggest point there."

Kittery nodded grimly, studying the gray, hundred-foot bluff. He slapped his holster. "Let's get to it."

Leaving their mounts in a copse of mesquite, the trio watched from the edge of the trees but saw no sign of life around the bluff. "Is there a way for a horse to get up on top?" Kittery asked.

Baine nodded. "The back side. I've ridden up there to get cows before. But it ain't too much easier goin' than this side."

Gazing at the knob, Kittery made out a notch in the upper portion of it, a place that should be accessible for a man climbing up from below. He pointed it out to the others. "I don't wanna give 'im any escape route. Give me twenty minutes to climb up through there. You two find a way up on the back side—two different ways, if you can, and we'll meet on top. You see a man that looks like a skunk, I suggest no quarter."

On top of Gila Knob, Tawn Wespin stirred cornmeal gruel in a little skillet and gazed out over the heat-parched desert. His makeshift camp was situated beneath the biggest oak tree he could find, but it offered little protection from the battering sun that struck at him almost as brutally from the ground as it did from the sky. He had taken off his hat, grown weary of removing and replacing it to sweep the sweat from his forehead.

The hat-sized fire guttered under his skillet, its flames almost invisible against the white of the caliche. The gruel was turning a nice brown, but the smell of it didn't do much for his appetite. He didn't have any sugar or blackstrap, and it was going to taste like mud. Besides, it was too hot to eat hot food. What he needed was some of that vanilla-flavored ice cream he had devoured in San Antonio. That would set this day off right. Or better yet, some of that creek ice he had sampled up in the Santa Ritas in January.

He shifted his rump to the edge of the rock, letting the air reach his sweat-soaked pants so they could dry a little. It was time to leave this furnace of a territory. But where else could he go? Any place up north would still be in snow. He was fortunate enough to be here out of the cold and wet and shouldn't complain. Other than Texas and New Mexico and Arizona, he had never been anywhere. But when he picked up the money for this killing he would buy himself a new outfit and head north as spring came on. Cool green fir boughs made the best kind of shade.

The sudden clatter of rocks toward the front of the knob drew his startled attention. Leaping up and grabbing for his Sharps carbine, he spilled his gruel and swore. What was down below? An animal, probably, but . . .

As he reached the rim of the knob and looked down, he was shocked to see a man with an unholstered pistol trying to conceal himself in a pile of rocks. Wespin raised and cocked his rifle. Whoever this was, he was too close.

The sudden crack of a twig behind Wespin startled him and caused him to snap off a wild shot. He heard the *ping* off rock as he whirled around to face a new threat. A man with a pistol half-raised stood thirty feet away from him. Wespin's eyes flew wide, and he flung away his single-shot rifle and clawed for the pistol in his waistband.

The other man dropped into a crouch and fired three times. One of the bullets sank into Wespin's shoulder, spinning him around. The second missed, and the third left its mark under his arm, piercing his chest cavity. Wespin's backward lunge sent him over the edge of the knob, and he crashed to the rocks and scrub brush below.

Cotton Baine, the gunman, ran forward as Adam Beck came through the oaks behind him, and the two of them came together where the knob dropped off into space. Below, Wespin lay on the point of his neck, asleep forever in a dubious cushion of catclaw. Kittery, gun drawn, was climbing up to check on him, but from where Baine and Beck stood it was obvious there was no need.

As soon as Kittery, Baine and Beck entered the dusty main street of Castor, with Wespin already stiffening over his saddle behind them, Kittery knew something was wrong here. The town was far too alive for that time of day, when the disagreeable heat should have driven most of the citizens to their homes. The trio made their way up the thoroughfare, stared after by onlookers.

James Price spotted them from the jailhouse doorway and hurried out to meet them. "Randall's gone."

"Gone?" Kittery said, anger rising up inside him. "Gone how?"

"We don't know, but we've searched every place in town. His horse is here, but he doesn't seem to be." The hangman glanced over at Wespin's corpse. "Well, well. Your first triumph. And hopefully not your last." In a quieter voice, he leaned in toward Kittery as he climbed down from the black and said, "Come to the jail, Kittery. We need to talk."

They carried the outlaw's body to a cell, where they left it lying while the three hunters and Price returned to the office. Price turned to Kittery, and his mood was subdued. "I'm afraid we committed a serious error in judgment, my friends. Out on the street I was talking for whatever curious ears might be listening. But here is the truth: Randall didn't just escape. We left the keys where he could reach them, thinking perhaps he would make an escape and head for whoever his source here in town is. That was the plan. And, well . . . it backfired."

"Backfired?"

"Well, like I said, his horse is still here, and as far as we can tell there's no other horse missing from anywhere. I find it hard to believe even a man as foolish as Shorty would take off into the desert without a horse. He wouldn't last a day."

"So where would he be?" Beck asked.

"Answer that, and you'll solve our problem," Price said. "I feel like a fool for not having someone watch him closer. But in the first place I really didn't think he would take the bait. And in the second place I had no doubt that if he did it would be to run to Baraga or his informant. I was wrong on both counts."

"Well, he'll turn up," Baine said. "He's prob'ly waitin' for dark, to slither out on his belly."

"Where's the sheriff during all this?" Kittery asked. "I thought he was workin' the day shift."

"Day shift?" Price looked puzzled. "You didn't know? He and Beth left for Prescott."

Kittery was stunned to silence. "Gone to Prescott? What do you mean?"

"They went to visit their children and Luke's brother, Travis."

Kittery brooded. Gone to Prescott. He couldn't believe they hadn't told him. Prescott was a good trip from there, so they would be gone for a while. He was surprised one of them hadn't said something to him. But he had had other things on his mind. They had to have known it.

Kittery was hurt and angry at not being informed of the Vancouvers' departure. Now he was doubly angry that Price and Tarandon would try their foolish plan at all without consulting him and Beck, who had been the ones responsible for putting Shorty in jail in the first place. It was his firm opinion that Raines wouldn't be dead now without the little man carrying word about him to Baraga. For that, Shorty deserved to die. Now, Price and Tarandon's foolishness could serve to send Shorty on his way scot-free.

Kittery got up, not in any mood to listen to Price try to excuse his blunder any further. "I'm gonna go take a nap. If anybody finds Shorty, let me know."

"We'll do that," Price said. "Tarandon's still out tearing the town apart, so if he's here we'll ferret him out. I guess I'll head home for a while. That's where I'll be if you need me."

Price walked out and down the street before Kittery had stepped out on the porch. He turned to Baine. "Are there any guns missin' from here?"

Baine shrugged. "Uh . . . not that I know of. The rifles are all chained up and . . . well, I don't know. Price didn't say."

Nodding, Kittery picked his hat off its hook and stepped outside, back into the sun's onslaught. He started across the street, noticing Miles Tarandon stepping out of the dry goods store but not interested in greeting the deputy. He was still fifteen feet from the front door of the boarding house when he heard a window slide open, and he looked up. Momentary confusion made his step falter. Wasn't that— Which window was the one to his room? He was stunned to realize the face that stared at him was indeed leaning out the window of *his room*. He was more surprised to see the man's arm come into view and bring a small pistol to bear.

But he was shocked most of all when his startled brain recognized the pinched, white-haired face of Shorty Randall.

Chapter Fifteen
The Kiss

With an oath, Tappan Kittery threw himself sideways as Shorty Randall's shot cracked like the slamming of a metal door. Kittery reached for his Remington as he fell into an awkward crouch. He wanted cover, but there was none to be had.

A shot from Miles Tarandon's direction smashed glass near Shorty's head. The little man flinched but didn't draw back. Though fear filled his eyes, his determined face leaned out the window farther. He steadied himself with one hand on the windowsill. His right arm and the pistol drew a straight line, his sights on Kittery.

Everything seemed to stand still. Kittery had what seemed an eternity to realize that in the position he was in his pistol was locked against his hip. He dove to the left in time to see Miles Tarandon, running, slip on something near the mercantile's hitching rail. Tarandon caught himself with one hand and with the other leveled his Winchester carbine at the window and fired again.

Above, Shorty grunted in pain as Kittery tugged his pistol free of the leather's grasp. He and Shorty fired together. He felt the little gun's bullet *zing* past his head as he fired again.

For the briefest of seconds, Shorty Randall balanced on the windowsill. Then his supporting arm buckled, and he fell forward out the window, doing a somersault in mid-air. A thin cloud of dust lifted from the porch as the little man landed on its edge, then rolled off into the street. A wisp of smoke drifted past Kittery, then faded away.

Shaken, Kittery pushed to his feet and aimed his gun at the fallen spy as he stepped closer. Miles Tarandon had just come to his feet and brushed irritably at his muddied coattail, swearing, as he walked closer to Kittery, the carbine hanging at his side.

Kittery sucked in a deep breath when he looked into Shorty's lifeless eyes and saw the blood all over his shirtfront. Trembling, he holstered the Remington and looked over at Tarandon. "Pretty good shootin', for a one-handed rifleman," he said.

Tarandon swatted again at his coattail and cast a sour look down at the green scum on one of his boots. "Yeah! It was luck I hit 'im at all. There oughtta be a law against horses crappin' on the street."

Kittery chuckled. When he looked down at the manure on Tarandon's boot and pants and fancy frock coat and remembered the picture of him slipping and going down, the urge struck him to keep on laughing, and he had to choke it down. It was the emotion of a man who had just survived near-death, but he didn't think the townfolk would understand his standing over the dead man and laughing.

After one last look at Shorty, Kittery turned and almost ran over Tania McBride. "I heard the guns," she gasped. "I knew you . . . I was afraid something happened to you."

Kittery turned as he heard a woman speak his name. Maria Greene was standing there looking in disgust at the dead man. When she realized Kittery had turned to her, she forced her eyes to meet his. "He dropped this on the porch." She held out a little derringer, fully engraved on nickel plating, with ivory grips.

Kittery took the weapon and turned it over in his hand. He looked up to thank the woman, but she was already moving away, hurrying across the street to her husband's store. Miles Tarandon had been looking over the body of Shorty Randall, and now he walked over and stopped in front of Kittery and the girl. He looked down at the derringer, then looked after Mrs. Greene.

"Funny she'd give that to you, with me standin' right there with a badge."

Kittery shrugged. "You want it?"

"No, you keep it. It was meant for you anyway." With a little smirk at his own attempted humor, Tarandon sauntered off, the Winchester hanging at his side. Even after he had saved his life, Kittery couldn't help thinking he didn't care much for the arrogant lawman.

Doctor Adcock Hale was kneeling over the body now, and he had the shirt open wide to look at the pale, narrow chest. There were three bullet holes within the space of twelve inches in the torso.

Kittery looked away distastefully. "This is no place for you, Tania." He slipped the derringer into his vest pocket and took her elbow. "Let's get out of here."

Later, Kittery sat in the café with Tania at his elbow and Cotton Baine and Adam Beck taking up the other seats. When he had finished the last of his meal, he withdrew the nickel-plated derringer from his pocket and placed it in the middle of the table, looking at Baine and Beck to catch their reactions. Beck's look of interest drew his attention.

"Ever seen that before?"

Baine picked it up before Beck could and ran a finger along its smooth white grip. There was a strange look in his eyes. "Looks familiar, Tap, but I've seen a thousand like it."

Kittery nodded. "Same with me."

While Baine was looking at the weapon, Adam Beck reached into his own pocket and withdrew a weapon almost identical to it, except that his was blued and had walnut grips. Kittery took it from him when he offered it and glanced from one weapon to the other, then up at Beck. "Same gun. Stevens. Forty-one caliber. You must know somethin'."

"I know *some*thing," the manhunter said. "Don't know if it'll help. But that gun's new. Been on the market less than a year. As you can see, I own the poor man's version. But when I bought this one they told me a little about it. They only made twenty of them in silver plating. That's a rarity."

Kittery had raised his water glass, but that revelation made him put it down. "So we oughtta be able to track it to the store it was shipped to, within reason. And maybe the man that bought it. It's a cinch Shorty didn't have it in the cell. You know—you searched 'im with me. Somebody gave 'im that gun—maybe his inside man."

Adam Beck nodded. "I wondered why you were so curious. And you're right about tracking it. Even mine is rare. They only made two hundred of it."

Cotton Baine said suddenly, "I'll take this and try to find out who bought it." He started to slip the gun into his vest pocket.

Before he could let go of the gun, though, his eyes met Kittery's. They held for a moment while Kittery tried to decide if his friend was joking. He realized he wasn't. "Thanks, Cot, but I'll take care of it. I wouldn't mind havin' that when this is over."

Baine cleared his throat and looked down at the gun. The look in his eyes showed he was struggling for an argument. At last, he passed the gun back to Kittery. "You'd prob'ly be better at that investigation stuff anyway. I'm no Pinkerton." He plunked a tiny gold coin down on the tabletop and stood up. "I'll be seein' you boys later. Miss." He tipped his hat to Tania, turned, and walked out of the room.

There was a moment of uncomfortable silence, and at last Kittery shrugged. "I hope I didn't offend poor old Cot. He must've wanted this pretty bad."

Beck answered with a laugh. "Yeah. Maybe it was his."

Kittery and Tania laughed at that, but Kittery's laughter died quickly. He looked after Baine again, musing, then shook his head and stood up, dropping the derringer into his pocket. "I better be goin', too. After my little scare I could use a drink—and a nap."

They had packed the dead man away when Kittery and Tania stepped back out on the street. Other than tight little pockets of people talking and gesturing now and then toward the hotel, the town had returned to normal, and Cotton Baine was nowhere in sight. Many eyes turned to Kittery and the woman, and many lips stopped moving.

Ignoring the attention, Kittery walked Tania across the street to her father's sorrel, which she had ridden to town and tied in front of the bank. "I guess I'll see you later," he said, his eyes flickering toward the cantina.

Tania clutched his sleeve. "Please don't go. It's such a beautiful afternoon. I'd love to ride with you."

Kittery thought of the cantina again. He wanted another word with Cotton about the gun, and he wanted a drink. But his taste for it vanished. "Sure, Tania. I'll ride with you. Let's go get Satan."

"Oh, I don't think we need *him* along," said the girl. When he paused, confused, she said, "Oh! You mean your horse!"

Kittery caught the twinkle in her eye, and they laughed together. "All right. No Satan today. We'll just bring that black horse of mine."

They rode east out of town, toward Huerfano Butte. The land into which they climbed was cattle country. Bunch grass abounded here, although it had been overgrazed many years ago by the vast herds of the Spaniards. Numerous clumps of Spanish bayonet jabbed their slivers at the sky, and copses of mesquite harbored occasional bunches of ribby longhorn cattle. Once, near a muddy waterhole, they startled four mule deer from a stand of trees.

After a time, movement became apparent on the distant horizon. A herd of cattle or horses was moving their way, and one of the figures stood much taller than the rest—a lone herder. Two hundred yards away Kittery recognized Jed Reilly, the cowboy with a hatred for sheepherders. The puncher's mount slowed to a dutiful walk by the time he reached them. The range horses he had been driving veered to the sides, tossing their heads and blowing, glad for the respite.

The big man reached down and eased the Remington loose in his holster. He glanced over at Tania, but she was watching Reilly.

The cowboy tipped his hat to the woman, his eyes going to Kittery. The big man watched him coolly, making note of his gun belt and the holster full of Remington pistol. It looked like he had been to see the sheriff.

"Well, well," said the puncher. "I said maybe we'd meet again."

Reilly's face was straight, but in a moment it showed the beginnings of a smile. It was obvious he was trying to read Kittery's reaction to his presence. At last, the smile broke across his face. He glanced nervously at Tania, then back at Kittery.

"I guess you did. Howdy, Reilly," replied Kittery. His right hand rested on his thigh.

"I'm glad to see you," Reilly went on. "I ain't a good drinker, Mister— I mean *Captain*. I wanted to talk to you in town there, but before I got a chance I was drunk again. I'd like to change what happened between us—straighten things out, if I could. I ain't got no enemies, Captain. I sure don't need you to be the first." His eyes flickered as he spoke.

"Rest easy, Reilly. I'm not as lucky as you—I have plenty of enemies. I don't need to add you to the list."

The puncher urged his horse closer and tugged off his right glove, holding his hand toward Kittery. Kittery shook it and was impressed by the warm, firm grip.

"Glad to know you this way, Reilly. Sorry I hit you so hard."

Reilly's hand went to a large discoloration on his cheek, and he shrugged the apology away. "Heck, I deserved that and more. I got a lot of good story-tellin' out of my face, anyway."

"Sounds like me," Kittery said with a chuckle. "People 've been tellin' stories about my face long as I can remember. Comedies, mostly."

Reilly laughed. "I doubt that, Captain. I've heard people talk. They look up to you."

Kittery and Tania turned around and rode back a ways with Reilly until he had to turn south toward his and his brother's R Slash R ranch, at the bottom of Melendrez Pass, in the Santa Ritas. There they said goodbye to the puncher and wended their way back toward town.

They had ridden for several miles in silence when Tania turned to Kittery. "Can I ask you something?"

"Sure, Tania. Most anything."

She gave a broad smile, her eye corners wrinkling up. Then her face went serious. "What do you plan to do? I mean, do you plan to stay long in Castor? After what happened to your friend, I mean."

He waited for her to say more while he tried to form an answer. When she kept her silence, he cleared his throat. "Tania, I haven't had a whole lot of time to think about it," he told her, and that was a lie. "I've made some good friends here. I'll say that. But there isn't a lot more to keep me." He wanted to tell her how much he hoped she would go with him if he had reason to leave. But the words stuck somewhere in the back of his throat, and they rode on without talking. Only the horse hooves and saddle leather, jingling bit chains and the wind spoke to them.

Arriving in Castor, Kittery accompanied the young woman back to her house. When he made as if to leave, she said, "Couldn't you come in for a few minutes? Please. Father would be sorry he missed you."

Kittery shrugged, again losing his taste for cantina refreshments against the competition of Tania's company. "Sure, I can step in and say hello."

They led their horses to the corral, and Tania watched as Kittery unsaddled the sorrel and turned him loose with the gray. He watered the black and loosened his cinch, then tied him near the porch with half a bucket of oats.

On entering the house, the smell of roast hit them full in the face. Kittery's mouth began to water, but he looked over uncomfortably at Tania. There were three places set at the table, and he hadn't been invited.

"I prob'ly should be on my way," he said. "Looks like you have company." He glanced about, searching, in spite of himself, for a man's hat or coat. He found only McBride's.

Before Tania could reply, McBride walked from the back hall and waved them toward the table. "You're right on time, you two. Good evening, Tappan. Have a seat, why don't you?"

Confused, Kittery looked again at the three table settings, then over at Tania. She was smiling craftily at him. "So you . . . Now I see." He let out a laugh and looked over at McBride, who gazed at him expectantly.

"I must have missed something."

Kittery looked at Tania again and shook his head. "Just your daughter's idea of a surprise party. I've been hornswoggled. Didn't know I was comin' to dinner."

McBride looked at Tania reprovingly. "Now, Tania! What if Tappan had other plans?"

Kittery shook his head and held up both hands. "Don't worry, Rand. I would've broke 'em for this."

McBride beamed. "We'll take that as a compliment, Tappan. Please. Have a seat."

The banker served them from a pot full of venison roast, potatoes, carrots and turnips, with onions laced over the top of it. He sliced off a big chunk of bread for each of them, which he buttered and leaned against their plates. Then he filled three glasses nearly to overflowing with deep red wine.

When they were seated, McBride bowed his head. "Gracious Father. Bless this table and this food. Bless this house and those who enter here. Especially bless this fine new friend who has found our home. Keep us in thy protective hand. Amen."

During the meal, McBride kept looking at Kittery's shoulder. Finally, he asked about it.

"It's feelin' pretty good, Rand. It was a little sore after I played the violin the other night, but it's comin' on good." Later, he said, "Speakin' of my shoulder, I think it's about good enough to shoot a rifle again. I think I'll try to find a nice fat buck tomorrow. You sure do it justice."

McBride thanked him with a laugh. "It's the only thing Tania's mother was ever able to teach this old goat."

They retired to the sitting area after supper and had coffee. When the hour grew late, the three of them stood in the semi-darkness of the front porch. McBride shook his new friend's hand, said good night and turned inside. Tania and Kittery stood alone, and he looked down into her dark, mysterious eyes that gazed back, not at him, it seemed, but into him. Tappan Kittery had never thought much about death since the trauma of the war. But he thought about it now. Death meant losing this girl. And affiliation with the Castor Vigilantes certainly put him in direct danger of violent death. He wanted to tell this girl how he felt about her, but . . . It wasn't fair. His life could end at any time.

Bending down, he brushed his lips across her cheek, then stepped from the porch and to his horse. Tania dogged his tracks. He could feel her behind him as he tightened his latigo. After stalling as long as he could, he turned to face her. She stood there silently, and here away from the house light he could hardly see her face. But he could tell she wasn't smiling. She stared up at him, and it was almost as if he could feel the heat emanate from her face and body. But of course that was heat from his own. The girl took a deep breath, as if about to speak, but she remained quiet. What in tarnation did she want him to do?

Tappan Kittery had his back to the horse. When Tania pressed forward, there was no place for him to go but up against the saddle. And there was no give to Satan. He planted his feet and eased into his master with as much force as he felt against him. Tania's body pushed against Kittery's, and he sought in vain to see those dark eyes as her arms closed about the small of his back. He could see there was no escape, and he didn't care. Hearing his pulse pound against the inside of his ears, he bent and kissed Tania on the lips. It was his only way out of this, he told himself. He was forced to it.

They stood that way for a long time, folded in each other's arms. It hadn't been much of a kiss, but it had satisfied her, it seemed. She squeezed him to her, and he was surprised by her strength.

She leaned back her head and whispered, "I love being with you, Tappan." His breath caught in his throat, and he hugged her closer, not knowing how to respond. He fancied he could feel the blood flow through her body. His hands reached up and caressed her long, black hair, his lips finding hers again in a kiss almost savage in its intensity. Her hands roved his back, seeking for a better position, a position that would let her pull him closer against her. But they could be no closer.

At last, breathless, he grasped her shoulders and eased her from him. He wanted to turn and go, but he couldn't. With one hand, she reached up to the back of his neck and pulled him down toward her lips again. Those lips were so full, so warm and soft and yielding, yet so full of passion, and demanding, too, if he didn't respond quite right. Her hungry kiss made his desire well up inside him, made him almost say he loved her. But he didn't know quite how, so again he said it with his kiss of passion.

His mind reeling, he put his fingers against her lips when she pulled back her head to look at him. He pushed back from her and squeezed her shoulders, probably a little too hard. "I best go, Tania,"

She smiled and raised her chin, lifting her hand to press two fingers to his mouth, the way he had done to her. "You be careful," she said. She turned and walked to the porch, and he couldn't pull his eyes away from her. She opened the door, sending a rectangular blossom of light across the yard. Then she disappeared inside.

Clutching the saddle horn, he tried for the stirrup and missed, then again. Cursing, then grinning at himself, he practically jumped into the saddle. He turned and looked at the window and saw Tania watching him. Lifted his hat to her, he trotted down the lane to the ridge road.

Chapter Sixteen
Vengeance Isn't Mine

Nine o'clock the following morning found him in the hills, the nose of the black pointed toward Castor. He held the lead ropes of two rented mules, and on the back of each dangled the carcass of a mule deer, their heads and forelegs discarded. Riding to Kingsley's café, he skinned out one of the animals, cut it into quarters with a saw Bart Kingsley loaned him, and sold the meat to the chef for a ten-dollar gold piece. Next, he left the unloaded mule at the livery corrals and skinned out the second deer, cutting off a quarter. He wrapped it in white cloth and rode up to the McBride place to leave it with Tania.

Returning to Main Street, he ran into Zeff Perry. The white-haired rancher offered him seven dollars for the remaining three quarters of his venison, and the deal was struck. As simple, and almost as accidental, as stepping in a cowpie, Tappan Kittery had entered into business as a professional hunter.

Kittery was wiping down the stallion at his stall in the livery when the Stevens derringer in his vest pocket thumped him in the ribs as he straightened up. That got him thinking about it again, and the first person that came to his mind was Cotton Baine. He had spent a lot of time thinking about him since the incident in the cantina. He knew Baine couldn't be the man who armed Shorty Randall, but why had he acted so strange about the derringer? Or had it been Kittery's imagination? From out of nowhere a thought struck him. The first place he ought to check for information was Greene's Dry Goods. And then there was Albert Hagar's mercantile, although most of their weapons were second hand.

After turning the stud loose, Kittery walked to Greene's. The doorbells tinkled as he shut the door behind him. The obese Greene turned from dusting his shelves, and his eyes became sullen when he recognized his customer. "Captain. What can I help you with?" He tried to force a smile, but it was a sorry attempt.

Not wanting to waste time on any niceties with a man of Greene's caliber, Kittery pulled the derringer from his pocket. "Do you know anything about this weapon?"

Greene stared at the gun for a moment, expressionless. He made a show of swiping at a swirl of dust on the glass counter. "Just a moment," the storeowner said. He pulled a big book from under his counter and started flipping through records, stopping briefly on what Kittery thought he recognized as a Stevens Company flyer. Greene looked up from the book when he closed it, but he didn't meet Kittery's eyes.

"Where did you get that?"

"Well, I pulled it off Shorty Randall yesterday after he tried to kill me with it. The thing we don't know is, where did he get it?"

"So nobody has any idea? You've shown it to everyone?"

Kittery nodded, wondering why Greene was so full of these questions.

"Well, I can't help you. I've never seen that gun," Greene said. "Anything else?"

Kittery shook his head. "No. You were more than helpful." With that, he turned around and stepped back out on the porch.

He crossed to Hagar's mercantile and was surprised to find the Negro, Solomon Hart, at work behind the counter. "How do, Mistuh Kit'ry," greeted Hart. The black man seemed genuinely pleased to see him.

Kittery asked him about the derringer and got the same answer as he had at Greene's. They spoke for a few minutes more about the little gun, but since Hagar's didn't seem to have any new gun catalogs he was no help in looking up any information on it. Kittery really hadn't harbored any hope for Hagar's anyway. The gun was probably too new and rare to have made its way to a gun store that sold mostly used guns. Saying goodbye to the black man, Kittery went back outside.

Brooding, he walked along the street. He wished Cotton Baine were around somewhere. He needed a smoke, and Baine was always handy with the makings. He sure wouldn't go back in Greene's and buy them.

Kittery had a sudden thought, and he headed toward the south of town, where James Price was supposed to have a little gunsmith shop. Price not only knew guns, but he was a vigilante. He would want to help Kittery, and Kittery could share with him his reasons for being so curious about the derringer.

Fortune smiled, and Price was home at his shop. He took a great interest in the derringer, and they were able to find it in one of his catalogs. Every bit of information Beck had presented about the derringers was verified. This one was extremely rare, one of twenty made. Price looked up over his reading spectacles. "It shouldn't be very hard to track that," he said. "Not hard at all."

"What do we do?"

"I'll send a wire or two, Kittery. I'll take care of it and let you know. I can't promise it'll be fast, but I'll write in my capacity as a hangman and see if it hurries them up at all."

Kittery nodded. "It's better than anything I had, Price. Whatever you can do I'll greatly appreciate."

"My pleasure," replied Price. "And sorry about Shorty's escape. I heard all about the shooting. I'm glad you and Miles pulled it off."

After Kittery left Price's, he walked down the hill and stopped behind the livery stable, plucking a splinter from a post as he set his heel on the bottom rail and felt the sun warming his shoulders. He picked his teeth with the splinter, again wishing for one of Baine's cigarettes. He didn't smoke much, but right now it would do him good.

On a whim, he leaped the corral and walked among the horses and mules to the back door of the livery. He went inside and saddled Satan for the second time that morning. There was one thing that would take away his cares better than any of Cotton Baine's cigarettes ever could. Or maybe she would just give him different cares. Either way, he had to see Tania McBride again.

As he rode up the ridge road, light clouds paced the expanse overhead, and a cool breeze drove in from the north. Kittery tied the stallion in the yard and looked around. He had seen Rand McBride's horse tied behind the bank, so he knew the man wasn't here. With that knowledge, and feeling a little mischievous, Kittery stole into the house.

The house seemed empty, but he could hear Tania singing on the far side of the closed hallway door. He tiptoed over and opened the door soundlessly. Stepping down the hall, he found another door, and from behind it came Tania's voice. He smiled to himself as he eased open the door. It opened into a spacious bedroom furnished with a dressing table, a large bureau and settee, besides the bed and night table. Tania perched on the arm of the settee, looking out a window that overlooked the back yard—a jumble of brush and broken boulders. For a moment, he watched her brush her glossy black hair, its tresses tumbling nearly to the middle of her back. She started to put the hair above her head, the rounded cut of her light blue bodice revealing the blemish-free olive skin of her back. Kittery was taken aback by her utter beauty and grace, the perfect curve of her back gliding into shapely hips which the tautness of her dress fabric did little to hide.

"You sure have beautiful hair."

Tania whirled at the unexpected sound of his voice. Her hair swung around across her breasts and across her cheek with the momentum of her turn. She put a hand to her chest. "Oh! I didn't hear you come in. You shouldn't come up on a woman that way." She started to self-consciously gather her hair again.

Tappan stopped laughing. "I apologize. I just had to see what your bedroom looked like." Tania blushed, and he laughed again. "I came to see if you'd be willing to accompany me on another ride. The day couldn't be much nicer."

"That would be wonderful!" she said, beaming as her embarrassment fled. "We could make a picnic of it, if you'd like."

"Sounds even better. You'll need more time, then. When would you like to go?"

She thought for a moment. "I can have a basket ready in an hour. Come back then. Or stay, if you'd like."

"I'll stay." He walked to a bookshelf built into the left wall while she went about her preparations. The first book that caught his attention was *The Holy Bible*. Thanks to his mother, whose religious beliefs had run deep, he had grown to enjoy reading it. It held treasures from wise men like a man could find in few other places.

Time swept by, and the hour had soon come and gone. Tania walked in from the other room and came to stand behind him, resting her hands on his shoulders. He looked up at her, and she smiled. "We can go when you'd like."

They rode down to the bank to tell McBride they were leaving, then took a route close to the one they had ridden the previous day, only this time they turned south a little sooner than where Jed Reilly had left them the day before. It was a fifteen-mile ride by the time they had made it partway up Casa Blanca Canyon (later to be known as Madera Canyon). The grassland fell behind, and white oaks grew here in little stands, intermixed with sycamore and cottonwood, along the draws. A lively stream wound down through the rocks, and beside it in one place they found a grassy nook, shaded by large trees.

At the bank of the stream, they watched a whitetail doe and fawn drink, browse a while, and then amble up into the oak trees on the hillside. After they had gone, Kittery and Tania got off their horses and took off their bridles, laying them across the saddles. Loosening their cinches, they turned them loose to graze.

Kittery had spotted a trout waving its tail beneath the bank in one deep pool. He dug into his saddlebag and came up with a spool of string and several small hooks caught in a piece of felt. With a stick, he poked into the stream bank, finding it to be heavy with clay. No worms there, so he settled for the several grasshoppers he could catch. He hid some of them under his hat for safe keeping, then left the hat, washed his hands, and went to sit on the blanket beside Tania.

Looking hungrily over the spread of edibles, he spied a bowl of gingersnaps. "Special occasion, huh?" he asked with a grin.

The girl returned a faint smile. "For me it is."

"And for me."

The delicious food was no surprise to Kittery. He expected it of this woman, even though the picnic had been spur-of-the-moment. When he was through, hardly a crumb remained. Then, while Tania lay down on the blanket and stared up at the sun-dappled leaves, he rigged up a slapdash fishing pole with a branch of desert willow and went to the water's edge. A feisty one-pound trout sucked down the first kicking grasshopper, and half-pounders were the rule after that. His stick had five of them on it after only twenty minutes. Holding it up, he grinned over at Tania. "Supper." She smiled back.

After scrubbing his hands again with sand from the creek, he sat on the blanket and leaned against the smooth bark of one of the sycamores. Tania sat beside him and gazed at the gurgling water and the shiny smooth, brown, black, white and red stones over which it ran. He, in his turn, gazed at her.

Tania had managed to put her hair in braids above her head again, tied with a blue ribbon. Reaching out, Kittery untied the ribbon while Tania watched him. As the ribbon fell away, she shook her head and ran her fingers through the braids, unraveling them. The hair fell loose about her shoulders, and Kittery smoothed it with his fingers.

The woman looked up to meet his eyes, then, and he tried to smile. It wasn't much of an attempt, though, for his heart was hammering too hard to smile. Putting his hand on the back of Tania's neck, Kittery drew her face to his and kissed her.

When she leaned back from him, he grinned and let his head fall back, smacking the tree trunk. "I should have you know, girl—I brought Satan with me today."

Tania giggled. "I could tell."

As if he had heard his name, the black horse blew through his nostrils, glancing over at them and shaking his head violently. His mane ended up in his eyes, and they laughed.

"Can I ask you something?" Tania said.

"Anything you like."

"I want to know about your life. You're always so busy asking about everyone else. What was it like when *you* were growing up?"

Kittery sighed and gazed at the girl, her face soft with shadow and raven tresses cascading down over her breasts. Why did she hold him in her clutches so?

"I usually don't see much point in dredgin' up the past. Seems like to me, if you know a person as they are now, that's what counts. Not whatever they done before. But if you want to know more, I'll tell you. Some of it ain't pretty."

Tania's only response was to continue searching his eyes.

"I was born and raised in Carolina—*North* Carolina, in the Great Smoky Mountains. We were poor. We grew everything we ate, or bartered or hunted for it. My pa, he wasn't much to speak of. He ran off when I was pretty young. But Ma, she was a big, strong woman. She claimed I got my size from her. I learned everything I knew from her or my big brother, Jake. And some sisters, too, but they had bridges of their own to build. Jake and Ma taught me together. How to shoot a rifle, climb a tree, track a wounded critter. They taught me how to handle a wagon, and how to boss a team of mules, pullin' my plow.

"Ma was gentle with all us kids, though I remember times she tried her best not to be. She said bein' too nice to us would bring our character down, and we'd all turn out to be weaklings. So she had us out in the fields young, shootin' rifles taller than we were and walkin' behind plows that were heavier. Even my sisters could shoot.

"Ma, she was pretty smart when it come to plannin' things out. She decided early to spend extra time on each of us, teachin' us one certain task really good so everyone could do their part and have a special chore only they could handle good. I was the plowhand.

"There were four of us boys and five girls. I was the youngest of the whole batch, except for little Annie. But then I was the biggest by the time I was fifteen. Prob'ly from usin' that plow. Abinidi was my oldest brother. Then Jake. He was my favorite. Kind of like the father that left me, I guess, since he was so much older. Ruth was next, and Sarah, Sharlotte, Derek, Eliza, myself—and little Annie.

"I remember Ma havin' us read the Bible every night by the light of tallow candles, before we went to bed. In fact, scripture-readin' is sort of where Abinidi and Jacob come by their names. Some man named Pratt came by one year, the way Mama told it, sellin' a little book called *The Book of Mormon*. Abinidi and Jacob were both in it. Jacob, he was a wise prophet, and Abinidi was a brave one who died by fire before he'd give up his ideals."

Kittery looked at Tania once in a while and had the urge to smile. She had her arms folded and was leaning forward, listening to him like a little kid listening to her grandpap.

"Anyway," he went on, "things weren't good for real long. We had some neighbors named Casey who were always lookin' for trouble with us Kitterys. Well, us and everybody else. They had more folks in their family, and they thought they should have the biggest, strongest boys on the mountain. Only problem was, all of us boys but Derek were bigger.

"Noah and Zeke, the two youngest of the lot, found little Annie out in the woods one day, pickin' some blackberries for a shortcake she was plannin' to make. Annie was about fourteen then, I guess, and she knew how to fight. But they were too big for her. Somehow she got stabbed with her own knife."

It pained Kittery to tell about that day. He almost relived it every time he thought about it. But Tania had asked about his life, and how could she ever know him without hearing everything that had shaped his personality?

"I think Annie dyin' was on accident. Those boys didn't mean to kill her. But they sure meant to do somethin' against God's laws. So the boys set out after 'em. They never caught 'em, though. The war come up two days later.

"Abinidi and Jacob took off to fight in the war. Derek and me stayed behind to look after the place and the girls and Ma. Well, partly that, and partly because Abinidi made us promise to take care of Noah and Zeke Casey. I was sixteen, and Derek was nineteen. We both itched to go off to the war and fight the Yanks." He laughed at that name, knowing he had become one. "We were huntin' a bear Derek had shot and let get away. We found the bear, all right. It and four of the Casey boys standin' over it. I held my rifle on 'em while Derri started in on old Noah. He did 'im up real good, too. Didn't kill 'im, though. And I don't think he intended to. Even though Abinidi wanted 'em killed, me and Derri had been brought up readin' the Ten Commandments, and we sure weren't itchin' to go to hell.

"Next thing I knew, before Derri could get started on Zeke, old man Casey showed up with a rifle." Kittery sighed and looked down at the palms of his hands. "That was the first man I ever killed, Tania. I don't remember that I ever cried about it. Guess I should've. Maybe it would've made me a different man now."

"Have you cried about killing before?" the girl asked him.

"I did in the war, yeah. I cried when I was walkin' through the battlefield at Gettysburg. I didn't fight there, but we showed up two days later. There were dead horses and men in piles so thick you couldn't count 'em in a month. Blood and flies everywhere. And even worse, there were men so thick in the hospitals, some with pretty small wounds, and those doctors were cuttin' their legs and arms off faster than you could say 'Kentucky.'" Thinking about it, a chill ran up Kittery's spine.

"So what happened with the Caseys?" Tania's words brought him back to his story, and he spread his fingers, watching his hands as if they were about to perform tricks.

"Well, I emptied my rifle when I killed Dred Casey. Then Zeke, Hadley and Trick come at us lookin' for real blood. Hadley grabbed his pa's rifle, and he should have shot me. He should've, but he didn't. He shot Derek instead. Trick, he was the biggest and the oldest of 'em. He stood about three inches shorter than me, and a good twenty pounds lighter. But he was some fighter, and he had experience on me, too. I had also gave 'im somethin' real to fight about, I guess. He jumped me and near had me beat before I got the knife out of my belt and killed 'im. He was only the second man I'd ever killed, and both in one day. But Trick was the hardest. It's a pretty awful feelin' bein' that close to somebody—close enough to feel a blade slide into their body.

"I've always regretted bein' forced to kill Trick. He was the only decent man in that family. He wouldn't even have been in the fight if it hadn't been for Noah and Zeke. He used to be Jake's best friend, before the trouble.

"Anyhow, by then Hadley had his pa's rifle loaded up again. But before he could do any shootin', Ma and the girls showed up, and Hadley was dead before he could touch the trigger. Ma always was a crack shot.

"Noah was still there, all beat up. And Zeke, too. But the rest of the Casey boys had gone up to war. We let Zeke go, draggin' Noah away by himself. I remember he was cryin', and I wondered why I wasn't. I didn't even cry when we went over to Derek and found out he was dead, too."

"So Zeke and Noah started the whole fight and got away with it?" Tania asked.

"Well, no. We heard a rumor a few months later that Pa had come back to the mountain. And that's when all the trouble with the Caseys ended. I swear it was none of my doin', but they found Zeke and Noah hangin' from a walnut tree down by the old gristmill. We always figured Pa did it for little Annie.

"When we got word that Abbi—that's what we called Abinidi—had got himself killed at Bull Run, I went off to the war myself. I joined the Rebs. Didn't want anyone tellin' me how I had to think. Then, when Gettysburg ended we found out Jake had been hurt pretty bad there, and he just up an' disappeared. I've never heard from him since, and he only wrote the girls twice. Maybe it didn't matter a whole lot. When he found out I'd gone over to the Yanks, he wrote to the girls that I was no longer any brother of his." It almost tore his heart in two to recall the day his sister Eliza told him that. He had gone off into the woods by himself only to find out even the war had not wrung all the tears from him. Jake was like his father, and his rejection made Tappan cry plenty.

"Why did you change sides?" Tania asked.

"That's a hard question. A hard question for sure. Mostly, I got thinkin' about how strong this big old country could be, if it ever got over its differences. I decided I didn't want any part of breakin' it up.

"I guess I ended up where I belonged, anyway. They made me a captain within a year after switchin' sides. 'Course, it was mostly boys and old men I was in charge of. And I was a good six inches taller than most of 'em, too. They figured giants make the best leaders, I guess."

"Did you ever fight again, on the north side?"

"I was involved in a little skirmish in Atlanta, but nothin' big, no. I started my fightin' back up when the war was over. I reenlisted as a sergeant in the regular army. Came west to fight Cochise."

"Why didn't you stay with the Army? My father has friends who did, and they retired pretty well off."

"I had some disagreements. That's the gist of it. I couldn't abide by how the Indians weren't given any rights at all. Yeah, they did a lot of things they never should have. But we were paradin' around like we were saints, and we were doin' the same things they were—wreakin' our vengeance on people who hadn't done anything because the guilty ones were too cagey for us to catch. Stuffin' the Indians on the most worthless pieces of land we could find. Pretty shameless dealings. I couldn't take the politics of it. I landed a lieutenant on his butt one day, and I landed myself in the guardhouse. So here I am."

"What have you done since?"

"Odd jobs. I worked on a couple ranches—one with Cotton Baine. In Prescott, on one big ranch, I bought and trained horses for sale to the army. Then I drove a freight wagon and loaded freight for Tully and Ochoa—you've seen the big green and yellow wagons."

"They say you hunt bounties for a living, though."

Kittery laughed. "Nobody hunts bounties for a livin', Tania. They'd starve to death. You might luck into somethin' once in a while, but it's not often. Yeah, I'm huntin' a man right now who has a bounty on 'im. The money will be helpful, but I'm huntin' this man because he likes to kill women and children. He killed an entire family up in the Santa Cruz valley. Butchered them and their livestock, too. Just for fun."

Tania looked down, shaking her head. He wished he hadn't said that last part to her, but it was good she knew. He wanted her to know he was hunting Ned Crawford for something more than just money.

"What happened to your sisters?" Tania asked.

Kittery thought for a moment. "I couldn't tell you for sure. Ma died during the war, and they all drifted away from the home place. Ruth and Sarah, they were big girls like Ma. When they were younger they didn't have any boys chasin' after 'em—though a few did when they got older. But I heard they moved west together to buy a farm. And Sharlotte—now, she was real pretty, but strong like the rest of us. She sort of took after Pa. He was a handsome man when he wanted to look the part. Anyway, Sharlotte, she got married to some boy from the valley right after I left to fight. I never did see her again, but I heard her husband died in the war and left her with a son. Eliza's the only one I've seen since the war ended. I saw her with her husband last year in Nebraska. She got married to an army officer, of all things."

Tania smiled and sat there nodding for a moment, searching Kittery's eyes. "Thanks for telling me about yourself, Tappan. I was just . . . aching to know."

"Yeah, well, now maybe you'll know why I'm not always the most happy-go-lucky feller around."

"You're happy enough for me. And I plan to make you far happier."

She was looking at Kittery like she wanted him to kiss her, but instead he pushed up from the ground and held his hand down to her. She looked up at him, confused, and finally took his hand, coming to her feet.

"What's wrong?"

"You're father trusted us, Tania, givin' us his blessings to come out here on this ride alone. As it is, they'll be talkin' around town. I'd just as soon it wasn't true."

Tania nodded and smiled. "You're right. Let's take that devil horse of yours and go home."

Every night that week Tappan Kittery took supper with the McBrides. He and Tania grew closer each day. Castor remained a tranquil place to live, and talk of the killings became rare. To all eyes, the little settlement had slipped back into its peaceful state of dormancy, like in some fairy tale. The cool spell that had crept down from the mountains hung on tenaciously. These pleasant days were the way April in Arizona should be. Cotton Baine continued searching for a job, so he was seldom in town anymore. But when Kittery did see him, the relationship between the two of them was the same as ever. The Desperados had not been heard from since the death of Shorty Randall, and Miles Tarandon walked his rounds without so much as the arrest of a drunk.

It was early one morning when Kittery made the discovery. He was cleaning the Stevens derringer and had removed the grips to get at any grime that might have collected beneath them. In the morning light, two initials carved on the inside of the grip were very plain: C.B. Kittery mused over that. He kept repeating them over and over in his head, but at the moment they meant nothing. Deciding to see if Price had learned anything yet, he went downstairs and up to Price's gunsmith shop. Price was home, and he greeted Kittery with a nod as he opened the door.

"I was thinking I should go up and see you," said the hangman. "I'm glad you came over. I got a wire back for you."

"Good. What'd it say?"

"That little weapon went to Stephens Hardware, up in Prescott, eight months ago. I still have to wire Prescott to see if they remember who they sold it to, but we're narrowing the gap."

Kittery stood there speechless. Price turned his head to one side curiously, expecting him to say *something*. "That's good, Price. Good," he stammered. "Let me know." Without another word or a proper farewell, he turned and walked back toward town, feeling sick in the pit of his stomach.

The initials C.B. had not been enough. Now he found the gun had been purchased in Prescott within the last eight months. The initials matched Cotton Baine. And Cotton Baine had ridden down from Prescott six months ago.

Kittery was too sick to let Maria Greene or anyone else see him on the street. He had to get out of town, so he went to the livery stable and entered through the back door, hoping to avoid Jarob Hawkins's prying questions. He threw the saddle on Satan and was tightening the cinch when he heard a familiar voice at the corral behind the building. It was Cotton Baine! He could hear Jarob Hawkins's gravelly voice in conversation with him.

A bad feeling welled up inside Kittery. He had to know what was going on, and he had to know now. He had already let his mind wander way too far, and he didn't like suspecting his best friend of treachery. He had to know.

Leading the stallion out the back, he stopped a few feet away from Baine as the blond cowboy and Hawkins turned their attention on him.

"Sorry to interrupt. I'd like to talk to you, Cotton."

The cowboy glanced over at the one-legged Hawkins, who took the hint and shrugged, hopping away to where his pitchfork leaned against the back wall of the building. Taking it up and using it as a crutch, he went past them into the livery.

Kittery motioned with his head for Baine to follow him to the far side of the corral. When they had leaned up against the top pole, Kittery said, "Roll a smoke for yourself."

Cotton shot him a queer glance, then shrugged, pulling out the makings. He offered them to Kittery, who passed. When the cowboy had touched a match to his smoke and the first cloud snaked beneath the brim of his hat to dissipate, Kittery turned his eyes on him.

"I have to know about that Stevens derringer."

Baine chewed on the blunt question for a moment. He had never been much of a poker player, and Kittery read in his eyes the moment his friend realized the futility of telling a lie. "I bought that gun in Prescott, Tap. Special ordered it from the factory through Stephens store."

Kittery was silent and scared. He waited for Baine to tell him he had given the gun to Shorty Randall. He was remembering the night his friend walked into the jail with dinner for Shorty. It had struck them all as odd, but they had laughed it off. He remembered joking about a gun on the plate.

Baine's laugh surprised him. "You think I gave the gun to Shorty?"

Kittery, caught in his thoughts, had to laugh, too. "Well . . . it's not what I'd like to think."

"Then don't." Baine took out his cigarette and threw it into the dust in disgust, his face taking on a drawn look as his eyes scanned the houses along the ridge. He turned back to Kittery. "I lost that gun in a poker game three months ago."

Kittery's surge of new hope was overtaken by cynicism. "Yeah? To who?"

"I'd rather not say."

"Even if it's the man who gave Shorty that gun?"

"It can't be him, Tap. He wouldn't do that. I gotta talk to 'im first. I just got to."

"You talk to him then," said Kittery. "But by thunder you tell me now who it was in case you wind up dead."

Cotton Baine blew out a deep breath. "It was Zeff Perry."

When the words registered, Kittery felt a huge relief wash over him, for he believed his friend. But the relief was trodden down by the thought that Zeff Perry, a respected rancher in Pima County, was Baraga's inside man. It was a sobering thought.

Luke and Beth Vancouver returned from Prescott on a balmy Saturday afternoon. Luke steered the buckboard up in front of Kingsley's café and checked the plodding of his two weary bays. Although bags had started to form under her bloodshot eyes due to lack of sleep, Beth looked beautiful in a yellow cotton dress that clung to her torso. She wore a white shawl, and a fringed parasol kept her face in shade. The lawman wore a dark gray suit, now covered with a thick coat of dust, and a black string tie.

After he had climbed down and squatted to stretch his legs, Vancouver helped Beth off the seat. They stepped into the shade of the café together, brushing the dust from their clothes. When they espied Kittery and Tania McBride at one of the corner tables, they came over to them.

It was a happy reunion. Its one grim moment came when Kittery had to be the one to tell Vancouver about the violence that had beset the town in his absence. After learning the circumstances of Shorty Randall's death, the lawman seemed to accept it, and the four friends fell to talking and enjoying their meal.

Late that night, after Kittery had left Tania at her front porch, he sat alone in his room. The stars were a wash of sparkling diamonds across the black satin sky as he gazed down on the street. Everyone had long since gone to their homes. Emptiness greeted the big man's pondering gaze, and an occasional dry breeze played at the curtains and the hairs on his bare chest, cool against the sweat that sparkled on him.

Up on the ridge, a single lamp shed its beam into the shadows, but besides Kittery's it was the only one. He realized after a moment that the light was Tania's. Was she looking out her window at *his* light, thinking of *him?* Was her heart as full of longing as his was? A shadowy form passed in front of the window. In a moment, the lamp went out, and there was only blackness.

A cool breeze wound its way along the street, finding Kittery's window and fluttering the curtain, cooling his damp forehead. His Bible lay open on the bed, and the same breath of wind turned some of its pages.

After a last thoughtful moment, he turned from the window and sat down on the bed. He lifted the Bible without giving it any conscious thought and began to read. He had read over a page, paying no attention to the words, for his thoughts were still on Tania. But his interest in the book returned, and he found himself backtracking to find what had captured his attention. What he saw made his stomach tighten up, and his heart begin to beat soddenly. He read it again, and was overcome with the sensation that these words were put there for him. This wasn't where he had been reading earlier. He had been in the Gospels, and this book was Romans. One verse kept drawing his attention, and that was the one that made a quiver run through his body. Any thoughts of Tania and their night's goodbye had flown from his mind.

He read once again, in the twelfth chapter of Romans, the nineteenth verse. "Dearly beloved, avenge not yourselves, but rather give place unto wrath: for it is written, Vengeance is mine; I will repay, saith the Lord."

The memories were distant now, but he recalled the night his mother had read what must have been that same verse to him and his brothers after Zeke and Noah Casey killed little Annie. They cut him to the heart now, as they had then. He prided himself in believing the teachings of the Bible, but this passage made him feel no better than it had all those years ago. One of his best friends, like little Annie, had been murdered—mutilated! And those who did it still walked the Earth, free men, doing as they pleased, breathing and enjoying life and taking the lives of more innocent people every day. How could a man with any pride be expected to let that go? Sure, vengeance was the Lord's. He promised to repay. But when? *When?* And how many more lives would have to end in smoke before Baraga and his men were gone?

Glaring at the black book where it now lay on the bed, he clenched his fists in anger. *Vengeance is mine; I will repay.* He remembered his sweet young sister's death. How many innocent lives might have been destroyed by Noah and Zeke Casey if it hadn't been for his father's revenge against them? They would have gone on living in their sins, like Baraga and his heathens did now.

In a rage, Kittery swept the Bible off the bed with his hand, sending it slamming into the wall. For the first time since the marshal's death, his eyes welled up with tears, blinding him for a moment. He looked over at the Bible through dim eyes and went and picked it up, smoothing the dust off and returning it to the bed. Looking up at the ceiling, he whispered, "When, God?"

Chapter Seventeen
Ambush!

All that night it seemed to Tappan Kittery that his heart and very soul were torn apart by the words of Holy Scripture running non-stop through his mind. His mother had taught him to put all his trust in the Bible, for it never lied. But at the moment it seemed if he didn't take care of the Desperados, no one would. They would survive to be old men, living off the sweat and hardships of others. They would continue to terrorize the territory, murdering and plundering at will. The struggle within him was tremendous. It pitted his mind between thoughts of instant justice and the eternal truth of the Bible. His mind burned in confusion.

But as light started to grow in the eastern sky, the painful decision came to him. In his mind he saw his mother smile. He was done with Castor, for he was done with this crusade. Two men had paid for the marshal's death—Shorty and Wespin. Perhaps that was enough. No longer could he hunt the rest of them, because his assurance that it was the right thing to do was gone. Perhaps he would stop bounty hunting altogether. He was sorry to leave Castor, to leave Baine and the Vancouvers and all the friends he had made. But that was how it had to be. To stay would mean to take the chance of becoming obsessed all over again, and he couldn't let that happen. Not after being served such a powerful warning.

Most of all, he hated the thought that he was breaking his promise to Joe Raines to see all of the Desperados dead. But in his mind the words remained: "Vengeance is mine. I will repay." Someday . . . Then so it would be.

When the sun broke over the mountains, Tappan Kittery fell asleep at last.

He yawned awake once more when the sun was large and yellow and high in the sky. He sponged the grime and sweat off his body, shaved, dressed and combed his hair, then went to the street. He wore the same outfit he had bought for Joe Raines's funeral—a black broadcloth suit, white, striped shirt and cravat. Something about his decision of the night before marked this a special day—a turning point in his life. He stood for a moment in the shade of the awning, then strode next door to Cardona's cantina. He leaned up against the long, pockmarked bar.

"Give me your very best, amigo." He smiled at Mario Cardona.

"Ver' good." The Mexican reached beneath the bar and pulled out a tall black bottle of Scotch whisky, pouring three fingers of the dark amber liquid into a shot glass. "Señor Adam Beck was here before," he volunteered. "He wish to see you."

"Gracias."

"It is you' birthday?" Cardona gestured with a hand to indicate Kittery's attire.

"No, no," Kittery replied with a chuckle. "But it feels like it could be. It's a good day."

"In this case, I will make it better for you. The drink, it is *gratuis*—no charge." Cardona beamed, revealing his broken teeth. "Is good when customer feels good. You would like a cigar, also?"

"No, thanks, amigo. I bought this the other day," he replied, pulling a black pipe and tobacco out of his coat pocket. The pipe looked identical to the one Raines had smoked, and Kittery felt no shame in admitting to himself he had bought it in his friend's memory.

When he had the pipe lit and was puffing on it, smelling up the room with its acrid-sweet fumes, he looked over at Cardona. He caught the Mexican staring at him with an amiable gleam in his eyes, and he smiled. "Thanks again for the drink, my friend. Say, where is Adam Beck now, do you know?"

"I think he is eating at the café. But he will be here soon. Señor Baine has left him his room. He goes away ver' soon—perhaps you hear, no?"

Kittery paused and lowered his drink. "No, I didn't hear. Goes where?"

"To his *trabajo*—his . . . work, you know? Right now he is with the Señor Beck."

Kittery gulped down the remains of his drink and turned away from the bar, stepping outside and to the café. He found Baine, Beck and Jed Reilly seated together at a table. All of them greeted him amicably.

"Cardona tells me you were lookin' for me, Adam."

Beck shrugged. "Oh, yeah. Well, just to tell you I'll be around town for a while if you'd like to get together for any rides out toward the Baboquivaris."

Kittery nodded peremptorily and turned to Baine, looking at him for a long moment. "I hear you're leavin' town," he said.

"Yeah. Jed and his brother found a place for me on their spread. I reckon I'll still be seein' you on Saturday nights, though."

"I reckon you better." Kittery forced a smile and clapped his friend on the shoulder. "Well, if I don't see you before you go, I'll wish you good luck." He glanced across the table at Jed Reilly. "Good to see you again, Reilly."

"Call me Jed."

"All right, Jed," Kittery said with a smile. His throat tight, he turned back to Cotton. "So did you talk to 'im?"

Baine's face flushed. "I didn't. I ain't had time yet."

Kittery sighed and nodded at the others, then went back outside.

Standing on the porch looking along the street, Kittery grimaced. The way things appeared, he would never see his friend Baine again, once the puncher rode out for the ranch. It was as if he was losing another best friend. Not that Baine's finding work had anything to do with that. Kittery never planned to come back to Castor again either way. In fact, he was having stupid notions of somehow trying to track down his brother Jake. Now wouldn't that be something, to see him again after all those years? Who could tell—maybe Jake had forgiven him for becoming a Yank.

Puffing on the pipe, Kittery went to the mercantile at the end of the street, across from the dry goods store. Lafayette Bacon, one of the vigilantes who was also Zeff Perry's ranch hand, said hello to him at the door. Inside, Albert Hagar, the silver-haired owner of the warehouse and this business, stood behind the counter. The sixty-year-old was one of the original founders of Castor who had come in from West Virginia with the Vancouvers, McBrides, Doctor Hale and Sabala.

"Can I help you, friend?" Hagar asked.

"You might." Kittery unbuckled his gun belt and holstered Remington and pushed them across the counter. "I'd like to sell this."

Hagar scrutinized the revolver. "Prime condition," he decided at last, flipping open the little leather case attached to the belt and lifting out the loaded spare cylinder. He let it roll across his palm and nodded, looking back up at Kittery. "I assume you're seeking to upgrade."

"No, in fact I'm not. I just need some travelin' money. Headin' out of the country, and I won't need it much, where I'm headed."

Lafayette Bacon overheard Kittery's words and whipped his head around as if to make sure who had spoken. After watching Kittery for a moment, he departed.

"Won't need it much, huh?" Hagar responded with a chuckle. "Sounds like a place I'd like to go. Well, the only thing about it is cap and ball weapons are falling out of favor. The gunman nowadays wants everything simplified, you know. I can sell a good used cartridge gun for twice the price, and there are those of them going for as little as five dollars, depending on condition and make. Still, it's a good gun. For it and the holster, I can give you ten dollars. Sorry I can't do better."

Kittery hefted and pondered the revolver. He fingered the smooth surface of the cylinder and down the length of the octagonal barrel with a touch of nostalgic regret. The pistol had given him years of faithful service since the war. Now it would hang at someone else's side.

"I'll take that offer, Mr. Hagar."

Kittery pocketed the gold eagle the old man produced. When he stepped out on the porch, he saw Cotton Baine and Jed Reilly headed toward him at a fast walk.

"Is it true what I hear?" queried Baine as they came together. "You sold yer gun and yer leavin' town?"

"News travels fast."

"We need you here, Tap," Baine protested. "You forgettin' about yer friend?"

"Not forgettin', but I'm tryin'. I did a lot of thinkin' last night, Cot. Someday I'll tell you about it."

"Tap . . ." Baine looked into Kittery's eyes, and his shoulders sagged. "We still have money comin' fer Wespin. You'll hang around to collect that, won't you?"

"It's yours, Cot. You made the shot."

Baine sighed, and his eyes dropped. "Well . . . so long, partner." He set his jaw, turned and walked off toward the dry goods store, with Reilly trailing after.

Later, Kittery stepped from the hotel. It was late afternoon, and although only April the heat that typified Arizona was crowding down on the desert town. He wore his old clothes again, his gray and white striped pants and the dark blue pullover shirt with a patch in its left shoulder where Beth Vancouver had mended it after his stabbing.

Adam Beck spotted him from the shade of the cantina's doorway and came over to step onto the walkway. "Baine's lookin' to see you. He's leavin'."

A strange feeling of loneliness came over Kittery. This was the last time he would see his good friend. It could not be just another parting.

He and Beck stepped to the cantina, where Baine's line-back dun was tied at the rail. As they were starting through the swinging doors, Baine met them. He grinned and nodded, moving to his horse to tighten the cinch. For a drawn-out moment, no one spoke. Then Baine turned and looked up at Kittery from where he had been fiddling with his saddlestrings.

"You know that banker's daughter has took a perty serious likin' to you, Tap. Yer leavin' will break her heart."

"I'll ask her to come with me."

"Tap, you fiddlefoot. I s'pose there's not a thing in the world I c'd say to make you stay, is there? There's still another bend in the river you got to ride around."

"Yeah, that's it. You know me too well."

"The Vancouvers won't be happy to hear yer leavin', either. I s'pose you'll ask them to go with you, too."

Kittery laughed. "They'll survive without me."

"Yeah, I guess they will. The way his wife looks at you, the sheriff may be glad to see you go." Baine forced a grin. "Well, old boy . . . seems you an' me've known each other forever. Been what you'd call best friends, huh? But . . . I know all good things have to end. Who knows, maybe someday . . ."

He thrust out his hand, which Kittery grasped and held. "Best friends, Cot." The big man forced a smile, his throat tight. He was losing another brother. "We always were, always will be. Don't let those long, lonely nights bring you down. Somewhere out there's a woman waitin' for you. Don't make her wait forever."

"Only woman waitin' for me's a cow," Baine said with a grin. His face went serious, and the two of them embraced, separating and avoiding each other's eyes.

"Write me a letter, Tap," said Cotton Baine in a quiet voice, and he swung a leg over his saddle. Reining the dun around, he lifted a hand in farewell. He met Jed Reilly at the dry goods store, and together they rode into the desert stillness.

For a while, Kittery and Beck watched their retreating dust cloud. Then the bigger man turned and said in a husky voice, "Join me for a drink?"

Without reply, Beck pushed through the swinging doors and went to sit down at his accustomed table, where a bottle and empty glass still waited. Kittery took the seat across from him, looking at a dog-eared deck of cards that lay with four cards separate from the deck, face up. "Your game?" He looked up at Beck.

The words drew Beck's attention back to the big man. "Uh . . . no. That's Cotton's."

Kittery picked up the four separate cards and looked them over. An ace, a king, a queen and a jack—all diamonds. One ten short of a royal flush. He showed Beck the cards. "Looks like Cot was drawin' a pat hand. Wonder why he quit."

Beck chuckled. "Said somethin' about quittin' while he was ahead."

Kittery smiled wryly and tugged a card out of the center of the deck, turning it face up on the pile of four. It was a deuce of spades.

Beck glanced down at the card as he struck a match with his thumbnail and held it to the tip of a fat cigar. "Some people would say that's bad luck."

Kittery grunted and shoved the cards back into the deck. "Some people are superstitious."

An hour had passed since Cotton Baine rode away. Kittery lay on his bed, sweat dampening his sheets. He had tried to sleep, but he couldn't, so he stared at the ceiling. His mind kept drifting to Tania McBride. What was he going to say to her? How could he tell her he had decided to leave Arizona? And did he dare ask her to come with him? He had nothing to offer her, not even the simple necessities of life. He had no work, no property—nothing a young married couple would need except love, and that might start to leave a bad taste in the mouth after the belly went empty for four or five days.

His instincts told him the girl would say yes if he asked her to go. But was it fair to ask? She didn't really know him. For that matter, he didn't know her as well as he would have liked. But he didn't live in an era when men could afford the luxury of getting to know a woman well before asking for her hand. There were at least ten men for every woman out there, and with Tania's attributes it was a miracle she wasn't already married. The most important thing was Kittery had never before felt the love he did for Tania. He had never had the chance. Army posts and mining camps weren't famous for putting men in contact with available, virtuous women. He would be making a grave mistake saying goodbye to this one. It didn't take a very intelligent man to know he would never find another like her.

The big question was, would Tania be able to leave her father so suddenly? Perhaps not, but he had to ask, all the same.

He stood before her that night in the darkness of her yard, his hat in his hands. He had turned down her offer of supper, unable to work up an appetite. He had come to the house after he was sure supper would be over and the dishes cleared away.

It hadn't been as hard as he had imagined, telling her of his imminent departure. But now came the silence, the awful waiting . . . the pain.

Finally, the woman looked up. "Why, Tappan? Please tell me why you have to go now."

"I can't tell you now, Tania. You'll have to trust me when I say I have to go. It's going to take me a while to know how to tell you what happened."

Tania swallowed, glancing around the yard with tear-filled eyes. The look in her eyes was that of someone who had just realized they didn't know where they were. "There's so much to do, Tappan. So much to think about. Papa will be all alone then. I'm the only one he has to care for him."

"I know, Tania. I thought about that a long time. It won't be easy. But . . . I'll understand. However you decide, I promise you it'll be all right. I'll wait for your decision."

He raised a trembling hand to her neck and leaned forward to brush her parted lips with his, not bringing his body closer. "Good night, Tania." He made his way down the walk on foot and trudged down the ridge road with his throat too tight to swallow.

The answer came at breakfast next morning. It came in a way Kittery could never have foreseen, in a way that told him God had a hand in the outcome. Luke Vancouver located Kittery in the café to bring him the news that Rand McBride had fallen while he and Tania were out taking the horses for a sunrise jaunt. His horse had shied at a flushing covey of quail, losing its footing on a loose slope. Going down under the horse, McBride had broken a leg, and one arm in two places.

Kittery thanked Vancouver for coming, paid Kingsley for the meal and left it half-eaten, hurrying to the McBride place. Doctor Adcock Hale let him inside, and they walked into the parlor. "He's fine, Captain," Hale answered Kittery's question of concern. "Well, at least it won't kill him. But he won't be getting out of bed for a while."

Kittery's heart tightened as he looked at Rand's bedroom door. "Is Tania in there?"

"Well, yes she is," Doctor Hale replied. "But, Captain, I—" He stopped.

"What is it, Doc?"

Hale looked at him, searching his eyes. He shook his head and turned away, waving him on. "Sorry, boy," Kittery heard him say under his breath.

"Thanks, Doc."

Drawing a deep breath, Kittery knocked, and when he heard Tania's voice he stepped inside. A thick yellow blanket hung over the window, but sunlight struggled through to light the room.

Tania sat beside the bed, holding her sleeping father's hand. Her eyelids were puffy and red from crying. She looked up when Kittery entered but didn't make any move to stand. They stared at each other, unable to speak for a time. Kittery broke the silence, not quite able to meet her eyes. "It's all right, Tania. It's all right. I told you I'd understand."

Tania's eyes were rimmed with tears, and they began to run down her cheeks. She stood awkwardly, releasing her father's hand. Stepping to Kittery, she grabbed his sleeves. "Do you—Do you have to go now? It isn't fair to me, not now."

"Tania . . . I have to go. I don't expect you to understand, but it's out of my hands. There's somethin' bad that will happen if I stay."

"Oh, Tappan. I'm sorry. I'm sorry," she whispered through her tears, wiping at her face with her sleeve. "I can't say any more than that."

Kittery's hands hung loose at his sides, and he found no words that would mean anything. He wouldn't cry. He felt like it, but he wasn't going to show her that weakness. And to protect himself, he must leave now.

"I'm sorry, too, Tania. I hope your father will be all right. I'll be on my way," he said. He looked at McBride. "Tell him I said goodbye."

Tania clutched at his sleeve again as he turned around. "No, Tappan," she said with a quiet sob. "You can stay for a day or two. It won't be long. You know I love you." She took his wrists in her strong hands. "Please."

"I can't stay," he said stubbornly. "I've stayed too long already. Don't make it hard, Tania. Maybe I'll come back for you someday." Even as he said it, he somehow knew he wouldn't. He could never come back here. Once he rode away, he would keep on riding. But he felt obliged to say something to ease the girl's pain.

Again, he turned, and Tania let go of his wrists, her hands going limp. He hoped she hadn't seen the tears come to his eyes.

"Tappan, please . . ."

He paused with his fingers on the doorknob, but only for an instant. He stepped through and away, moving past Doctor Hale with his eyes fixed straight forward. "G'bye, Doc," he said, low-voiced.

He let himself out before Tania could decide to come after him and strode briskly down the ridge road, his heart filled with loneliness. He fought down the boyish urge to run back to town. He kept his eyes straight ahead.

Behind him, Tania rushed to the porch, and through tear-filled eyes she watched him go. She opened her mouth as if to call out but held her silence and supported herself against an awning post. Then there was only the stillness, just her heart pounding in her head and the sound of an oriole singing from its perch in the catclaw thicket.

Doctor Hale stood behind her, and he laid a gentle hand on her shoulder. "He'll come back, honey."

"No . . . He won't."

It was early that same morning that Cotton Baine reined his horse away from the squat gray buildings of the Reilly ranch. He had simple orders: ride south and west into the foothills of the Santa Ritas; check any stock and keep track of their location; get to know the general lay of the country. He rode through the half-light, and when he threw a glance backward, the dim yellow glow from a window shone like a struggling beacon in all the vastness of the broken hills.

As it grew light, he scanned the grassland. Bunchgrass dotted the low-lying hills, accented now and then by a mesquite or the stiff, sharp leaves and towering stalk of century plants. Next to the trees, he always saw the tramping grounds of cattle, littered with abundant droppings and dry, bleached bones. But he saw no life, so he rode on.

Early in the afternoon, the sun glaring overhead, a wide draw opened up before him, gaping its olive-colored innards to the sun. A wide and sandy creekbed came sweeping in from the left, then made a bend and continued on away from him. Its path had undercut the ground where Baine sat his horse so that it dropped down four feet or so to the rock- and gravel-strewn creekbed. On Baine's side and beyond the opposite cutbank, trees bunched together, stretching their bony limbs to the sky as if to beg relief from the sun that ruled their arid land. He had seen fifty or sixty head of longhorn cattle earlier in the day, but they had been widely scattered. Here, more than any place he had seen, seemed a likely sanctuary for a larger herd.

Dropping over the lip of the cutbank, Baine descended into the draw. Man and horse were engulfed by the shadows of sycamore and cottonwood and ash trees. Ahead, Baine caught a flicker of movement, made out a brindle patch of hide. He grinned. Now, *he* was a cowboy. He had pegged this place, sure enough.

But in a moment more the high-pitched neigh of a horse surprised him. He had gone out of sight of the patch of hide, but the noise came from that direction.

He followed the sterile streambed in the direction from which the sound had come. When he spotted it, he was taken by surprise once more. This horse was not just loose range stock. The stocky animal, a light chestnut in color, closely resembling palomino, wore an old Texas saddle and a horsehair bridle. It had a *mecate*—lead rope—lashed to a cottonwood limb. It was the shadows from the cottonwood leaves that had given the horse's hide the brindle appearance.

"Howdy," Baine called into the stillness. His answer was the rustle of the grass and leaves, a stomp of the chestnut's hoof. His dun nickered. "Whoever you are, I don't mean yuh no harm. Just come on out where I c'n see yuh. You hurt?" He spoke more to break the silence than because he expected anyone to answer him.

As he called out, he nudged the dun farther up the wash toward the chestnut. He could smell the sunshine and the dead grass and the dusty gravel flung up by the dun's hooves. "Easy boy. I'm not here to hurt you." Throwing a nervous glance about, he pulled his Colt. This strange horse gave him a bad feeling.

Behind Cotton Baine, two pale eyes peered from the shelter of boulders balanced along the far bank. They took in every move the horseman made, gazing with the patient curiosity of a trained huntsman. The man to whom those eyes belonged knew the puncher had seen his horse and was puzzled.

He glided his rifle barrel along a notch in the granite boulder he leaned against. He cocked the long-barreled weapon and lined up the marks inside his telescopic sight along the cowboy's collarbone as he turned his dun broadside. The cowboy started his dun forward, holding his gun and looking everywhere but back.

The rifle barrel lunged back and up, spewing a heavy white cloud of smoke and filling the draw with a deafening roar.

Cotton Baine heard the crash of the rifle at the same instant the .45-70 slug shattered his collarbone and neck, clutching him from the saddle. The dun bolted, racing in panic down the creek bed. It carried splashes of blood on its neck and saddle.

Cowboy Cotton Baine lay dead among the stones, his fists clenched in final agony.

The killer stood motionless, the barrel of his rifle still resting in the vee of the boulder top. He wasn't particularly anxious to see the results of his deed, so he waited while the last of the smoke riffled away in the air. He rolled a smoke and lit it, drew deeply and blew two streamers of gray out his nostrils. He swatted a deer fly on his neck.

Finally, he slid down the embankment and moved toward the body, his moccasins making no sound on the rocks. He nodded with satisfaction as he turned Cotton Baine onto his back with a forceful shove of his moccasined foot. His shot had been true. The puncher's eyes were open to the sun, his shoulder and chest soaked in blood from the wound.

The killer stooped and went through the cowboy's pockets. He found a well-used jackknife and a half eagle—a five-dollar gold piece. He took the cowboy's pistol, too, along with the Mexican loop holster and cartridge belt full of shells. Turning, he climbed up the embankment to his horse. He looked after Baine's dun for a moment, then shrugged and mounted his own animal. Reining it around, he dropped over the bank and rode down the streambed.

Captain Tappan Kittery stepped into Jarob Hawkins's stable with bedroll and bulging saddlebags in one hand, his rifle and a gunnysack full of biscuits in the other. When he had saddled and bridled the stallion and started tying down the bedroll and saddlebags, he heard Hawkins approach from the shadows of his office, stumping along on his wooden leg.

"So it's true then, what they say." Kittery didn't turn around, so the old man continued. "Yore leavin' Castor. Seems a dang shame, to me."

"What's a shame?" Kittery turned half-around as he drew the saddlestrings tight.

"Seems to me a man like you wouldn't go leavin' his friends thisaway. Seems like with a friend like the marshal was t' you, you'd be wantin' to stay an' bring in them that killed 'im."

"Who am I to do that? That's the sheriff's job."

"Oh, that's bull! Don't talk to me like a imbecile, Cap'n. I don't deserve that. Don't you think I know about you an' Price an' all them? Hell, I was the one got the vigilantes t'gether."

Kittery spun around and stared at the old man. "You're serious, aren't you?"

"Dead right I'm serious. I seen what was happenin', an' I went t' Price. The law couldn't handle it, Cap'n, an' you'll see it still can't. Leavin' now's a craven thing to do."

"You're pretty free with your opinions, Hawkins. I'd hoped to leave here peaceful."

"Then leave peaceful, Cap'n." Hawkins made a broad gesture with his arm toward the door. "An' ther's many a young fella hereabouts that'll take yore place with the banker's daughter." He winced at his own words and turned away sheepishly. "Dang it, Cap'n. Sorry for that. Good travelin' t' you."

Kittery had turned to stare out into the street. Now he lowered his head. "Thanks, Hawkins. You take care of yourself and this town. So long."

He swung aboard and cantered out and up the street. He had almost passed the point of no return, the turn-off to the ridge road, but at the last possible moment he turned and rode between the dry goods store and the water tower and made his way up past the McBride place to the Vancouver homestead. He dismounted and tied the black in the shade at the side of the house.

Beth opened the door at his knock. She greeted him with a forced smile. He could see she had heard the news. He was glad, in a way, for he hadn't relished telling her.

"Can you come in, Tappan?" She smiled uncertainly.

"I just rode up to say goodbye," he said. "Reckon Tania told you I was leavin'."

"Yes. Yes, she did." She searched his eyes and glanced behind her at the door. "Are you sure— Couldn't we talk for a while?" she asked, giving him another pleading smile. "I'll make some coffee—sit down and I'll be right back, okay?"

She disappeared into the kitchen, leaving no room for argument. Kittery found himself stepping inside and dropping into the chair he had adopted. He cursed himself now for coming up here. It would have been easier to leave without saying goodbye. But he had to see Beth again. And if he admitted it to himself, he was stalling. Perhaps he was hoping something would happen to make him stay, but he knew he couldn't. He had to be rid of this place forever, before he heard any more about Baraga's Eight.

Beth returned several minutes later. "The coffee will be a minute." She sat down across from him, resting her hand on the tabletop. She began to massage that hand with the other one, and for a moment, neither spoke.

He shifted in his chair. "I just came to say goodbye, ma'am."

Beth raised her eyes in surprise. "*Ma'am?* We know each other too well for that, Tappan Kittery."

"Beth," he corrected himself. "I figured ma'am was easier."

"Perhaps it would be. But are goodbyes between friends supposed to be easy? I have yet to see it. And I expect this one to be least of all, because we both know you won't be back. Why are you leaving us?"

Kittery's face twisted. "It's not you I'm leavin'. Why do people make things into somethin' they ain't? I'm leavin' this land—not anyone in it. Not even the land, really, but some things that haunt me here."

"You won't see us again. To me that says you're leaving *us.*"

"I came to say goodbye, Beth," he repeated gruffly. "My mind's already made—no movin' around that fact. I was sort of hopin' we could part happy."

"How do you expect us to part happy, Tappan? I love you."

Kittery had opened his mouth to speak, but he clamped it shut as he sank back in his chair. "Pardon?"

She laughed, embarrassed. "Oh, Tappan. I mean *we* love you. We don't want you to go."

"There's more to this than you can see. You can't tell me to stay when you don't comprehend the situation."

"Listen. I don't care what the situation is." She rose, turning her back to him. "I just know a person can't go on moving, always. There has to come a place and time to settle down, and if you keep telling yourself 'next time', you'll never find it. You must take a stand." She spun on him. "You have to face yourself, Tappan. I don't care if it's you you're running from or someone else. If you don't stop and face it, you'll never be happy."

This time Kittery remained silent for a long, pondering moment. He knew she was right. He *was* running—from himself. Running because he was afraid of what he might become here—a hardened killer. To be true to himself and his friends, he should stay here and find a way to bring peace to this land without breaking God's laws or man's. But his mind wasn't going to be changed now. He pushed up from the chair and clamped on his hat.

"I'd better be goin'."

Her face fell with disappointment, but she smiled. "Your coffee should be ready. I can't drink it all. Stay for a minute—please. I'll be nice."

Kittery sighed and returned her smile. "I'll stay for a minute."

Beth departed then returned to pour two cups full. "You made Tania cry." It was as if the thought had just occurred to her. "I've never seen her cry like that since she lost her mother. She was serious about a boy once—a boy by the name of

Alan Peck, who's up riding with General Custer's cavalry. But she's never really been in love before you."

"Same for me," he admitted. "She's the last person I'd want to hurt, but after what's happened to Rand—you wanna talk about a sign from God! Now she can't rightly go . . . and I can't stay."

"And there's nothing I can do to stop you."

"Close to nothin'."

Beth's face brightened cautiously. "So there is *some*thing?"

What he intended to say was that if all the Desperados disappeared he would stay. "It's only a dream, Beth—only a dream."

He turned to the widow, sipping his coffee too fast and burning his throat. Inside, something told him he should stay. But he shoved that thought aside and swallowed his coffee down to the dregs. He turned back to Beth and set his cup down. "I'd best be movin'. Night's not far."

"I . . . All right. You'll leave no matter what I say. That's plain. I wish you'd come back some day, still. Come and find Tania again. You don't have any idea what your leaving is doing to her."

"I believe I do. My heart is full, too. But don't tell her that. I want her to forget me."

"That's stupid. She never will. None of us will."

"Well, you can try."

The big man reached the door, then felt Beth's hand gentle on his arm and turned halfway about. There were tears in her eyes.

"Tappan Kittery, don't you leave us," she whispered. It was the most pitiful face he could have imagined her putting on. She put her arms around him, and her face pressed against his chest.

"Don't be sad, Beth." He patted her back and gave her a squeeze. "Prob'ly one day we'll meet again."

When he tried to step back, Beth wouldn't let him, so he held her longer. She looked up at him, searching his eyes. He pressed her away from him and hurried outside. He climbed on Satan and brought him back around to the front of the house. The black stood broadside to Beth, and Kittery sat hunched over the saddle, his hand resting on the horn. He watched the woman, who stood in the doorway half-shaded by one of the pine beams protruding from the front of the house.

Kittery tried to smile, but it was a sad effort. "A man'll ride a lot of trails in his life when he lives like I do—a lot which aren't of his choosing. But a lot of trails tend to lead back to where they started. The one I'm on now, maybe it'll lead back here."

Beth forced a smile to match his, and tears flooded her eyes. She lifted a hand, and the big man broke away and moved down the slope. He didn't look back again.

Armand Gonzales unloaded and loaded freight at Albert Hagar's warehouse. That evening, he was stepping outside to gather red peppers from the string that always hung from the rafter ends of his adobe hut. His little wife was cooking *huevos* and a big pot of *frijoles,* and nothing made them taste better than red peppers.

Gonzales heard the horse hooves drumming toward him, and at first he was scared. Everyone knew about Juan Torres being shot. It had been done right under their noses. And Raoul Mendoza had been stabbed in his own home. They assumed it was because Mendoza and Torres had been friends with Gustavo Ferrar, but that didn't make any of the other warehouse workers feel any safer. For all they knew, there was some vendetta against all Mexicans.

But when the horse came into sight, it was riderless. Curious, Gonzales went into the yard, waving his arms to slow the horse down as it neared him. He managed to grab the reins at the bridle. They were knotted together the way a roper would do, so the horse hadn't tripped over them. Gonzales brought the tired animal to a halt. The first thing he noticed was blood splattered on its saddle and on its mane. Yelling out to his wife what he had found, Gonzales flung himself onto the horse and rode at a gallop to find the sheriff.

Vancouver received the Mexican's news grimly. He stepped from the office and swung his eyes to the hitching posts at the front of the cantina. He wasn't surprised to see Kittery's black stud still tied there, hip-shot and head hanging, half-asleep.

Thanking Gonzales, Vancouver trotted across the street and pushed through the batwing doors of the cantina, half expecting to see Kittery passed out drunk at a table. Instead, the big man had his hands on a deck of cards, his eyes fixed on the tabletop. The brown bottle before him had hardly been touched. The rest of the room was quiet, with only two men at the bar.

Sheriff Vancouver walked up to Kittery's table, wondering if the bottle might be his second, and if so how he was still conscious. "Still here."

Kittery snapped out of his trance and shot his glance up at the speaker. When he recognized Vancouver, he scowled and looked past him at the swinging doors. He shoved his chair back and towered to his feet.

"I was about to leave. Figured I'd wait till it got cool. Besides, it's safer to travel at night."

He started to brush past Vancouver, but the sheriff caught his arm. "Tappan, Cotton Baine's horse just trotted into town. There's blood on it."

Kittery whirled on the lawman. For a moment, they stared at each other without speaking. Then Kittery strode outside and to the stallion, Vancouver fast on his heels.

"Tap, hold up!" The sheriff's words caught the big man with his foot in the stirrup. "You don't know where the Reilly ranch is, do you? I'll ride with you."

They drew into the ranch yard of the R Slash R an hour later, peering at the lantern light cast from the house. Tying their jaded mounts, they stepped onto the porch. Vancouver pounded on the door.

It opened, and Joe Reilly squinted into the shadows, his hair hanging in his eyes. "Hello, Sheriff. What in tarnation brings you out this time of night?"

"Trouble," Vancouver answered grimly. He and Kittery entered at Reilly's invitation and stood in the kitchen holding their hats. "Where's the last place you saw Cotton Baine?"

"Right here—just this morning. What's happened?"

"We don't know. His horse showed up near town tonight with blood on it."

Reilly looked alarmed and let out a curse. "I told my brother Cotton should've been back by now. Well, you can't do much in the dark."

"No, but we plan to get an early start come morning. Mind if we stay here the night?"

"No. Please do. Please, make yourselves to home. You c'n sleep by the stove," Joe Reilly offered.

The pair wasted no time with small talk. They spread their blankets, pulled off their boots, and rolled into bed.

The smell of coffee reached Kittery in the first cold hours of dawn. Rolling over, he slapped on his hat and tugged his boots on, moving to the stove. The heat had driven him away during the night, but now he needed all of it he could get. Despite the warm glow of the cast iron, he could feel the chill creep in from outside. He roused Luke Vancouver, and the sheriff rolled over and came up without hesitation, going to gather his gear.

Jed Reilly stepped from the bedroom wearing shotgun chaps, a large scarf and a heavy denim jacket. Kittery could tell by the straw and manure on his boots he had already been out.

"Cold this morning?" Vancouver asked.

Reilly dipped his chin. "Yep. I saddled your hosses. Figgered you'd want an early start. I also saddled a third one, in case . . ."

Vancouver nodded understanding. "Thanks, Jed. We may need it if we find him."

Wolfing down a plate of cold beef, they chased it with a pot of black coffee and stepped out into the darkness. Stars still dangled loose across the sky like stray dogies.

Saddle leather was cold from the desert night, and Kittery felt a shiver go up his spine as he settled into his. Vancouver swung aboard Spade, his bay, and they rode out of the yard, waving farewell to Joe and Jed, who stood on the porch.

By nine o'clock, the heat drove down, and they rode in uncomfortable silence, scanning the country. By eleven, their backs, chests and sides were soaked with sweat. Kittery wiped at the beads that kept gathering on his cheeks and tickling his whiskered jaw.

Several times during the morning, they came upon Baine's tracks. He had been searching for cattle as he rode, swinging in wide circles to cover more ground. So the searchers, keeping in a straight line, cut his sign only occasionally. They rode ahead following the Reillys' information where they had told Baine to go.

It was near noon when they spied the old dry watercourse. They followed the dun's tracks down into its jaws.

As they rode in, a sense of foreboding loomed over Kittery. He had only to look over at Vancouver to see he felt the same dark sensation. The wind whispered eerie through the treetops overhead, and the shadows on the ground danced back and forth as if to the tune of some strange, distant flute. All was silent but the wind, the saddles, and horse hooves clattering against the rounded stones. The horses rolled their eyes. Listening into the silence, beyond the noises of their travel, Kittery fancied hearing the desert goblins and fairies frolic in the grass and rocks, laughing taunting laughs, watching him and the lawman go forward to what horror awaited them. Somehow, Kittery sensed they had reached the trail's end.

Then it was there. The broken body of Cotton Baine lay before them in the rocks of the creekbed.

Kittery and Vancouver dismounted, moving to the corpse. The vultures hadn't discovered it yet, here in the protection of the trees. Kittery saw where a big slug had smashed Baine's collarbone and broken his neck. There was a pool of blood where he had lit before rolling over.

From the sign, they saw that the killer wore moccasins and that he was far bigger than the average man. They also found a .45-70 caliber rifle shell by the rocks. They found where the killer had mounted a big-hoofed, heavy horse and ridden down the wash, leaving an easy trail.

Kittery read these signs in silence. The big man hadn't really allowed himself to think yet, at least not beyond trying to form a mental image of the murderer. He hadn't yet stopped to realize how much ruthless hate had built inside his chest.

Kittery's own voice almost startled him, after the long minutes of grim quiet. "One of us has to take Cotton back to town, Luke. I know he was my friend, but I'll leave that to you. It's one to one now—me and the dead man ridin' ahead of me."

"The dead—" Vancouver stopped himself, grasping Kittery's meaning. "I know how you feel, Tap, but I'm the law. I'm the one who should have to go."

"No." Kittery's voice left no room for argument. "Sorry, Luke. We're too close to Mexico. This man could make it there, and you'd have to turn back. Besides, you might show him some mercy if you catch up. I won't."

"What if you don't find him?"

"I'll find him. I'll find him if it has to be in hell."

Vancouver didn't want to go back. Had Baine been *his* friend, he would have buried him here and gone on the hunt. But Kittery was right. He was a lawman, and he believed in his badge. If he caught the killer alive he would have to bring him in alive. If Kittery had other intentions, Luke Vancouver had no desire to be there. He didn't know if he could stop another killing even if he felt like trying.

Together, they loaded the body on the back of the spare horse. They mounted, and Vancouver turned to the big man. He unbuckled the nickel-plated Colt Peacemaker from around his waist and held it out to Kittery.

"That carbine carries one shell. It isn't going to be enough, Tap. You'll need this."

Kittery took the belt with a grim nod. "I'll have to sling it on the saddle horn," he said. "No way somethin' that'd hang on your skinny little butt would fit mine." He tried to smile, but he couldn't. He looked up with the last ounce of friendliness and humanity he felt inside him. "Thanks, my friend."

Then he reined Satan about and headed down the rocky wash. He knew that somewhere ahead, either he or the murderer of Cotton Baine would breathe his last.

Chapter Eighteen
One Down: "Now You Will Keel Them All"

Cotton Baine's murderer left an easy trail. He puttered along the route, halting here to sit beneath a tree and smoke a cigarette while he rested his horse, stopping there to climb into the rocks and scan the country. Once, he took the time to stalk and kill a big buck mule deer. Kittery came upon ravens picking at its entrails and the front half of its body and severed head, its nubs of antlers encased in velvet.

It was plain to see that with every step Kittery's horse took he narrowed the gap between him and the killer. He knew he had to stay alert. It was already proven this man liked to kill from ambush. Always, Kittery had to be aware of the chance that the man might take a pause to check his backtrail. The man was relaxed now, but Kittery wasn't fool enough to think he would remain that way. Jake had taught him better, hunting dozens of black bears with him through country where relaxing at the wrong moment could mean death or grave injury.

He discovered the killer's camp three hours after leaving his own the next morning, but he didn't believe he was three hours behind. He had been on the trail early, and it appeared the man he followed had taken his ease before leaving. Now and then a flame licked up among the coals of his fire, and there were quail bones lying around that were still moist. It looked like the man had killed three of them nearby that morning, judging from the fresh guts and tail feathers he found. He had eaten them all down to the bones. He also found most of the mule deer's hindquarters, discarded nearby in the brush, where the birds had yet to find it. This was a man who didn't mind squandering game.

As the sun rolled past its zenith, Kittery knew he was closing in. He had ridden at a good, steady pace, and the killer was still moseying along. Kittery drew in at a ribbon of stream to water the black and himself and to replenish his canteen. He made himself a sandwich of a biscuit and some dried-up beef, washed it down with the cool water, then rode on his way.

What he followed now appeared to be an ancient game or Indian trail, probably turned into a cattle trail now. It was a well-used highway leading uphill from the stream. He followed it down into flatter country, with occasional outcrops of rock breaking out of the ground to form their battered battalions around him.

Kittery soon found himself paralleling a dry wash. The ground sloped off to his left to drop down to it and a dry meadow where the ancient creekbed meandered, still a place for run-off, when the summer monsoons set in. On its banks stood massive, dying sycamores, sentinels over the land.

Up ahead, towering columns of basalt reared up, heaving their shoulders a hundred feet into the sky. He stared at the formations in wonder as his thoughts turned around his friend's killer.

The man was big—very big. He knew that and the fact that he wore moccasins. Perhaps he was a renegade Apache—some had been known to get that big, Mangas Coloradas being the best example. But it would be odd for one to be traveling alone unless for some reason he had been expelled from the tribe for an evil against his own people. He wondered if the man was a habitual killer or someone who had seen an opportunity for gain and crossed the line.

Kittery scoured the rocks with his eyes, studying every outcrop capable of concealing a man and horse—or just a man. He slipped back into his youth for a moment, in the rugged Smoky Mountains. He had played the hunter then, searching out the smallest clue his prey might have left on the forest floor. Little had he known then that one day it would pay off this way.

The sudden, sharp but muffled sound of a pistol pounded somewhere up ahead, and he jerked Satan to a halt. The black pricked his ears toward the sound. Kittery scanned the rocks and brush but saw nothing. He was guessing the shot had been fired at game. Whatever the reason, it might have saved his life. He nudged the black on at a slow walk.

They had gone on for five minutes when a rattling of rocks somewhere off to the right brought them once more to a halt. Though the sound seemed to come from very near, he couldn't pinpoint its exact location. After several minutes it hadn't repeated itself.

The terrain had flattened out. In place of the cliffs, brushy hills rolled off toward the Cerro Colorado Mountains, in the north, and the Las Guijas Mountains, whose base was yet seven miles ahead. Only to his immediate right was there any hint of the type of landscape he had just traveled through. This was a huge mass of basalt forty feet high that fell straight off at its summit. A tangled mass of thorny vegetation surrounded its base, with individual clumps of prickly pear as tall as a man's waist, and some catclaw bushes nine or ten feet tall.

He scanned the intermingled catclaw and cactus from where he sat the saddle. He searched the boulders nature had chiseled from the cliff and dropped in that place to rest. A man and horse could hide in cover like that—perhaps several men. Yet he could see no sign that anyone had entered there.

Nervous, he urged Satan into the dubious shelter of the brush at trailside, unholstering Vancouver's long-barreled Colt as he did so. He resented the sheen of the silver plating, but the engraving on barrel and cylinder helped to break that up, at least at a distance. If the shot he had heard came from the killer, the pistol's shiny finish wouldn't matter anyway. He was too close for its sheen to give him away. It was the two hundred and fifty-grain slugs bedded in their steel chambers that counted. Following basic safety precautions, he normally loaded only five chambers of the revolver, leaving the one under the hammer prudently empty, to prevent an accidental discharge. Today he carried six.

The afternoon bore down, making sweat trickle down Kittery's sides and back and from under the brim of his black hat. Nowhere was there the cry of a hawk, nor a thrasher nor a wren. Only the cicadas made their unnerving racket in the grass. But somewhere in that forest of brush and broken rock waited death . . .

The sudden neigh of a horse sounded clear in the afternoon. It rang from the brush where Kittery had trained his eye. Kittery gave the black three sound thumps on the side of his neck to caution him against answering. The horse looked back at him out of the corner of his eye and blew his irritation out his nostrils. But he didn't make a call to the other horse.

Kittery's heart battered against his chest. His right hand clutched the .45, sweating until it felt like the grips were coated with lard. He heard a sound like gravel grinding beneath a hoof. His grip tightened on the gun. Another minute crawled by. He heard the stirring again, then again and again, each time closer to the trail. He stared at the spot where he expected the horse to emerge. His head pounded against his sweatband. He blinked against the sweat that trickled into his eyes and off his chin. By now his shirt was soaked, as well as the seat of his pants, and his throat felt like a sandy draw. He could smell the musk of his saddle and the sweaty horse and his own man scent.

He swore under his breath. Inadvertently, he had moved Satan onto an anthill at trail's edge. The insects had found their way up the horse's legs and onto his own, creeping up over his boots and onto his pants. Satan stomped a hoof, and he swore again. He thought he had heard the other horse move at the same time. If he hadn't, the man would surely have heard the pounding hoof.

An ant managed to find its way inside one of Kittery's boots. As he felt the stinging sensation at the back of his calf, he lowered his eyes. In that fleeting moment, when he glanced down then back up, the other horse lunged forward.

There, reining to a halt in the center of the trail, was a huge man atop a light chestnut horse. His sandy-colored hair flowed long and slightly wavy at the ends, down over his shoulders, and his beard ruffled as he jerked his head and stared in surprise at Tappan Kittery. A man of such size could never be mistaken.

Rico Wells, the Desperado they called Big Samson.

The outlaw's surprise flashed across his face. His mouth curled into a snarl of despair and anger. He had his rifle, a big model '73 Springfield, across the flat-topped horn of his Texas saddle. It was pointed in Kittery's general direction, but too far left. For a moment, Kittery felt frozen in place. And then, with a curse of defiance, Wells lurched low in the saddle. He swung his rifle to bear on Kittery. He jammed his heels into the chestnut's flanks.

Kittery allowed for the killer's change in position. He snapped off two shots as he hauled back on the reins and wheeled the black to the left. Wells, thrown off by the explosive lunge of his mount, fired and missed. Kittery fired again. He didn't know where his bullet struck, but it made Wells lose his reins. The chestnut with the flowing cream-colored mane and tail bolted forward again with a squeal of fright. But its rider left the saddle. He toppled backward to land with a bone-jarring impact on his right shoulder.

Most men would have been stunned by the fall alone enough to stay down. But the huge outlaw surged to his feet, hatless, his hair spider-webbed across his face. Blood flowed down the front of the man's white silk shirt from two wounds. Kittery yelled at his horse to be still. He aimed the Colt again, furious and frightened to see Wells still alive.

In the fall, Wells had lost his rifle, but it didn't matter. It carried one shell, and Kittery would never let him load again. The outlaw's right arm hung useless at his side. Although it was set for a right-handed draw, Wells's left hand clawed for a Colt pistol behind the waistband of his fringed buckskin trousers. It cost him too much valuable time.

As Wells jerked his Colt free, Kittery looked into his blue, wolf-savage eyes, and shot him in the throat. He shot him again in the chest. Wells's eyes went wide when he realized he could get no air. He dropped his gun and clutched at his throat. Then he pitched forward on his face into the patchy grass and struggled fiercely to his death.

Kittery managed to calm the black down, and he slid over his side, hanging onto the saddle horn for a long moment before thinking to reload. The sixth cartridge still remained in the Colt, but for Rico Wells it didn't feel like enough. With trembling hands, he thumbed five cartridges from Vancouver's belt. He put the gun on half cock and clicked its cylinder around as he slid them home.

Then, leaving Satan with one rein trailing on the ground, he walked to Rico Wells and crouched to shove him onto his back. The blue eyes looked dark with the expansion of his pupils and didn't constrict even with the sun bright against them. Kittery looked at the big man's right hand, realizing why he hadn't drawn with it. His third shot had shattered it.

Nearby lay a Gambel's quail Wells had dropped, probably the reason for the shot that warned Kittery and saved his life. He stood over Wells and at last took a ragged breath, then let it gust out through his nose.

The dead man's dark beard ruffled, and the ends of his long, yellow hair lifted with the warm summer breeze.

The next morning, the black stud's and the chestnut's hooves made plopping sounds in the dust as Kittery walked them briskly up to Jarob Hawkins's barn. Sagging out of the saddle, he led the horses inside.

Hawkins gave him an uncertain smile and a nod, his eyes flickering beyond him to the body of Rico Wells, draped over the chestnut. "Howdy, Cap'n."

Kittery returned the nod. "Where's Luke Vancouver."

"Home. 'Sleep, I 'magine. But he's 'spectin' you."

Thanking the hostler, Kittery left the black and chestnut in his care and crossed to Hagar's mercantile. "Two days ago I sold you a Remington pistol for ten bucks," he said to the white-haired man. "I have reason to buy it back."

Hagar nodded. "I was afraid when you told me about that place where you wouldn't need it that it didn't exist," the old man said.

Kittery shook his head, looking away.

"Well, all right," Hagar said, after a moment's study of Cotton Baine's Colt, hung up behind the waistband of Kittery's striped trousers. "It's yours." He reached below the hardwood counter, came up with the weapon, and placed it before Kittery. It had been cleaned and oiled. "You ought to consider a conversion kit on that thing, if you plan to hang onto it. Cap and ball doesn't cut it anymore for a man who does the work you do."

Kittery knew he was right, but he was in no mood to talk. "How much?"

Hagar's eyes flickered. "Well, I paid you ten dollars two days ago. I wouldn't ask more."

Kittery placed a gold eagle in the man's palm. As the old man was putting it in his cash drawer, Kittery took a coin purse and emptied five twenty cent silver pieces on the countertop. "That's for the oil."

"Oh, well . . ." Hagar scanned the coins. "I didn't use any dollar on oil."

"Twenty cent pieces are a nuisance anyway, and you aren't in business to clean people's guns for free," said Kittery, buckling on the familiar old pistol as he walked outside. It was good to have a belt that fit him again.

From the mercantile, he walked to the cantina, cutting the dust from his throat with a shot of rye whisky. When he returned to the street, Vancouver's horse tied at the jail was the first thing he saw.

As he strode toward it, Vancouver's voice came from the doorway of the livery. Kittery walked toward him feeling like every last ounce of strength and human kindness had been sapped from his being. He hadn't shaved, his clothes were coated with dust, and he must look like a railroad hobo. He could feel the grating of sweat and grime every time he moved his legs or arms.

"Good to have you back, Tap," said Vancouver, stopping a few feet away.

Kittery grunted, forcing a half-hearted smile. He liked the sheriff, and it warmed him inside to hear him calling him *Tap*. "You were snoopin'," he said. "Did you see the man? Rico Wells?"

Vancouver nodded. "Figured that was him. Shot full of holes."

"Uh-huh. He thought he couldn't die. He was headed home like a tortoise with an ingrown toenail."

"Looks like he almost *couldn't* die," Vancouver said.

"Your gun saved my life, Luke. I'll always owe you for that."

Shrugging that off, Luke smiled. "I don't know if you realize it, but you're two thousand dollars better off. It'll take a while to collect. I've learned the government is almighty happy to see their criminals brought to justice, but something about an outlaw's demise shrinks their purses up mighty sudden."

News of the reward didn't cheer Kittery much at the moment. He nodded, too tired to think. "That'll help."

Vancouver sought more, but Kittery didn't embellish. "Cotton's things are in my office, what little there was. Unless you know how to reach his family, I think you ought to have them."

Kittery nodded again and tapped Baine's pistol, snug behind his gun belt. "I've got the most important of 'em here."

They walked together to the stable, where Vancouver collected his pistol and belt. Kittery offered to clean it, but Vancouver refused. "I don't get to use it much, myself. Cleaning it will get my hands used to the feel of it again."

After arranging for Wells to be taken away, Vancouver left, and Kittery retired to the cantina. Perhaps partly from the whisky he was pouring down, partly from fatigue, he soon found his heart thudding heavy inside, making him short of breath. He stopped drinking and let his mind whirl through the last two weeks of his life. He had gone from a man on the side of the law, to a blood-seeking vigilante trying to decide how to keep the woman he loved safe, to a man with passive intentions, to a killer who didn't know if he cared about anything in this world anymore. The world sure didn't care about *his* feelings.

Tappan Kittery had a job to do, of his own choice: to protect the innocent citizens of Arizona from further death and destruction by the Desperados Eight—or the Seven, if that was how the papers would refer to them now. This time he would not be detoured or detained. Nothing could sway him, ever again. Not love, fear of death or even needing to fulfill his own happiness. His two best friends were already dead. Now the Desperados must be killed—not chased out, not brought in. Killed. His goal was to protect the innocent, but his drive was to see seven more men die.

His other goal was to leave his Bible shut until it was all over.

He walked to the dry goods store with black clouds hunching together across the sky and dust skittering down the street. Little sheets of light danced up inside the clouds, and now and then a bright thread would twist and curl, then spear its way down toward the mountains. The far away sound that came rambling several seconds later was like horses running in a distant canyon. By the time he reemerged fifteen minutes later, outfitted with supplies, the wind had grown to half of gale force. It gusted by, hurling dust and small debris and bringing with it the pungent smell of the desert and of rain. The blue sky was nearly gone, and rains bulged at the pockets of dirty cloud that settled low over the valley. One of the clouds hurled a giant arm of bright white light at the side of the Santa Ritas, and the crack of thunder rolled like a horde of berserkers through the town. No one moved along the street. Even the warehouse packers had taken refuge.

And the rains came down.

Kittery ducked beneath the shelter of the hotel's awning and watched the water pour down in sheets. The force of the wind sometimes drove it horizontally. Thousands of tiny drops spattered in brief confusion across the parched boards of the porch, sinking in at first to leave small, dark splotches as sign of their passing. The rain whipped now and then against his face, and he let it. It felt cool, cleansing. Sheet lightning danced across the clouds, bright streaks parried one another across the sides of the dark mountains, and some long, rolling white whips curled and lashed back upward into the gloom of the sky that had spawned them.

In the loneliness of the deluged street, Kittery leaned against the hotel's wall and let himself drift into thought. He knew what must be done now. He would find James Price and join himself once more to the vigilantes, if they would have him. They had admitted they needed him. But if such were no longer the case, he would find Baraga alone. And why not? He had always worked well alone. Besides, he was never alone; Jacob was always beside him. And if the vigilantes *did* need him, he would see to it that every able-bodied man in that part of Arizona who had any inkling of a desire for justice took on the oath of the vigilante.

It didn't matter to Kittery how the Desperados died. At that moment, he would just as soon the killing all be done by himself. But he had to be realistic. No matter who killed them, it only mattered that they die. Rico Wells was gone, and he was Baine's killer, so that was good. But they had all ganged up on Joe Raines, and his promise to the marshal was still left unfulfilled. Kittery had no intention of easing up on the rest of the band. Baine's killing had only served to acutely remind him he had to drive the others to ground at the earliest possible chance.

Even in his state of impatience, Kittery was prudent enough to realize he had to have money. Sure, the two thousand from Wells would be good, but he didn't have it yet. He had to have some cash before that, and that meant taking up hunting in earnest for at least a few weeks.

His thoughts turned involuntarily to Tania. Perhaps he should go to her, maybe ask about her father. The desire to see her was keen. Yet if she had any common sense she would cut a wide swath around him. He was a man with death written all over him. So he would make it easy for her. If she wanted to avoid him, it was her choice. If she wished to see him, she knew where he could be found.

All the riding, the thinking, the vengeful fantasies—all of it took its toll on Tappan Kittery, and the bed upstairs cried out for him.

The rain lessened, the wind stilled. Thunder grumbled in the distant north, and heat flickered in the purple clouds. A few large raindrops continued to dot the street, but they came sparingly, disappearing into a hundred tiny rivers of brown that already laced its length. The air smelled fresh and clean, renewed in the wake of the storm. Kittery breathed deep of the fragrance from the soil and the hills. Ironically, his own storm had just begun.

Maria Greene sat at her desk, her soft dark hands shuffling through a sheaf of papers. Kittery stopped before her and nodded.

"I am glad you have come back. You left your Holy Bible in your room."

"I meant to," he said.

Maria nodded and gazed up at him with bold frankness. "I knew you would avenge Señor Baine."

"You know about that?"

"Yes. An' your job is done now, no? You will go from here?"

Kittery shook his head. "One down."

Her eyes skimmed the Springfield rifle in his hand. "I know wha' you think, Captain Kittery. Now you will keel them all."

Kittery paused for a moment, studying the dark contours of Maria's face, approving of the slender, supple neck that disappeared inside her red silk blouse. He chose his words for their impact. "Not long ago, here in Castor, my two best friends were livin'. Now they aren't."

Smiling her understanding, Maria told him, "You must of course follow the will of your heart. It tells you what to do—what is right and what is wrong."

"Ma'am," Kittery drawled, "I . . . You seem to understand the way a man feels. It's easy to talk to you. I'd be pleased for us to speak more sometime." As he spoke, he knew it was the liquor working on him. Tania was the only woman he needed to see, and she understood him as well as anyone.

Maria smiled and lowered her eyes. When she looked up and spoke again, the subject had been dismissed. "You will want your same room, Captain?"

"I'd like it if it's there."

"It is. An' for you it is one dollar."

Kittery smiled, this time not needing to force it. He placed two tiny gold dollars on the counter. "Two dollars. You think about what I said, ma'am—about talkin'."

"I will. An' you, Captain. You think about your Holy Bible. I have it here." She pulled the thick black book from a drawer and held it up.

"Ma'am . . . you better keep that for now."

The room was the closest thing to a home Kittery knew, of late. It was somehow like the greeting of an old friend he never thought he would see again. He left Rico Wells's rifle and his supplies on the bed and returned to the stable to recover his saddlebags and bedroll and his own Springfield from its scabbard.

On the street once more, a sight at the dry goods store hitching rail spun him around. Tania McBride's gray mare stood there hip-shot. The urge to rush to the girl hit him like a rock. She was so near, yet so far away. But he knew more than anything else he couldn't drag that girl into the maelstrom that was about to hit southern Arizona. And he would be at the very heart of it.

Lowering his head, he turned back to the hotel and went inside. The sad eyes that followed him from Greene's window went unnoticed. As the big man disappeared from sight, a single tear rolled down Tania McBride's cheek. She had had to hear from someone else that her man had come back. She had hoped he would come to see her. But now she wondered if she had lost him forever.

Kittery sat on the bed and tugged off his boots, his feet damp and weary. The long hours in the saddle and the warmth of Cardona's whisky worked their mystical spell. He fell asleep minutes after his head touched the pillow.

Later, he started awake, cold sweat standing out on his brow. He had had a nightmare, of what he didn't remember. His only memory was that Jake was in it. Outside, the sun had gone. The sky had cleared, and stars licked bright in the sky, for there was no moon. A lamp burned in the jail when Kittery crossed to the window and looked down. A yellow light stretched warmly across the street from the cantina's doorway. Horses lazed in front of the cantina and hotel and Kingsley's café.

Kittery was drawn by a sudden, strange urge to go down and mingle in the cantina. Something told him to go partake of strong drink, to play some cards, to joke and laugh. But he couldn't. He no longer fit into that crowd, if he ever had. In a way, the death of Cotton Baine had changed not only his life, but his nature. Only when the last of the Desperados died did he feel any hope of returning to his old self. He realized that with a profound regret.

Somewhere a coyote yapped mournfully at the sky. Before all went quiet, three more of them had joined the chorus, all from different points. Kittery tried in vain to seek them out on the darkened hills.

A corner of the big man's mouth turned up in a lopsided smile. "I know how you feel, boys," he said to the prairie wolves.

Drawing the curtains, he turned from the window. Now he would sleep.

When he awoke once more, the world lay clothed in sheets of gray. The air in the room was cool, thanks to the brief rainstorm, and it smelled of dampened desert. He went to the window and pushed aside the curtain to look out. A whisper of wind strolled mischievously along the street, ruffling a freighter's shirtsleeves as he drove his high-boarded Tully and Ochoa freight wagon into it. The broad yellow wheels ground through mud, the horses' hooves flung it up with each step and trace chains jingled an early morning wakeup. The doors of the warehouse creaked in closing, and someone hollered out in Spanish. These were the sounds of the dawn.

In the chilly half-dark, Kittery dressed and went down to the street. A few lamps burned—in the jail and in some of the houses along the ridge—and wood smoke curled and coughed from the chimneys into the cool, still air.

Jarob Hawkins tapped his pipe against a timber, watching Kittery smooth every wrinkle out of the saddle blanket before he threw his Denver saddle on Satan. Hawkins had been around horses all his life. He appreciated a man who cared for them as Kittery did.

Kittery made a deal with Hawkins and bought a big gray mule and a smaller bay one to pack supplies and game. When all was ready, he rode to Price's gunshop on the southern edge of town.

"Hello, Kittery," Price greeted him at the door. "We had a feeling you would decide to stay. What can I do for you this morning?"

"Well, I *am* leavin' town. But only for a couple weeks or so. If you'll take me back in the vigilantes, I'm gonna go recruit some help in Tucson. We need to do this right. When I come back, I'll be with you for keeps."

James Price studied him for a long time. At last, he nodded. "I've made a point and a rule that anyone who is asked to join us must be approved first by me, Kittery. You're one of the very few men I would ever break that rule for. Recruit who you will." He gave him a satisfied smile. "And we'll be waiting for your return. Good luck.""There's a couple more things, Price," said Kittery. The hangman looked at him questioningly, and he unholstered his Remington pistol. "I've changed a lot in the last few weeks. There's somethin' else I'd like to change now that I'll be in the business of killin'. Put a conversion on this old gent, would you?"

"Sure. What was the other thing?"

"The Stephens derringer. I know who bought it in Prescott, and he's since lost it in a poker game. I reckon I better tell you who won it. Just in case I don't make it back."

Chapter Nineteen
The Creature

When he rode down the street, Tappan Kittery had to force himself not to look up at the ridge road. But he couldn't force himself not to think of Tania. Her thoughts rode with him for miles away from Castor.

With the sun revealing its upper fringes above the eastern hills, Kittery downed a mule deer buck half hidden in low-lying scrub brush. When he went toward it, he jumped a second, larger one, and killed it, too. Gutting them and removing their heads, he cut them in half and lashed them onto the pack mules. As he rode on, he scanned the hills for any sign of other travelers.

The sun forsook its curtain of hills and climbed high into the sky. Kittery rode into Tucson and sold six quarters of his venison between the Shoo Fly restaurant and the Cosmopolitan. He also spoke with a few trusted friends and asked them discreetly to spread the word about the vigilantes.

One man mentioned a jaguar up in the Santa Catalinas that carried a two hundred-dollar reward for its hide, and Kittery took that as a sign where he should go. Glad to leave town, he headed back into the desert.

In the late evening, the saguaro cactus-studded desert floor began giving way to the oaks and sycamore and juniper of Tanque Verde Canyon, west and north of the Rincon Mountains. Here, the faint but fragrant scent of mesquite wood smoke telegraphed a warning as he rode down a rocky draw. Ahead, a horse called out.

The cow camp lay around the next bend, where the canyon steepened sharply, struggling to flank Agua Caliente Hill. It was a two-bit affair, three men standing spraddle-legged about a too-large, smoky fire, drinking coffee and smoking cigarettes. Saddles hung on rocks, surrounded by a disarray of saddle blankets, bedrolls, lariats and holstered pistols. Three horses stood weary-lipped and three-legged in a slipshod rope corral, heads drooped. Farther away, three more watched Kittery approach with their eyes, ears and nostrils picking apart everything about him. After a brief study of the camp, Kittery judged it for the peaceable cow camp it was. He allowed the black to clatter through a pile of rocks as he drew within eighty yards. The three men turned about in surprise and watched Kittery come in to fifty feet away.

"Mind if I light? I could use a rest."

A lanky, mid-forties puncher with a generous, sweeping mustache gave him a welcoming wave. When Kittery dropped from the saddle, the older puncher stepped forward and thrust out a weathered hand.

"Name's Riley Brand. Have yerself a cup."

"Tappan Kittery." The big man nodded, shaking hands all around. "An' I'm obliged."

"Shore," said Brand. "Dark's comin'. Yer liable to run into a sneck or somethin'."

Kittery had to fight off a chuckle at the man's accent. He had run into a few "snecks" lately, some of the two-legged kind.

The other two men were both young, no older than twenty or twenty-one. Fernando was a lithe Mexican, smooth in his movements, with an easy, friendly smile. Jethro was rougher, a little belly-fat, with broad shoulders and chest, a straggly beard and a shock of flaxen hair poking out from under his black hat. Kittery noticed the hat not only because most cowboys wore white or light gray, but mostly because he wore a hatband made of rattlesnake skin. In all Kittery's days, he had only seen two other men do that. Most people wanted nothing whatever to do with snakes—*or* snecks.

"Good deer thar," said Brand. "You wouldn't wanna spare some of 'im, don't s'pose. Fresh food's bin kinda scarce."

Kittery glanced over at the deer and shrugged. "Sure. I can't use it all. Be a shame to waste it."

The deal ended by Riley Brand forcing two dollars on Kittery, and Fernando carving off three quarters of what remained. He lashed the remaining hunk under his tarp, then returned to the fire to squat on his heels.

"Maybe you boys could help me out." He let his gaze drift over the three. "I'm up from Castor, an' I hear tell about some jaguar they claim is marauding in these mountains. You wouldn't know anything about where they last saw it, would you?"

"Hell, meester," Fernando said with a grin. He waved to indicate the surrounding hills. "Thees hills are full of animals. Bear, wolf, cat—anything you weesh. That big *gato*, he moves aroun', you know. All of them do. Eet's like following the wind. But there ees a man up here—a hunter. You wan' to hunt, you shou' meet thees man."

Kittery nodded. "Sounds like that'd be a good beginnin'."

They talked until well after dark, eating chunks of venison as they came out of the flames. Kittery ate his half-raw, but all of the cowboys, typical for their breed, liked their meat done almost to a crisp. When they had filled their bellies and learned all the news either party had to share, they rolled into their blankets and slept.

Dawn came, and Kittery awoke to the sound of low voices. He heard the amiable voice of Jethro hailing a newcomer, calling him "Creature." Rolling over, he laid back the blankets and reached for his boots. Shaking them out, he tugged them on. He rose and picked up his hat, stepping to where Brand, Jethro and the new man stood around a flickering, frying pan-sized fire.

A hunter or trapper, by the look of his clothes, the new man carried a long knife and two Colt Navies tucked into his strap leather belt. A big brown mule stood beside him, and a long-barreled Sharps rifle hung from a sling across the saddle.

The new man had his head lowered so his hat concealed his face. He looked up as Kittery approached, and he held out a hand. "Mornin', mister. Name's Judd Creech. The boys tell me yer here on a huntin' expedition."

As Creech lifted his face, the firelight flickered across it, casting an eerie yellow glow. The reason for Jethro's calling this man "Creature" struck Kittery sharply. His entire countenance was plastered with a mass of scar tissue. His beard grew only in scant, unblemished areas, and there it grew untamed.

"That's right," Kittery finally managed to say, shamed by his involuntary pause. "I'll hunt anything that'll land me a stake. I'm Tappan Kittery." They shook hands.

"Wal, if yer huntin' cougar, yuh couldn't be talkin' to a better man. I invented that game."

"That's what I'm after. That, and maybe this jaguar I hear tell about."

Creech guffawed. "That'll be the day we see *him* around here again. He warn't stupid enough to stay."

Kittery shrugged. "Then perhaps a bear or a wolf or two."

"Now yer makin' sense. An' Mister Kit'ry, I gen'rally make a habit of ridin' alone. But I wouldn't mind havin' someone to listen to fer a spell—maybe fill me in on goin's-on down below. Seein's you ain't none too familiar with the country, what say yuh fall in with me?"

Kittery grinned and flashed his eyes across the others, then looked back at Creech. "Well, Mr. Creech, I make a habit of ridin' alone, too. But I appreciate the offer. I think I'll take you up on it; it's a big country to be alone in."

Dawn had cowered away, and when the sun threatened camp, with its bold light ringing up behind the mountains, Kittery saddled the black and repacked his mules. After a breakfast of venison and sourdough cakes, Kittery bade so long to the cow crew. He and the trapper rode north at a steady clip, flanking Agua Caliente Hill.

It was a rocky, brush-studded land, cut through by gullies and sharp-lipped canyons. The going became maddeningly tedious. They made a mere fifteen miles, as the crow flies, before coming to a halt and making camp in Buehman Canyon, in the foothills of the Santa Catalinas. Blue shadows cupped themselves like water in the cuts and washes along the mountain ridges, softening their rugged edges. On the higher rocks played a soft, golden light.

Later, as the cook fire began to die low, Creature leaned back against his bedding. With his chipped teeth he picked the remains of venison from a bone. "Kit'ry?" the trapper gruffed. His voice was the only sound but that of the wind in the rocks and the crackling of the mesquite fire. "You ever hunted these hills?"

"Nope. I've only passed through," the big man admitted, his gaze taking in the stars.

"'Course y'ain't. 'Course y'ain't. It's a lonesome sort o' land, Kit'ry. A man l'arns t' be more alone than he's ever been afore. Yuh l'arn out here that you are yer onliest friend." Creech sighed, tossing the bone off into the brush.

Kittery thought of Cotton Baine. Then his mind glanced across Joe Raines. "It's better that way," he said. "If you believe you're the only friend you have, you know there'll be no one mourning you when you die. And you know you'll never have to mourn over anyone else. What does it add up to? You'll never have to mourn, unless Saint Pete lets you hang around long enough to mourn for yourself."

He watched the pits of the trapper's eyes as he stared at him through the darkness. "It's true, what yuh say. Never figgered it that way. Still, I'm thinkin' some folks'd give anythin' for a chance t' mourn, even if it was over the loss of a good friend. At least then they'd know they had a friend one time—somebody to enjoy whilst he was livin'."

Kittery nodded grudgingly. "I've heard there's two sides to everything in this life. No good without the bad."

"Maybe it's t'other way around," mused Creech.

Early the following day they came in sight of Creech's cabin. As they rode in, Creech turned to Kittery and said, "Yer soon t' meet my friend."

The friend Creech referred to was a dog—a huge and black and wild-looking mongrel with a long, heavy scar running down the side of his nose and into his lip. When they rode into Creech's yard he appeared from behind the cabin and trotted forward, eyeing Kittery and his animals warily. Not once did he bark, but his keen eyes and nose missed nothing.

"That's Griz," said Creech with a proud smile.

From that time forth, the three were constant companions. They hunted the broken rims and timbered slopes, searching for cougars and whatever else would earn them a dollar. Creech gave Kittery several deer hides he had tanned, and then, of a night, the two would sit by the fire fashioning a rough suit to fit the big man. "Them Arizona thorns'll rip up the clothes yer w'arin', Pilgrim. Now you won't have to go huntin' jaguars nekked." He roared with laughter.

They spent hours around the fire, spinning yarns and recalling memories. And then one night the trapper sat back in his hide-covered chair, chewing on the end of a pipe. He gazed at Kittery. "How's come yuh never asked me why m' face is like this, Pilgrim?"

Kittery looked up at him, shrugging after a moment and taking a tentative sip of his coffee. "Figured it was your business."

"Well, I wanna tell yuh. Might's well know. Ever'body wonders."

Creech got up without waiting for Kittery to reply. He went to a shelf, drew out an old tin box and came back to take his seat, the box across his knees. The big man sat in respectful silence as the other man opened the box to reveal its contents. He showed an old tintype portrait of a handsome family—a man, his wife and two boys. He picked up a gleaming bone-handled knife, and then, last of all, a string of long, amber-colored claws. Kittery recognized them as those of a grizzly bear. He remembered the day he had asked the trapper if he ever went after bear. Creech had told him no.

The story unfolded. One night, an angry grizzly had raged into Creech's unprotected home, killing his entire family. He had found them upon returning from his traplines. One of his sons was partially eaten. The other was sprawled across the shattered table, and his wife lay down near the creek where the bear had dragged her.

For two days, Creech tracked the guilty bruin, through creeks and forests, and across talus slopes where a normal man should have lost all sign. He found the bear resting among some blown-down timber. Or rather, the bear found him. It charged.

In the terrific combat which ensued, Creech lost his rifle without firing a shot. Then he lost both pistols, managing only to put one bullet in the beast. All he had left after that was the bone-handled knife in the box. But when the battle ended, the bear lay with the knife deep in its heart. Creech lay mutilated and nearly dead beside it, his face a mass of bloody flesh. If not for some of Cochise's warriors, who found him and nursed him back to health out of respect for his bravery, Judd Creech would have died.

"You ever figure out why that thing went on the rampage?" Kittery asked after a long moment of silence. The silence that came afterward was even longer.

"That's the worst part, Pilgrim. That's the part that almost kills me. I didn't know that griz were wounded when I was trackin' 'im. A bear's blood gets all matted up in his fur an' sometimes don't even git t' the ground."

Kittery met Creech's gaze, swallowing.

"Guess yuh got an idea what I'm a-gonna say, don't yuh, Pilgrim? I's the one shot that bear that same mornin'. Shot 'im in the shoulder an' let 'im git away in the trees before I c'd reload. It was me that kilt my family, Kit'ry. Nobody but me. I've saw hundreds of bears since then, some as close as ten yards. But I ain't never shot at one since then. They wanna kill me, then I'm fa'r game."

Kittery sat silent, watching the yellow sparks dance from log to log in the open fireplace. Now he knew the awful story behind Creech's scars. The man called Creature had been a handsome man once. (In fact, he had looked strikingly like Jacob might have turned out looking at an older age, which seemed almost funny, since Jacob was a very handsome man, and there were people who now might have described Creech as looking "hideous.") And Creech had had a fine-looking family. Now all that was gone, replaced by a one-room dirt-floor cabin, a mule and a lop-eared dog named Griz. How could one man endure such tragedy? How could Creech have gone through all that and still be sane? How could he still want to live? Kittery sometimes didn't comprehend the human will to survive. He himself had suffered. He had lost two friends, and he grieved for them. But his losses were nothing—not when compared to Creech's.

The day came when Kittery had been away from Castor for three weeks, and he knew he must return. His feelings were bittersweet. He had come to love this life. He felt close to this hard-bitten mountain man. But the life was lonelier than he had imagined.

One cold dawn, Kittery saddled Satan and loaded his goods and his hides on the backs of his mules. Creech lay in his bed until he figured Kittery was ready. Then he rolled over, stretched and stumbled outside, blinking against the growing light.

"I shorely hate t' be seein' yuh go, Pilgrim. I have to admit it," Creech said after a moment.

Kittery drew the stallion's cinch tight. "Yeah, me too. But I've gotta be movin'. I've been gone too long from . . . from my real job."

"What is yer real job, anyway?"

"It'd take too long to explain." Kittery averted his eyes, then looked back. "But it's somethin' like your story about the grizzly. You come to Castor some day and ask about me. They'll tell you about my real job."

Creech studied his partner for a moment. "I'll do that. Wal, so long, Pilgrim . . . Kit'ry." He held out his hand, and Kittery shook it warmly.

"So long, Judd. You take care of yourself."

Creech smiled. "I alluz do. You already know that. You come back some day. We'll find that jaguar." With an informal salute, he turned back inside.

Kittery let the stallion amble out, Griz following along. Within a quarter mile, the dog turned back home. When Kittery turned back to see him go Creech was standing inside the doorway of his cabin, watching him. The mountain man turned back inside.

Kittery gazed at the surrounding land, pondering the life of Judd Creech. The ultimate survivor—that was Creech, who lived all alone in this rugged, lonely, sometimes violent land. He thought of himself as a survivor, too, but doubted his own strength and ability to make it out here alone. Creech had deep wounds, not so much physical as emotional. Perhaps this dictated his chosen life. As for Kittery, he was still in the early phases of the kind of reforming that had turned Creech into a hermit. He still had to kill his grizzly.

But then what? Anyone looking at his life might believe he had already chosen a path like Creech's. He had forsaken friendship, in a way—particularly that of Tania McBride. But now he knew he didn't want to live his life alone, despite the potential pain that comes with any close relationship. Right now, he resembled Creech, in his smoky buckskins, his grimy skin and his beard. But that was as far as he wanted the resemblance to go. If Tania would have him, he would go back to her, in spite of the chances of his dying a violent death. She could make that choice herself. But a hermit's life was not for him. He would enjoy life, enjoy his friends, while he still had the chance. Tappan Kittery would remain a vigilante until the land had been cleansed. But killing would not become his life. He knew his brother Jake would have approved of his decision.

Chapter Twenty
The Falling Out

Riding Satan and leading the two mules, Tappan Kittery plodded beneath the swinging sign that read, CASTOR, around noon the following day. The bay mule carried a headless deer as its only cargo.

Riding into the stable, Kittery was grateful for the shade. His shirt was soaked with sweat, and he had almost been stuck to the saddle, and the fierce sun had been baking the exposed parts of his bearded face with a vengeance. He had felt the temperature rise from somewhere in the seventies in the high country the day before, to close to a hundred down here. He almost regretted coming back.

He hung the deer from the rafters inside and sawed off twenty pounds of the venison. Some of the horses fidgeted in their stalls and blew out their nostrils, shaking their heads like they weren't too sure they liked the smell of the deer blood. He wrapped the meat in a clean white cloth and rode up to the Vancouvers' front door.

Beth, her yellow hair hanging loose and lovely on her shoulders, answered the door at his knock. Her eyes glowed with surprised delight as she saw him. He removed his hat, and the dark hair fell across his forehead. His combed beard lay in course curls on his face and neck. He wore a dark blue and black checkered, loose-fitting cotton shirt and black trousers tucked into his fourteen-inch cavalry boots.

The big man smiled. "Hello, Beth."

"Hello, Tappan! You got back to the lowlands just in time for the summer heatwave."

"Yeah, it's quite a change." He smiled at Beth. "You sure look pretty this afternoon. It's good to see you."

Beth laughed, embarrassed. "Oh, I haven't even touched my hair today. But thanks for the compliment. I can't tell you how good it is to see you, too. Can you come in?"

"Well . . . yes, ma'am, I guess I can, but I can't stay long. Mostly, I just came to bring you this." He hefted his package, now partially soaked in blood. "Shot a nice one on the way down from Tucson this mornin'."

She thanked him for his thoughtfulness. Leading him into the kitchen, she placed the package on the counter. They returned to the main room, and Beth poured two cups of coffee. She sat at the table, and her guest plopped down in his accustomed chair. For a few moments, they made idle talk, but neither one mentioned Tania McBride. Kittery was bursting with the need to hear about her, but he wasn't going to let on right now that it made any difference to him. For his pride, it was better to wait.

The conversation came around to Cotton Baine. "I wanted to tell you how sorry I was to hear about your friend Mr. Baine. I'm happy you've stayed in Castor, Tappan. I really am. I just wish it could have been us you stayed for and not . . . not that."

"Forget it. I'm used to death. I've been around it all my life. It doesn't matter much anymore."

Beth frowned and sipped her coffee. She searched Kittery's eyes, trying to find the truth there. But she already knew the truth. This man was full of pain.

"We brought supper to Randall and Tania McBride the day after you left. I can't tell you how sorry they were to hear you had left without having a chance to say goodbye. Tappan . . ." Beth paused, searching his eyes. Finally, her eyes flickered away from him, and a long sigh escaped her lips. "You haven't . . . you haven't been over there. Have you?" When he shook his head, she quietly said, "Oh. I . . . I think Randall wants to talk to you."

What about Tania? Kittery wondered. Does she want to talk to me, too? There was something odd about Beth's voice, something odd about the way she couldn't meet his eyes. Had something happened to change things while he was gone?

"We're glad you're home," Beth said again.

Kittery forced a smile. "Thanks, Beth. I wish I could say I was glad to be here. I wish things were different." He stood up. "Thanks for the coffee. I'd better be goin'."

He turned to leave, but Beth's hand on his sleeve stopped him, and he turned back. "Tappan? Go and see Randall before you go into town. Please."

He had to pass the McBride place on the way down the ridge road, so he pulled in there minutes later, glancing at the neat, clean-swept flagstone path. The red and yellow flowers on the windowsill blazed back at the sun that tried and failed to wilt their beauty.

In his right hand, he carried a huge vase with red, black and yellow Navajo designs painted on it. He shifted it to his left hand to knock.

"Tappan, my boy!" came the surprised greeting as McBride opened the door. "Good to see you, lad! We've missed you."

McBride's right leg was in a cast, his right arm also in a cast and a sling. He leaned on homemade crutches.

"Good to be here, Rand. I'm glad to see you gettin' around so good."

"Oh, hell. A few broken bones won't keep *me* down. Come on in and have a seat." The older man glanced at the vase out of the corner of his eye, but his glance scanned past as if he hadn't noticed it. He showed Kittery to a seat, then sat down himself and crossed his feet out in front of him, glancing at the vase again.

Kittery shifted in his chair and let his eyes flicker about the room. As he looked back at the older man, he swallowed. "Oh! I brought this along for Tania." He forced a smile and hefted the vase in one hand. "Thought it might brighten her room. Rand . . . I'd like to see her. I owe her an apology."

McBride stared at Kittery for several seconds. At last, he cleared his throat.

"Uh . . . Tappan. You . . . I thought someone would have told you. Tania is gone."

The words struck Tappan Kittery like a blow. For a moment, they stared at each other. Kittery felt his face turn hot, and he set the vase back down. "Gone where?"

"She went to live with a sister of mine in Las Cruces." Kittery could only stare. "Tappan, she didn't think you'd be coming back. She couldn't stand living here alone with me anymore, I guess."

Kittery steeled himself and forced a thin smile. "Las Cruces, huh?" He cleared his throat and glanced down at the vase, trying to ease the sudden pain in his heart. "Well, that's okay. I, uh . . . I only meant this as a sort of token of friendship. I'll give it to you. Maybe it'll brighten *your* room."

McBride didn't look at the vase. "I'm sorry, son. Believe me, I tried to make her stay. I need her, too. She seemed to think she needed some excitement in her life. She said she was tired of living in a sinkhole like Castor, with nowhere to go and no one to do anything with."

Kittery forced a smile again and shrugged, trying to look calm. "Well, a man can't hardly blame her for that. There isn't much excitement here, that's for certain. I guess I oughtta go and leave you to your peace and quiet." Kittery stood up.

"Wait." McBride held up his hand. "I wanted to tell you, since Tania's room will be empty now . . . there's a place for you—and a good, soft bed. I built it myself. The house isn't much, Tappan, but if it doesn't bother you to be in Tania's room, you're more than welcome to stay here."

Taken by surprise, Kittery had to pause and consider. It could prove very difficult living in a house where so many memories lived. On the other hand, perhaps living in a room that had been so much a part of Tania McBride would ease his pain. He looked at McBride, and the expression he read in the older man's eyes was almost a plea. Maybe they needed each other. Any small piece of Tania they could cling to.

"If you really don't mind the comp'ny, I'll take you up on that, Rand. Thanks for askin'."

Randall McBride smiled up at him, a look of relief coming into his eyes. "Good, good. You can move in today."

Maria Greene's eyes seemed to fall when Kittery told her he was giving up his room. Her smile was swept away, and she looked down at her desk to shuffle some papers that didn't seem to need it. After a moment, she looked back up and forced a smile. "It is good tha' you have friends. The banker, he is a ver' good man; you will be very well taken care of—I am sure of it." She turned her attention back to the papers as she continued. "I am sorry tha' you were gone so long. It seemed the last time you came you desire' tha' we shou' talk."

She looked up, searching his eyes. Her disappointment was poorly hidden.

"I'd still like to come and talk to you, ma'am. Sometime soon. I haven't forgotten about it." He smiled. "I'll leave my things here until later, if that's all right."

As the sun oozed behind the purple-hazed Sierrita Mountains that evening, Kittery sat alone at his old table in the cantina. A half-empty bottle and glass sat before him. But his thoughts were on Las Cruces and Tania McBride. He knew without any doubt how much he had hurt the girl. Why else would she have left so suddenly? And she had told him how she dreaded cities. Not that Las Cruces was any San Francisco, but it was much bigger than Castor. But no, she hadn't left to see Las Cruces. She had left to hide from the memories floating like spirits through the streets of Castor.

He cursed himself and poured a glass full of whisky. Why had he been such a fool? All he would have had to do was visit Tania when he came back from hunting down Rico Wells. He had needed her, and they could have helped each other so much. Instead, he hadn't even told her of his plans. Of course she had assumed him gone for good. The only person he had told any different was James Price, and the hangman wasn't the type to go telling people every little thing he heard. Tappan Kittery had gone away to find Judd Creech and to rediscover the meaning of life. He had accomplished that, but too late.

But then, he owed no one. He had no ties—no responsibilities. He didn't have any deadlines to meet or anybody to please. But now he regretted his decision to leave things up to Tania. She had revealed her mind plainly.

Las Cruces, New Mexico. He hadn't been there for some time. But from what he remembered they had a theater, some fine restaurants, perhaps an opera. If nothing else, they would have the facilities for dancing. They would dance on the plaza, if there was no place else. One thing Las Cruces would have was plenty of bachelors looking for a mate—or just a good time. Those men would fall all over a dark-eyed country princess like Tania. The more he thought about her there in Las Cruces, living the high life, going out on the town with the young men, losing her sorrows in wine, dance and song, the more he drank. The more he drank, the surlier he became. His blood-shot eyes stared dimly. He didn't take notice of the Saturday night crowd that had begun to gather.

Within an hour, the room was abustle with cowhands, miners, and the occasional freighter or merchant. The Mexican laborers and farmers also made their appearance, most of them getting along well with the *gringos* whose ancestors had overrun their land. The revelers danced and sang, drank and played cards, hollering at one another across the tables. They joked and laughed, spilling their drinks and sweeping the evidence away with soiled shirtsleeves. Stifling blue tobacco smoke began to fill the room, writhing in snake-like swirls. Its odor and that of liquor and *cerveza* permeated the air.

Once, a half-drunk man slammed into Kittery's table, jostling his drink and spilling some. Kittery burned the man with a glare but made no move until the apologetic stranger tried to straighten the table. Kittery surged up, grasped the man's arm and hurled him away into the bar crowd. This incident, coupled with the malevolent look in the big man's eyes, assured him a table to himself. Though men soon waded three deep at the bar, no one drew near Kittery's lonely corner table.

An hour later, with the cantina's din at its height, a man came through the swinging doors. His eyes flitted about the room and lit at last upon Kittery. He slunk forward. Kittery recognized him as Dabney Trull, one of the few white men who worked at the warehouse, and also the man who had been hired to dig graves out at the cemetery. He was a vigilante. He stopped next to Kittery, spoke loudly through the uproar.

"Can I talk to you outside, Captain?"

Kittery squinted at him through the acrid swirls of smoke. He shrugged and pushed back his chair, and downing one more glass of whisky with a single tilt of his head he picked up the bottle and followed the man out.

Once outside, they stood in the alley. The other man glanced about, then leaned close and spoke in a whisper. "Don't talk, Captain. Just listen. You want the Desperados dead? Me too. But I hate a traitor worse than I do Baraga. I couldn't find James Price, so I came to you. I take it you know Zeff Perry by now? The rancher that owns the Double F Ranch? Well, I just saw him talkin' to two strangers fifteen minutes ago. And I'd give a dang good hoss if they wasn't Silverbeard Sloan and Crow Denton."

Kittery felt a chill go up his spine. His eyes turned hard. His skin began to tingle as the blood rose hot inside him. *Zeff Perry!* "Where are they?"

"Far as I could tell, Sloan and Denton rode out. Headed west. But Perry's inside the barroom. He's sittin' in a card game in the middle of the room."

A cold fury took hold of Kittery as he thought of Zeff Perry and the question of the mysterious Stephens derringer Shorty had used to try and kill him. The vigilantes had talked of a leak somewhere in the organization, of a man somewhere around Castor called the Mouse, who supplied information to Baraga's gang. But the only thing that could have laid suspicion on anyone was the Stephens derringer. The only thing, at least, until tonight. And the new information pointed to Zeff Perry as well . . .

Without another word, Kittery dropped the whisky bottle, whirled away and shoved back through the swinging doors as his informant faded into the deeper shadows. True to Trull's word, Zeff Perry was at the center table, at that moment shuffling a deck of cards. Other than Perry, the only man Kittery recognized at the table was a dark-eyed gambler who went only by the name of LaFitte. The man dressed as a riverboat gambler and wore a leather patch over his right eye, giving him the look of a pirate. Kittery heard Zeff Perry say to another man, "Draw, Frank." But Frank had no chance.

"No, Perry. *You* draw." Kittery dropped his hand to the butt of his Remington. "Stand up. Stand alone."

Perry's eyes flickered up in surprise. "What?"

Kittery was too furious to repeat his words. Instead, he leaned forward and reached between two of the players, grasping the table's edge. With a grunt, he heaved it over, showering cards and currency onto the floor and knocking Perry down. LaFitte stumbled backward, his dark eye flashing and filled with danger and his hand near his belly gun, unnoticed to Kittery. The other players scrambled away to the safety of the crowd.

Perry stood up and stepped away from Kittery. With no more warning, Kittery, looking wild-eyed and terrible in his towering bulk, closed the distance between him and Perry in two steps and smashed a fist to his jaw. The rancher sprawled back into the crowd, where rough hands caught him and helped him stand.

Like a hungry beast, Kittery lunged forward, grabbing Perry by the lapels of his vest. He ripped him out of the hands of his supporters and slammed a knee into his belly. A big fist battered Perry's cheek, and he went to his knees.

Once again, Kittery stepped in . . .

He felt a blow to the side of his head. It staggered him, and he turned in confusion to seek out his new opponent. Deputy Miles Tarandon stood in the middle of the clear space, legs braced and fists poised. Kittery moved quickly, and Tarandon flinched in surprise as the big man twisted at the waist and sent a roundhouse left that should have taken him down. But unhindered by alcohol Tarandon slung his body and ducked beneath the blow, then straightened back up to give three hard jabs to the big man's ribs, ripping the wind from him.

As Kittery straightened, he shot out a right hand that shocked Tarandon, catching him on the mouth and chin. He reeled back into the bar, scattering the patrons then recovering before Kittery could reach him again.

Calls and cheers of encouragement rose up in the room, most of them for Tarandon, because of Kittery's vicious, seemingly unprovoked attack on the rancher. But Tarandon, though fast and wiry, was no match for Kittery. His pugilistic blows had little effect on the raging bull that was Kittery at this moment.

The two met again, going toe to toe. Kittery absorbed five solid blows for the two of his that connected. The second of his was an arcing right. Tarandon took it on the temple and crumpled to the floor. He didn't rise again.

Kittery swung about. His blurred vision sought out Perry and found him amid the crowd, eyes wide. Kittery moved toward him, and the crowd parted. Perry was left to stand alone.

"I won't fight you, Captain," said Perry through his teeth. He rubbed his bruised cheek. "You're loco. Whatever you think I did, you're wrong."

Kittery stepped forward and found himself staring down the barrel of Perry's Remington.

"I don't know what's wrong with you, mister," Perry blurted, his eyes frozen in fear. "But you take one more step and you won't use your gun arm again—ever. You ain't gonna hit me again."

Even dead drunk, Kittery was no fool. Perry meant what he said, and he had the drop on him. One more step, and Kittery knew he would have a broken shoulder at the least. More likely, he would be dead. And no one would blame the rancher, for nobody knew him for the traitor he was.

Kittery lifted his hands away from his sides and took a step backward. His mind was groggy, but his words were clear enough. "I know about Denton and Sloan. I'd say it's time for you to sell your ranch and move on—before I catch you without that gun on me."

Perry's face twisted in confusion. "What're you talkin' about?"

"You heard."

Perry stared at him, his eyes flickering around at the crowd. A light appeared in his eyes and grew explosively. "You mean—You tryin' to tell me those two . . . That was Crow Denton and Silverbeard I was talkin' to?"

Kittery stared the rancher down, trying to read him. Was it possible Perry was genuinely surprised? The rancher had never seemed a talented man when it came to speaking. The part of a convincing liar didn't fit him. But he seemed to be taken aback. The sickening dread that he might have made an awful, brash mistake washed over Kittery like a wave. Had he accused the wrong man?

"If you . . . If you didn't know 'em, why'd they come to you?"

"Them two came lookin' for me claimin' they was after some horses I had for sale. I had notices around town advertisin' them horses. Figgered those two was legitimate."

Kittery's face was flushed and hot. He knew it had to be from the heat of the room. He let his groggy mind whirl in confusion, then forced it to clear—at least as much as half a bottle of whisky lets any man's mind clear.

"What about the Stephens derringer you gave Shorty Randall to kill me with?"

Perry stared at him. "I don't have no derringer, Stephens or any other kind."

"Silver plating with ivory grips. Does that ring your bell?" Kittery growled.

"What— I don't have that little gun no more. You talkin' 'bout the gun I won from Cotton Baine? Dang, man, I lost that thing the very same night!"

Kittery was stunned and a little sick to think the words might be true. "To who?"

"To Dabney Trull," Perry said.

Kittery's rage was ready to spill over. Not rage for Perry, but for Dabney Trull. What kind of game was the half-breed playing? "I got some bad information, Perry." He swung his eyes about the crowd but saw only blurred faces. No Dabney Trull. He looked at Perry again. "I'm sorry. There's not a thing more I can say."

With blood raging in his head, Kittery eyed the room again, feeling suddenly dizzy. He stooped to retrieve his hat and almost fell over. Then he stepped over the unconscious Tarandon and lumbered from the silent, smoky room. No one spoke for a long minute.

Back in his new room at McBride's, Kittery stared at the blank wall. He cursed himself and his brashness. He had drunk a lot of whisky that night, but the fight had served to clear his head somewhat. It was clear enough to know he had acted the fool. That tarantula juice did it to him every time he allowed himself to consume too much of it. And anymore, two glasses was too much.

In his mind, he ran over and over the fight. He swore out loud. He was a fool! A fool to let the word of one man bring him to a conclusion that could have such a lasting effect on the rest of his stay in Castor, and on his relationship with the vigilantes. But he knew there was a traitor in the vigilantes. It seemed pretty plain it was Dabney Trull. Why else would he have been so anxious to point a finger at Zeff Perry, if he didn't want to cover up his own trail? Now, more than ever, Kittery meant to bring down the traitor. He had to, if only to redeem himself in the eyes of the others. Dabney Trull . . . he had to find that half-breed.

But more than that, he had to find the Desperados. And for that quest he knew he would forsake all else. Tania McBride had gone away, just when he had come to realize how much he needed her. The only thing left in his life that mattered was seeing the Desperados pay for what they did to his friends. He would never rest until their blood soaked the Arizona ground.

In the cavern where the Desperados penned their mounts, Silverbeard Sloan turned in his stocky gray, wiping a sleeve across his sweaty cheeks. He dismounted and tugged off his saddle. Crow Denton stepped down and stretched his back, then unsaddled his own chestnut and went to the corner where they kept the oats and grain. When he turned to take some of each to the weary animals, Sloan was gone. He had overturned their saddle blankets to dry, taken his saddle and walked out.

In the Desperado Den, Sloan held his saddle over his shoulder, a can of saddle soap in his left hand, and stared at the fire pit in the cave's entrance. The coals appeared cold, and the cast iron oven beside them was empty but soiled. Used plates littered the floor. Feeling his stomach work up a ferocious growl, he tugged at his loose gun belt with his thumb. His lips opened and curled in a silent curse, and he turned back toward the canyon and threw down his saddle.

"Filthy pigs can't even save a man some food."

Sloan plopped down against his saddle, staring out across the canyon. He cursed again in anger, pushed himself to his feet, and turned back into the cave. His eyes met Slicker Sam Malone's and bore through him. He was furious, and his eyes went back to the Dutch oven. He shook his head.

It was then that something registered on him, something in Malone's eyes. He looked back at the skinny man, and his eyes narrowed. Malone's gaze swung away from him. With a growing wariness, Sloan looked farther back into the cavern to find Baraga watching him. So, too, were Doolin, Bishop and Morgan Dixon. Sloan looked back at Malone, who now refused to meet his eyes.

"Damn this place," Silverbeard growled, loud enough for everyone to hear. Then he turned away.

"What's the matter, boy? Hungry?"

The impersonal voice of Savage Diablo Baraga stopped Sloan in his tracks. *Boy?* Sloan turned back around. His legs were spread wide, and his hands dangled at his sides. Baraga remained seated, but he still made an impressive figure.

Sloan eyed the outlaw leader. "Hungry? Hell yes, I'm hungry."

"What did you bag us to eat?" Baraga asked. "One of the boys'll be happy to fry it up for you. You told us you were going hunting. You *did*—didn't you?"

"Yeah, we were huntin'." He was beginning to wish he had made a different kind of entrance. He had riled Baraga.

"Well, you weren't huntin' deer. What *were* you huntin'?"

Sloan shrugged. "We took a little ride into Castor. Thought maybe we'd find some excitement." He threw an accusing glance at Slicker Sam Malone. He was the only one they had told where they were going. But it was obvious Baraga knew it now, too.

Baraga thought over the information. "Uh-huh. Castor. Malone told us. Sloan, you don't go in there unless I tell you to. Not while you're ridin' with *this* bunch. I've told you that before. Mouse works through this band as a whole, not any one or two people in it. What did you think you were gonna get out of him? A little job on the side, perhaps?"

Sloan threw a murderous glance toward Sam Malone. Then he looked back at Baraga, his eyes insolent. "It was my own business. I'm tired of sittin' around. Tired of doin' nothin' all the time."

"That so? Well, you listen to me, Beard. As long as you choose to ride with us, whatever you do is *my* business. You *have* no business. This outfit holds together because it works as a team. If you can't accept that arrangement, I'll replace you—with someone with a brain. From now on you're gonna be like my dog. I'm your master. If I say something, I want it done. If I say to lie around, you'd better lie around—or get your carcass out of here. I started this outfit, and I don't intend to see it fall apart because of your stupidity. You understand me, compadre?"

Sloan's chest was full of fire. He couldn't move his eyes from Baraga, yet they were too blurred to see him clearly. When his vision cleared, he dared a glance at the others in the room, but he found no sympathy there.

Baraga shifted on his sandy seat. "Tryin' to line your own pockets, huh? You little weasel."

Sloan's entire body felt stiff, and his eyes widened and narrowed intermittently as Baraga's words bore through him. But Baraga didn't care.

"You made a mistake, Beard. Malone didn't want in on your scheme. His loyalty's with me. You want to line your pockets, maybe you oughtta go line 'em. And take the breed with you, if he likes you as a leader. What this bunch does, it will do together—as a military unit. I can't have men runnin' everywhere, doin' whatever they please and undermining what we've built down here. What do you think got Wells killed?"

Sloan's lips were pulled tight across his teeth, and his jaw muscles ached under the strain of clenching them. Baraga couldn't stop cutting him down in front of everyone. He had to talk and talk, never giving a man any chance to explain. He could have backed off, but Baraga didn't give him a chance. Now it was beyond that point.

"Well, I ain't never been a real part of your little band, anyway. Have I? Wells was the only one besides me who had guts enough to do what he wanted. And you!" His eyes swung to Malone. "I should kill you right now for openin' that fat mouth."

Baraga's icy-calm voice cut in. "And if you did, I would kill you."

The words broke something loose inside Silverbeard Sloan. He had been on the edge of something drastic. Now he was over that edge. No one could threaten Silverbeard Sloan. Whatever happened to him for the rest of his life depended on this very moment in time.

Sloan's lips began to move before he thought out his reply. "You ain't seen the day you could kill me, you old son of a bitch. If you think different, try." He turned to square his body against Baraga, his legs braced. He felt his breathing slow. It seemed to cease. His right hand gripped his holster.

Sloan heard Colt Bishop's calm, even voice like a whisper in the back of his head. "Kill me first, Sloan. Then kill him."

Sloan turned his eyes to the rear of the cavern. He began to quiver all over and willed it to stop. But it wouldn't. All his fantasies of gunning down Colt Bishop came back to him in a flood, and his eyes bored through the gunman, who was half-hidden in shadow like some creature of doom.

"I've waited for you a long time, Mr. Lightning-Hands." Sloan said the name with a sneer. "I never thought you'd have the guts to buck me. Stand up to me now. I'm done talkin'."

Colt Bishop sighed. Seeming almost lethargic in his movements, he eased up from his seat on the floor. He watched Sloan for any sudden move. At Bishop's full height, several inches taller than Sloan, he raised his eyes from the slighter gunman's hands to meet his gaze. His eyes didn't shift again.

"Sloan, you're a fast man. Mighty fast." Bishop kept his voice quiet. "On a lucky day for you, and a day when I was half in my grave, you might stand a chance at beatin' me shootin' at photographs of old men and families. But you won't live to see the day you could get a bullet in me. You've never been that lucky."

The blood pounded through the veins at Sloan's temples, making thin, purple twigs appear to pulse beneath his skin. His breathing came quickly now, heaving his narrow chest in and out. "You son of a—"

In a blur of movement, Colt Bishop clutched his pistol. In the same quarter second he brought it level. With his mouth open, Sloan's eyes appeared to follow the pistol's rise. It was as if he were too stunned to react. Then his own right hand closed on a Colt. It came halfway from its holster.

A single shot shattered the quiet of the grotto. Smoke billowed, and two hundred and fifty grains of lead slammed high into Sloan's chest, knocking him backwards to sprawl onto his back. Blood began to soak his chest and shoulder.

Crow Denton ran in from outside. As he saw Sloan lying on the ground, he came sliding to a halt. His eyes turned to find Colt Bishop standing there with his gun still leveled. The others were just then rising to their feet.

"What happened?" asked Denton, his voice showing as much emotion as it ever could.

"He challenged me," said Baraga coolly.

"So you set Bishop on him."

"I didn't *set* anybody anywhere."

Denton stared. "What about me?"

"Pick him up and get him out of here," said Baraga. "Then you make your choice. You ride with me, you'll act when I say."

Crow Denton nodded. "I'm tired of sittin' around gettin' soft. Sloan was right."

Baraga shrugged. "Take him out of here. I don't want to see you back."

As Baraga turned his back to the half-breed, Colt Bishop looked down at his side. Morgan Dixon's shotgun was cocked in his hand. The gunman gave a slight smile of satisfaction.

Crow Denton stooped, his face emotionless once more. He picked up Sloan's pistol and stuffed it down behind his gun belt. Then he turned toward his comrade, and for the first time he noticed the heaving of Sloan's chest. Without looking up at the others, he bent over his partner and pressed around the area of the wound, nodding to himself.

He looked up at last at Baraga and spoke without inflection. "He's alive. If you'll let me doctor him in here, he may stay that way."

Baraga stepped forward, the shotgun bores raised partway to Denton. "Perhaps he'll stay that way if you take him out of here, too. Or on my say-so he'll die now. He made his choice, Breed. And to me his life isn't worth as much as a lizard's, anyway."

Denton listened to Baraga with no expression and glanced around at the others. Baraga bent to look at Sloan's wound, then turned and eyed Bishop with a faint, grim smile. "Your aim could stand improvement."

Turning back to Denton, he let down the hammers of the shotgun and waved the barrel toward the canyon. "Saddle and ride, Breed."

Seven minutes later, Baraga listened to the clatter of hooves echo along the canyon. He didn't go out to see the two men off, but he assumed one of them would soon be a corpse. That was for the best, yet Savage Diablo Baraga had a moment to ponder the portent of this incident. There were now only five members of Baraga's mighty Eight. The newspapers must change the moniker they had given his band. And at the rate they were being dismantled, how long until none of them remained?

Chapter Twenty-One
Forbidden Threshold

People didn't often look once at Captain Tappan Kittery without turning to look again. He was a big man, bigger than most; a dark man, straight and strong and proud, and in his eyes shone a vengeance like a quiet, crackling fire.

Today, he wore his suit of buckskins and his dusty black hat snugged low on his forehead, shielding his eyes against the desert sun. He was not a man who dressed for style, as attested to by the plain-handled Remington .44 at his hip and the unadorned moccasins on his feet. He dressed and geared himself for the harsh life of the Sonoran desert of Arizona.

Kittery stood in front of Bartholomew Kingsley's café, savoring the flavor of black coffee lingering on his tongue. It was a luxury he feared he had taken too much for granted. He had a feeling he would be spending many days to come with no such luxuries, like today.

Today he rode toward the Baboquivaris—the lair of the Desperados Eight.

Kittery would ride alone and did not plan any attack of the outlaws. He didn't even care whether he found them. He only wanted to get a feel for the trail and for the lay of the land. Then when the time came to go after them he would have an edge he didn't have now. The Baboquivaris would be a harsh, barren range, and water would be precious. That was obvious even looking at them from a distance. But there was an oasis in there, a place where the Desperados could survive in comparative comfort, with plenty of food and water not only for themselves, but also for their horses. Kittery must find that oasis.

The task of hunting the Desperados Eight seemed easier than before. First of all, Rico Wells was dead. That had strengthened Kittery's confidence in himself, for Big Samson had been a formidable foe. Second, and more importantly, Tania McBride was gone. No longer would there be anyone who would grieve him in the event that . . . He stopped the thought. That kind of thinking would only weaken him when he stood against the Desperados.

Kittery clenched his jaw and stepped away from the café. He walked along the line of dusty, weathered buildings and stopped at Albert Hagar's mercantile store. Inside, he glanced over the weapons lined up on the shelf along the front of the counter. He motioned the white-haired Hagar over.

"I'll take a look at the big Smith and Wesson."

The pistol he had pointed out was the second model American, a pistol designed to be loaded by breaking it open behind the cylinder. This pivoted cylinder and barrel forward, and if done right it also pushed the empty casings out and away, leaving the weapon ready to reload. The American looked awkward at first glance, and being a .44 it didn't possess quite the knock down power of the Peacemaker and Schofield. Yet the American was a strong, well-balanced weapon, all the same; its predecessor, the first model American, had been employed briefly by the cavalry before the big .45 Colt's Peacemaker replaced it. The most attractive thing to Kittery was its second hand price.

Kittery didn't need a new pistol. He not only had his Remington, now converted to take cartridges, but he had also inherited Cotton Baine's Colt. Yet the search for the ideal sidearm seemed forever ongoing. Each had its good and bad points. And then . . . Heck, what reason did he need? He had been around long enough to figure himself out. Whenever troubles had come his way, it had been his answer to go out and buy another weapon. Most of them ended up sold or traded off, but it somehow eased his mind to spend his money. Maybe that was his feminine side peeking through.

The American was blued and scarred, gripped with plain walnut. For it and a box of Smith and Wesson .44 shells Hagar asked twelve dollars. On a whim, Kittery pulled out the Stephens derringer and laid it on the counter. "You have any use for that?"

Albert Hagar looked at the derringer, then leaned closer to peer at it, laughing. "Well, where in tarnation did you get that?"

Surprised by Hagar's reaction, Kittery looked at him for a moment. "It's a long story."

"It must be," Hagar said. "That thing must've changed hands more than the only woman in a mining camp! Excuse the expression. That's a rare weapon, Kittery. I bought that thing a while back only to sell it one hour later, and the man I bought it from said it had two owners before him!"

Kittery took a deep breath. "Who'd you buy it from?"

"From a half-breed that works around here. Name's Dabney Trull."

The big man almost smiled. He wanted to laugh. Here he would have had Dabney Trull hung for a traitor, and there went another theory, up in smoke. "Well who bought it after him?"

"The banker," said Hagar. "Rand McBride."

Tappan Kittery made the rest of his deal in shocked silence. He paid Hagar outright for the American and shells, since he had decided not to use the derringer in trade. After filling a basic supply list, Kittery walked numbly down to the livery stable and to the farthest stall in the back. He couldn't get his mind off Rand McBride. But then he was able to comfort himself with the thought that if the little gun had changed hands so many times it could change once more. But why would a man buy a gun only to turn around and sell it? It was completely different than winning one in a poker game.

In the stall Satan greeted him with a restless snort, showering his cheek with moisture. Wiping his face with his sleeve, Kittery patted the stallion's neck and swore at him gruffly but good-naturedly and led him out by a handful of mane. He haltered him and tied him by a horse hair mecate to the top board of the stall.

As he worked, he heard the *thump, step, thump, step* of the peg-legged Jarob

Hawkins approaching from his office. As usual, the old man's mind, as well as his vocal cords, was active. "You know, Captain, that Baraga's a hard nut t' crack. Them others, too. But seems t' me yer plannin' on goin' after them Desperados . . . alone. I 'member another fella did the same. Never come back, neither—not on his own. Funny, I took to him, too." The old man chuckled and waved a hand as if swatting a fly. "I s'pose that's none o' my bid'ness, though."

Kittery nodded with sad memories and swung aboard the stallion.

"One thing, Captain," the hostler bulled on. "The marshal was dressed a sight stylisher." He passed laughing eyes over Kittery's greasy buckskins.

In Kittery's head was the image of Joe Raines going to his death dressed in three-piece black broadcloth. He truly was *stylisher*, as Hawkins put it in his hillbilly English. Giving the old man a smile, he touched a finger to his hat brim and spurs to his horse. "So long, old man." He left the stable at a walk. Hawkins leaned on his crutch and watched the black horse disappear.

Kittery's first order of business was to purge his mind of any thoughts about Rand McBride and the derringer. He was riding in dangerous country out here. It wouldn't do to let his mind be taken up with anything but his survival.

From his previous trip into the badlands to bring back Joe Raines, Kittery knew where to find the first water seep. He pushed the stud at a hard canter to reach it, and there he watered up, filled his canteen and trotted on. The temperature was already over one hundred, and he was cursing the suit of buckskins. But they were tough, and they blended into the desert.

In the late afternoon, when it had dropped clear down to ninety-five degrees in the shade, he rode through the Peñitas Hills, past the spot where Joe Raines had died. It was quiet and still there, as if nothing had happened on those barren rocks. Only the ghost of a wind rustled through the clumps of creosote bush and mesquite and through the upthrust chunks of rock. But there were bloodstains still on the rocks.

He camped twenty-five miles away from Castor. His camp was fireless. He had not only the Desperados to watch for but renegade Apaches as well. They came to roam these arid wastes, either jumping their reservations or never going there at all—not that he could blame them. They had been jerked far away from their homelands and stuffed onto a piece of dead, disease-ridden ground that would barely support a burro. He had seen the reservation at San Carlos. It was a blotch on the face of humanity.

Cochise, the great chief, was gone. He had died two years before, at San Carlos, following a long stretch of peace that only his word or that of a man like him could have ever brought about. He had been a strong leader of the Chiricahua Apache and a bitter enemy of the white man after they had taken some of his family members hostage and then killed them when he proved unable to return prisoners some other Apaches had captured. He couldn't have had any control over those other Indians; Apaches followed their own hearts. But white men didn't understand that. Lieutenant Bascom, who had been responsible for that affair, had assumed Cochise could make the other Apaches do anything he told them. So he showed Cochise how strong he was. He showed him he would keep his word: he executed his prisoners. Cochise's legendary word, in exchange, was to kill ten white men for every Apache Bascom killed. Some believed he had.

But that was past. Cochise had surrendered, thanks to the efforts of one

Thomas Jeffords. And then there was peace.

The renegades who had replaced Cochise, however, were a different breed. They seldom had a friend among the whites. If they knew they had the strength of numbers on their side they would give no quarter. Like Cochise, they still respected courage and strength. But a man with those attributes proved the most challenging to torture, and to kill him brought glory.

In the morning, before departure, Kittery drew rawhide shoes from his gear and lashed them with thongs around the stallion's feet to muffle the beats of his hooves. For his own feet, he wore the Apache-style moccasins that reached to his thighs and then were folded down to below his knees and held in place with crisscrossed laces.

As he rode, Kittery again had to fight off thoughts of Rand McBride and the derringer that now rode in his saddlebag. How he would like to know this very moment who the last man was who owned that pistol before Shorty tried to kill him with it! But if it was indeed McBride, perhaps he didn't want to know. There had to be something else he could occupy his mind with besides that.

He had covered nearly twelve miles as the crow flies across the flat plain. The major vegetation remained creosote, cactus and thornbrush. Occasional patches of dried grass were all that remained of a dominant vegetation now nearly gone. Before the arrival of the Spanish, this valley had probably waved with bunch grass. The long-horned Spanish cattle had overgrazed it, however. The creosote had come in to take the place of the grass, with its roots nearly impervious to intrusions by other vegetation. They secreted a toxic substance that insured its dominance on any landscape where it found a purchase.

As the desert vegetation changed, taking on the spires of saguaro, the spikes of yucca, the spines of snake-limbed ocotillo, Kittery became more wary still. They had entered the foothills of the Baboquivaris, with the massive granite dome of the peak like a guard over the range. A well-trod trail ran to the north of it—the trail to the Desperado Den.

Sweat ran in rivulets down Kittery's cheeks and torso as he rode. It soaked the insides of his moccasins. The sun hammered down on his back and flung up at him from the desert gravel below. His eyes felt like hot pokers, and his head inside his black hat seemed ready to explode with the heat.

He saw the first clear horse tracks somewhere in the midmorning. He was riding through a coppice of mesquite, and it protected the trail from the wind. What Kittery read in the tracks when he stepped down to study them took him by surprise.

The tracks he had taken time to look at the previous day had been wind-whipped and soft. There was no telling which way they went. But these sets were clear, and of the tracks he saw here now, the newest went not toward the Den, but toward Castor!

There were two sets of tracks beneath the two that headed toward Castor, and as near as Kittery could tell the same horses had left them. So if Silverbeard Sloan and Denton had indeed been to Castor, and these were their tracks, near as he could tell they were headed toward Castor again. How had he missed meeting them on the trail?

Kittery stared toward Baboquivari Peak, which looked like a solid fireball in the sunlight. He wanted to make it near the Den on this trip, to see what was back there. But if Sloan and Denton were headed back toward Castor, did that mean their business was unfinished? If so, there was a chance he could catch them with their informant, the Mouse. The Desperado Den could wait.

Tappan Kittery whirled the black around and started back the way he had come. He tried to keep the freshest horse tracks in his sights, though the winds had dimmed them. The fact that he had missed meeting up with the two outlaws kept gnawing at his mind. Was there another trail?

That puzzle was answered a few hours later. If Kittery hadn't been watching the tracks so closely, he wouldn't have caught it. He crossed a rocky stretch of the trail he remembered passing the day before. It was a little dry wash with a bed of sandstone swept clean by wind. On its other side, the tracks he had been following disappeared.

Kittery looked both ways down the narrow little wash. One way seemed to peter out and disappear into the desert scrub. The other way deepened and widened out a little as it went before fading among a jumble of rocks farther along. On a whim, Kittery took the latter path into the wash, riding slightly south. It wasn't long before he caught the first indentation of a track in the sand. After that, he observed one now and then and picked out the white scars made by horseshoes on the stone bed. After a hundred yards or so, Kittery discovered where two horses had left the rocky wash and turned east again, now paralleling the trail Kittery had taken the day before.

A chill went up his spine. He must have passed very close to the two outlaws on his trip the day before. But after a time he began to realize the trail was veering farther and farther south of his own. If it continued that way, it was going to skirt Castor by quite a distance. It would lead through the ranch country between Castor and the Santa Rita Mountains.

Kittery camped that night in a hollow guarded by mesquite. Considering the abundance of fuel and the hidden location, he risked a small, hat-size fire and cooked up a batch of sugarless cornmeal gruel. Because he was hungry and tired of jerked venison, it tasted fine. But he found himself longing for one of Tania McBride's meals. And again, he stewed over Rand McBride . . .

The next day, early on, he found the outlaws' campsite, probably from two nights back. It was only six or seven miles past his own. He found where they had dug indentations out for their beds. He could see where the horses had been picketed and where the outlaws had built their fire. He discovered the direction of their departure.

And he chanced upon the blood-soaked bandage.

Sand half-buried the bandage near a clump of creosote. He pulled it loose and shook it, then held it up for a closer examination. It was a torso wrap, judging from the size of it, a makeshift affair made from a shirt. Whoever had worn this had lost a considerable amount of blood.

Aching with curiosity, Kittery followed a set of tracks around through the brush, finding where someone had stopped at a dead saguaro and cut loose four of its long, smooth ribs. He followed the tracks until they wound back around to where the horses had been tied. The next sign he saw was that of the horses' departure. Two pairs of drag marks flanked the horse tracks, the sign of a travois.

It didn't take a genius to guess that whoever had worn the blood-soaked bandage was weakening enough to no longer be able to ride. For one thing, the second set of horse tracks wasn't as deep as the day before, except at the front of the toe, where it was deeper. It wasn't being ridden but pulling a load. And what else would two outlaws need with a travois but to pull one of them along?

The trail skirted six miles to the south and east of Castor, then swung sharply north toward Tucson. Later, it began to move west, toward the Camino Real.

On cattle range now, Kittery nooned in a cow wallow shaded by stunted sycamores and desert willow. Dried cow chips littered the ground, and he contemplated a fire. But his thoughts outweighed his hunger. So he rested and pondered on his discoveries while Satan cropped grass and waited to move on.

He let the horse rest for half an hour, then gave him a drink from his hat. With a handful of jerked venison strips, he stepped into the saddle and pushed on, knowing with each passing minute he drew closer to the Camino Real. If the outlaws had joined that road their sign would be obliterated by the horse and wagon traffic.

And so it was. Fifteen minutes later, he watched the tracks disappear in the turmoil of sign on the wagon road. He sat his horse in the roadside brush, looking toward Tucson and frowning in discouragement. He turned the black north.

Three miles farther, the smell and bleating of sheep betrayed a sheep camp long before Kittery came within sight of the wagon. The vehicle was parked in a shallow dip between some gnarled mesquite. Kittery recognized Efraín Valesquez, the herder whose acquaintance he had made thanks to Jed Reilly. He sat on the wagon tongue reading from a thick black book, a rifle leaning nearby. The smell of his mesquite wood fire was strong and sweet in the dying heat of the desert air, but not strong enough to drown out the musky, dusty smell that hung over the flock of sheep. Riding through the midst of the flock, he scattered sheep this way and that.

Valesquez, hearing the commotion, groped for his rifle with one hand as he strove to finish one more sentence in his book. Kittery's grin widened at the humor of it, and he waved as the Mexican looked up.

"Can I come in?"

A huge grin broke across the Mexican's face. "Si! Come, amigo. It is good to see you."

Riding up to the wagon, Kittery dismounted stiffly. "Good to see you're alive. And careful," he added, indicating the rifle.

"Oh, sí." The Mexican smiled and shrugged, returning the rifle to its leaning position. "You also. I hear many story of you."

"I bet. Do you have coffee, by chance?"

"Sí. You know, I am ready soon to eat. You would join me? It is not much, but for an amigo . . ."

"I do believe I'll take that offer," Kittery said, sitting down on his heels. "I've been livin' on jerked deer meat for too long."

"Ah, sí. I know how you feel." The Mexican shook his head with a grin, waving at the flock. "When a sheep dies, it must not be wasted. If you cut it and dry it, it is much like one of your American buffalo."

The sheepherder scraped coals off the top of a Dutch oven and slid open the lid, leaving the coals to burn themselves out. He portioned a meager stew between his plate and the one he handed Kittery. Despite its scant ingredients and the gritty texture of the old mutton Kittery savored each morsel.

After several minutes he put down his fork and ran a sleeve across his whiskers. Indicating the nearby Spencer rifle with an inclination of his chin he asked, "You had any more trouble?"

"No, *mi* amigo. All has been well." The Mexican encircled his throat with his hand. "But the skin still remembers the feel of the reata."

"Oh. I thought maybe it was somethin' else."

"What is that?" asked Valesquez.

"Thought you might've expected trouble from a couple of hombres I'm lookin' for. You seen a man with silver-colored hair and a beard come by here? He travels with a half-breed Injun. One of 'em was hurt bad and ridin' a litter."

"Señor Kittery, it is one night past, a man rode by leading a horse that pulled a . . . a litter? A *cama*. The man who rode the horse was dark, as an Indian. I did no' see the other. He rode on the cama beneath blankets. He could be the man with silver hair, no?"

When Kittery explained who the two were, Valesquez nodded gravely. "Then I am ver' lucky I stayed from sight. And you, you hunt them because of you' two amigos they have kill'. I know of the story. I wish you *mucho* luck."

By now, both men had finished their stew. Breaking large chunks from a round loaf of Dutch oven bread, they mopped up what remained in their plates. The Mexican then cut a quarter-pound of cheese in two equal parts and gave half to Kittery. Last of all came a wineskin. The Mexican took the plump container, held its spout a foot from his mouth, squeezed and swallowed two mouthsful. He passed it to Kittery, and the big man imitated him as best he could, swallowing half and wiping the rest from his beard as the Mexican chuckled.

Kittery rose, dusted the back of his pants, and shifted the gun belt around on his waist. "Thanks much for the meal, friend. And the wine, too. It makes a good beard washer, anyway."

"Here." The Mexican smiled broadly, shoving the skin forth at arm's length. "Take it, an' you will learn. I have another."

Kittery smiled and took the skin as he turned to where the black cropped grass. "Gracias, my friend. You take care of yourself."

"As I always do, Señor Kittery. It is you I worry about. Vaya con Dios."

Kittery swung into his saddle and raised a hand as he trotted back toward Castor.

He had decided while speaking with the Mexican he wouldn't continue after Sloan and Denton. The chances of finding them were not good. The two outlaws could ride almost anywhere. There were roads in all directions, and before long they would throw away the travois and find passage in a wagon, which he could not track. He wanted the two of them dead, if only because they had taken part in killing Joe Raines. But they could wait. If they still rode with the Desperados, they would be back. Baraga and Doolin were the ones he really wanted, particularly Doolin, for carving Joe Raines up with his knife.

As Kittery rode back toward Castor, he glanced at the red globe sinking into the western mountains, knowing he was looking toward the Desperado Den. He had the sudden urge to cleanse his mind of the outlaws. He wanted to relax with a glass of whisky, maybe sink into a hot bath. He wanted to take off his buckskins, shave and slip into fresh clothes. Above all, he wanted to feel Tania's arm around him, smell the freshness of her hair against his face. He ached with missing her.

Yet he couldn't think of Tania without her father and the derringer taking over his mind, and those thoughts were killing him.

As the stallion came abreast the hotel, Kittery felt drawn to pull in, and he did. Perhaps it was weakness on his part, but the things he desired awaited him inside. They had a bathtub, a relaxing atmosphere, and it would be an easy thing to find some boy to run up the hill to McBride's and fetch him some clothes, along with a bottle of whisky. And then there was Maria Greene . . .

Kittery felt a strange heat rise in his chest, a sensation that seemed to squeeze some of the air from his lungs. He stepped off the black and tied him to the rail, then paused.

Why was he here? He kept telling himself it was to bathe, to relax. But was that it? Or was it to see Maria Greene? If that was the reason, he should leave now. But no, he wanted that bath. So why did his chest feel so compressed? Why did he feel like he was about to cross a threshold that would never let him return?

Chapter Twenty-Two
A Lonely Man

What was he doing here, stopped at this hotel? He didn't have a room here. There were other places he could go, places where his heart wouldn't feel like it had been shoved up into his windpipe. Was it Maria he had come to see?

Pushing the thoughts from his mind, he scanned the street and saw two Mexican boys playing near the sheriff's office. He motioned them over, and when they stopped before him he clasped them both on a shoulder.

He asked one, "Do you know where Señor McBride lives? The man who owns the bank?"

"Sí, señor."

"I want you to go there and tell Señor McBride to give you my boots and a clean shirt and pants. He'll know who sent you. Hurry and bring them here to the hotel." He could have ridden up the hill in no time and picked up the things himself, but with the horrible question of the Stephens derringer in his head he didn't think he was ready to face Rand McBride yet.

When the first boy had gone, he gave the other two bits and said, "You take my horse and water him, then take him to the stable and tell the man to give him some oats and hay. When he's done bring him straight back and tie him here."

"Sí, señor," said the boy.

"And—" Kittery reached into his pocket and drew out three gold dollars. "When you're done, go to the cantina and tell the bartender Captain Kittery would like a bottle."

Kittery removed his rifle from the scabbard and watched, grinning, as the boy doggedly tugged his reluctant charge down the street toward the water trough.

The boys, like most young Mexican boys, didn't mind taking orders from perfect strangers. They were accustomed to it and knew the reward that came afterward, with faithful service. It was a good arrangement on both sides.

After making certain the boy and horse made it to the stable on friendly terms, Kittery turned and took a deep breath, looking up at the hotel. He had told the boy to bring the black back to the hotel and tie it in order to make sure no one thought he was trying to hide something. Anyone could come to the hotel, after all.

Kittery took another gulp of fresh air and pushed into the dim-lit room. Maria Greene looked up with a smile of surprise when he shut the door and stopped at her desk.

"Hello, Mrs. Greene. Is your husband around?"

The bright look on the woman's face dimmed, but she forced a smile. "No, he is not. He stays at his store. You wish to see him?"

Kittery was aware of that bizarre arrangement, with husband and wife separated not by a space between beds but by an entire street. "No, ma'am," he replied. "No, I came to see you."

"To see me?"

"Yes, ma'am. I'd be pleased to have your company. But first I wondered if I'd be able to take a hot bath and shave. I've been on the road a while."

Maria smiled and nodded, getting to her feet to reveal the bulge of her abdomen, tight with its new life. "Sí, captain. You smell of dust—and smoke, too."

Kittery grinned lopsidedly, silently thanking Maria for her tact. He knew he smelled of more than that. She came around the desk and walked toward a room on the other side of the stairway, and he followed and stepped in after her.

"There are towels there," she pointed to a cabinet. "And a razor and strop. I have a helper now," she said with a proud smile. "I will have her bring you water—it is already warming."

"Her?" repeated Kittery.

"Sí," Maria said, smiling. "Do not worry—there is nothing Alfonsa has not seen."

"Great," Kittery replied sarcastically.

He laughed to himself when the Mexican woman arrived with his water. Alfonsa appeared to be past her sixties, wrinkled like linen long stuffed into a drawer, with coarse gray hair tied back in a bun, loose strands hanging on her forehead.

Kittery had his shirt and moccasins off. He was clad only in the buckskin trousers. He stood by while the strong old woman went in and out of the room, filling the tub to a level she judged acceptable. He thought she would leave then, but Alfonsa stood there watching him, looking from him to the water as if she expected him to climb on in with her watching.

He motioned her toward the door several times, but the old woman held her ground. At last, in desperation, he called out to Maria. He was thankful to be saved by her entrance.

Trying to keep from laughing, Maria gave the old woman an order in Spanish, pointing at the door and pushing her toward it, soothing her with soft words. Alfonsa went, but reluctantly. She chattered in Spanish all the way out. Kittery could still hear her muttering after Maria closed the door.

He sighed with relief when she was gone, trying to compose himself. He felt himself blush in Maria's presence, aware of his bare torso. He turned away, making a show of inspecting the water.

When he looked back, he saw Maria's hand was over her mouth, her eyes twinkling in mirth. Then she began to laugh. He watched in consternation, ready to call it a night before starting the bath. But there was something contagious about the woman's laughter and, despite himself, Kittery began to smile, recalling the old woman's mulish stance. Then he too was laughing, and neither could stop before they laughed so hard tears came to their eyes. It was the first time Kittery had seen Maria laugh and the first time he himself had done so in some time. It felt good.

They managed to regain a semblance of composure after several minutes. Kittery asked, "What did she want, anyway? She was done bringin' me water."

Maria stifled another laugh and shook her head. "You do not know? She has been this way since I brought her here, Captain. When a man is to take his bath, she thinks she must watch until he is safely in the water. She told me it is because she must see that the water is just right and that there is no accident getting in. But I think she just likes to see men."

"Wonderful," said Kittery. "Well, now that she's gone, I'm going to take my bath. All right?"

"All right, Captain. You can shave you'self, or I will shave you after you' bath. In two minutes I will bring you more water."

Kittery had nearly finished his bath when Maria stepped in with her hands on the shoulders of the two Mexican boys. "Someone to see you."

"Oh, good." Kittery smiled and motioned for the one boy to deposit his clothing near the wall, which he did. He asked the other boy, "How's my horse?"

The boy smiled. "Ver' fine. He waits outside."

"Good," Kittery said with a nod. "He likes you, I think. What is your name?"

"Pepe," was the boy's reply.

"And José," the other boy piped up.

Kittery turned to Maria. "In my buckskin shirt there's a pouch of coins. Could you give my friends two bits a piece?"

Maria complied, and the boy named José ran away smiling. Pepe walked over and leaned down to place the bottle of whisky near the foot of the big tin washtub. "Eef you wan' more help, you can always fin' me, señor. I am happy to help you, especial with that horse."

Kittery held his wet hand up to the boy, and the boy took it and shook hesitantly. "I am Tappan Kittery, Pepe. When I need you I'll call."

When Pepe had gone, Maria turned to Kittery with a warm smile. "There is a good heart in there some place, Captain."

"I don't know where," Kittery replied. "But I do know the water's gettin' cold." He indicated the door. "Do you mind?"

Maria looked at the door, then back at Kittery with a crafty glint in her eyes. She smiled and folded her arms. But under Kittery's stern look she shrugged, moving to the door. She turned there and looked at Kittery once more, then laughed. Now that she had started, she seemed to like laughing. It suited her. She was a lovely woman when laughter lit her black eyes.

"I cannot blame Alfonsa for wanting to watch you, Captain. You are a very pretty man."

Kittery chuckled, his cheeks hot. "Thanks. Now go."

"Who will gather up your clothing and hand you a towel?"

Kittery glanced over at the towel hanging across the room. "I can manage, thanks. I've done fine up till now."

"I shall help anyway," said Maria, and she reached out and took the towel, tossing it to him. With a smile, she stepped out, clicking the door shut.

Kittery's entire body seemed to breathe in the air through his clean pores as he dried himself, pulled on his wool pants and shrugged into the fresh-smelling cotton shirt. He shaved so his skin felt like that of a child, combed his hair, pulled on his boots, then stepped into the main room carrying his hat.

Maria sat behind her desk. She raised an eyebrow in frank appraisal when she saw him. "This is how I will always remember you, Captain. You look ver' fine."

"Thanks."

"It is late now," she pointed out as she stood. "And all of our guests are in. I will lock up until you leave, so we can talk in peace."

When Maria had bolted the front and back doors and drew the blinds, she picked up the kerosene lamp from her desk and moved toward her bedroom. Kittery hesitated, but when she looked back at him, he followed her. Maria placed the lamp on her bureau and sat down on the edge of her bed, casting a longing glance toward the empty bed where Emilita had slept.

"There is a chair, if you like," Maria said.

Kittery pulled the chair up and sat, flicking his eyes about the room. He was unsure now what he thought he and Maria Greene could talk about. It came to him now with perfect clarity that his visit to the hotel had had nothing to do with bathing. It had to do with female companionship, and with that knowledge in his head he told himself he should go. But he stayed.

They muddled through a few minutes talking about Arizona and life in Castor. When Kittery told the woman how he had spent his last two days, she grew quiet. Her eyes would not meet his. He changed the subject.

"Ma'am, there's somethin' I often wonder about. Especially tonight. You are a very—" He caught himself. "You aren't a disagreeable woman to look at. You're . . ." He felt his face redden, but in the crepuscular light of the one lamp he knew she couldn't see that. "You don't have to tell me anything about your life unless you want to. But why did you marry Thaddeus Greene?"

Dropping her eyes, she twisted her fingers together and watched their shadowed grace dance in the candlelight. "I do not know if you have the right to know these things, but . . . I will tell you. I will tell you because I am not some hard stone with no feelings. I too need someone to listen to me."

Kittery wanted to reach out and squeeze her hand. His heart was racing, and that compressed feeling was on his lungs again, making it hard for him to breathe. He was glad the room was so dark.

"You would ask me if I love Thaddeus Greene. I know you wonder," she said. "No, I do not. I have never been in love. But I do not want you to misunderstand. I mean I have never been in love *with* anyone. But I have loved. I loved my Emilita. I loved my papa. But I have never been in love."

They sat silent. Kittery was at a loss for words, but it didn't matter. He had no wish to disrupt Maria's search for her own way of expressing herself.

"When I was young I dreamed always that I would be in love one day. But I was forced to marry Greene—you know? I was a *niña* only fourteen years old. My family all was killed by Apaches one month before I met Greene. I was living off the streets. He came to me, he told me I was the most beautiful girl in all of Mexico, the most beautiful girl he had ever seen.

"He was not a handsome man. Look at him. He does not have a face that ever could have been handsome. But I began to believe he was, for he was so kind to me. Also, he had money. Everyone knew it. He rode a big white horse and wore a black suit with a tall black hat. He wanted to be the governor of Texas, and he lost most his money to try. But before that I think maybe he could have bought my whole little village of Martinez and all the people in it. Instead, he chose me, a dirty little Mexicana girl dressed in rags, with not even the money to buy shoes or food. I could go with him or I could stay an' become . . . something I would be ashamed to be. Yes, Captain, I think that man could have bought all of my village, but he didn't. He bought me."

For a long moment she was silent, studying the hands folded in her lap, the hands that had at last become still. She untangled them and caressed her protruding belly, not looking up. "I have never told anyone this. I hope it will remain between only us."

He searched her face for a long moment, moved by her story, and glad she had found a way out of following a course many young women found it impossible to avoid. "I would never tell a soul, ma'am . . . Maria." Embarrassed that he had allowed himself to speak her name without her permission, he dropped his eyes. "I'm sorry there couldn't have been someone else for you. I'm sorry you've never been in love."

The woman took a deep breath, raising her shoulders and letting them drop. "I am afraid for you, Captain."

"Thanks for that, Maria. But don't be."

"You know, once before I said a man must follow what his mind tells him. But sometimes, when he follows his mind, a man will die. You will—I feel it. You follow you' mind to a strange land where many cruel men wait to kill you. You once said that I understand the way a man feels. But you were wrong. I do not understand why a man would try to have himself killed when so many care for him. That is what you do, you know. Men like this Diablo Baraga, they keel without caring and without feeling. But you cannot understand these ways. I do no' think you could ever keel this way."

Kittery gazed long and hard into the eyes that watched him, so full of grief and disappointment from her past.

"Maria, you do understand the feelings of a man. You're right in what you say. I can't kill as Baraga does. That's what sets the two of us apart."

"That is what I say, Captain! You could not keel another unless first you gave him a chance at keeling *you*. Can you not see that? That is why you must die. You cannot give many men an equal chance to keel you and still live yourself. You must either die . . . or become like them. Cruel and uncaring about life like them."

Tears had welled up in the woman's eyes to match the emotion in her voice. Without realizing it, Kittery stood, and Maria stood with him. He took her into his arms and held her.

The woman raised her face to him, her lips parted. He took a deep breath and tried to push away, but she held on. He brushed a tear from her cheek and touched her there with his lips. She searched his eyes, drawing him back into her. But then, after a few moments, she relaxed and lowered her eyes.

Kittery took advantage and eased her from him, turned and stepped almost blindly from the room. Fumbling at the bolt on the door, he shoved it open and stepped outside into a world dark with night. He sucked in lungs full of fresh air, glad to be outside again.

Stepping to the stallion's side, he untied the reins and slipped them over the animal's neck, then swung up. As he crossed the street to pass through the alley between the blacksmith shop and the jail, he didn't feel the two pairs of eyes that followed him. Miles Tarandon owned one of them. He watched from the jail.

And from the dirty window of the dry goods store glared the hate-filled eyes of Thaddeus Greene.

Chapter Twenty-Three
The Cleansing

Tappan Kittery unsaddled the black stallion and dried him off with the upper side of his blanket, pushing harder than necessary as he strove to drive the guilt from his mind. He shouldn't have ever gone to see Maria Greene. She was too lovely a woman for a man to allow himself the luxury of talking to her in such private circumstances—especially a man so weakened by the recent loss of his true love. She was a married woman! He cursed himself, wishing he had better sense. The only thing he could be thankful for was that he hadn't let her draw him into kissing her. He couldn't have forgiven himself for that.

Turning the horse into McBride's corral, he stepped across the veranda and into the house. He could see a candlelight burning but hoped McBride had just left it on for him to see by. He wasn't in the mood to be friendly to him tonight.

His silence was to no avail. McBride sat awake in the main room, but he was not alone. With him was James Price.

Surprised, Kittery gave the hangman a nod. "I didn't see your horse."

"I walked up."

"Price is spreading the word around, Tappan," McBride offered. "There's a meeting at four o'clock tomorrow out at Perry's place."

Kittery nodded. "Better get some rest then. What's this meeting about?"

Price stood and clapped on his hat. "Like I said, Kittery, you can't be too careful. You come in the morning and find that out for yourself. Don't miss it."

At fifteen minutes to four, Kittery and Rand McBride sat the seat of a buckboard as it rocked through the dense black night—or *morning*, as Price had referred to it. Rand McBride was silent all the way out, and Kittery was just as glad. All he could think about was that derringer, and he had no idea how to bring it up. He wondered if he should feel guilty for bringing McBride with him out to this vigilante gathering. After all, what if he really was the traitor? He could be the death of them all.

The road to Perry's place was in pretty good shape, so Kittery, at the ribbons, could canter the horses a ways, walk them, then canter them some more. But even making good time this way they reached the ranch house ten minutes after the appointed hour. There were numerous other horses and rigs in sight, and a dim light shone from the house. A curl of smoke that appeared silvery in the starlight filtered from the huge stone chimney.

Kittery parked the wagon near one of the ranch's round pens, fifty yards away from the house. He wrapped the reins around a corral post and climbed down to help the half-crippled banker out of his seat. They crossed a porch that extended the length of the ranch house and opened a screen door to step inside. The inner door was already open.

A man Kittery didn't recognize jotted their names down on a pad of paper, and he and the banker stopped to glance around. Faces turned to them from around the fireplace, eyes staring from dark hollows, occasionally highlighted by the flickering fire. Kittery caught the sheen of gunmetal close by every figure. Most of the men were either squatted on their heels or seated Indian-style on several braided throw rugs. Six had arrived early enough to procure a chair.

Zeff Perry stepped from the shadows, a dark bruise on his left cheek. His jaw also showed the stains of a blow it had taken from Kittery's fist. Kittery felt a twinge of conscience. He had apologized for his grave error in judgment, and Perry hadn't been interested in pressing charges against a fellow vigilante, but Kittery couldn't forget his bad mistake.

"Sorry for the lack of chairs, boys," said the rancher. "You'll have to make do."

McBride nodded and started to sit down when one of the seated men got up and stopped him, offering him his chair. As for Kittery, he remained standing, wondering if there was anyone left here who would ever want to offer *him* a chair, after his show in the cantina the other night.

Kittery watched the white-haired rancher walk away, hoping he would forgive his rashness someday. As for Miles Tarandon's forgiveness, he knew he had just as well wish for a friendship with Baraga. The deputy's pride had been wounded, and he wasn't likely to forgive or forget. He, too, had declined to press charges. But that was *only* because Kittery was a vigilante. Tarandon hadn't spoken to Kittery since the incident.

When talk died down, James Price raised up from his horsehide chair and lit a tallow candle, which shed scanty light across the room. It was enough for Kittery to see Deputy Tarandon across from him, his face wreathed in bruises. He had to hold back a smile. He and Tarandon had always walked on thin crust together. Now they were sunk in the center of the pie.

James Price's eyes traveled over everyone in the room before calling in the door guard and beginning to speak. "Does anyone know where Thaddeus Greene is? He said he would be here."

No one answered for a moment. Then Tarandon spoke up, his voice sardonic. "If he has any sense, he's home guardin' his wife."

Kittery tensed and forced himself not to look over at the lawman. When he looked down at McBride, the older man was watching him, waiting for a response. That made it obvious. Word had already found its way around.

James Price brushed off the import of Tarandon's comment. "Well, he's never done much for this committee anyway, except to stir up dissension among us. We're better off without him. He was told about this meeting far in advance."

From the back, a voice rose. "We'll sure miss his glowing presence." This brought a trickle of laughter to the room.

James Price held back his smile. "Some of you already know why this meeting was arranged, but for those of you still in the dark—" he smiled at his own humor "—Zeff Perry has lost a few of his horses over this way in the past months. Since we all joined this committee together, I believe it's time we stuck up for one of its members. That's the first reason we're here—to plan strategy.

"Secondly, the committee has grown so much in the last short week that I'm afraid half of you don't know who's a member and who isn't. We'll take care of that today."

The hangman had each of the members stand and introduce himself, and Kittery was surprised at the faces he didn't know. It was pretty plain the lawlessness of Arizona had a lot of people upset. He had a feeling some things were soon going to change.

In his head, Kittery went over the names of the new members, wanting to recall everyone as best he could later. There was "Tick" Bell, a miner by occupation; Frederick Ridgon, another miner; "Tule" Simpton, with whom Kittery had hauled freight for Tucson's firm of Tully, Ochoa and Company; Dabney Trull, the half-breed Mexican who had dug Joe Raines's grave; and last, Jake Long, a ranch hand on Zeff Perry's Double F ranch.

When everyone had introduced himself, Price concluded that part of the meeting by welcoming Kittery "back to the fold." Then he returned to the main part of the meeting—deciding what to do about Zeff Perry's disappearing stock. Kittery noted that now Price's eyes kept returning to rest on him.

"We have two theories on who's in charge of the horse thieving operation up here. Perhaps both are based in fact. Kittery, you'll be interested in the first one: there have been two reported sightings of Ned Crawford in the last two weeks within a ten mile radius of this ranch house. They tell me you came to Castor hunting him."

Kittery's heart raced. He had nearly forgotten Ned Crawford after the deaths of his two friends, but the back-shooting woman-killer was always somewhere in the back of his mind. He never intended to totally forget him.

"The other theory," went on Price, "evolved when we found out about Sloan and Denton hanging around. It could be them, but it doesn't match their *modus operandi*—their usual way of doing things. They usually work on a much bigger scale. But you can go with either theory. Perhaps they're all in it together. They say Crawford works with the Desperados sometimes. Incidentally, Amayo Varandez was seen with Crawford, for those of you who know the name."

Kittery recognized the name as that of a notorious Mexican bandit worth more to the government than Crawford because he had killed three good lawmen in his career.

Price remained silent for a moment, and it was Kittery's voice that broke the silence. "I have a tidbit that might mean somethin' to you, Price. I just come from the Baboquivaris. I found the tracks of two horses goin' in and the same two comin' out within a short time. If it was Sloan and Denton's tracks goin' back from Castor, they're gone again. This time toward Tucson."

Price raised his eyebrows. "Very interesting, Kittery. Anything else?"

"Yeah. Something even more interestin'. I found a blood-soaked bandage and then ran into a sheepherder on the Camino Real that saw the breed. He was draggin' Sloan on a litter headed toward Tucson. As much blood as he lost, Sloan could be dead by now."

"Now, that *is* right interesting. Maybe the two of them had a falling out with Baraga. I see someone here has done his homework." Price seemed pleased. "Maybe we'll send someone up toward Tucson. But in the meantime there's Perry's problem. I take this news of Denton and Sloan to mean that our horse thief is likely to be Ned Crawford, since horses have come up missing as late as yesterday. Be that as it may, it must be stopped. What I want is to appoint a man who will be a good leader, who has experience commanding men, to choose and lead a force into the Sierrita Mountains. That's the rustlers' apparent refuge."

Price's eyes swung again to Kittery. "I think Captain Kittery is our man. He is by far the most zealous among us, as you have seen . . ." He glanced at Perry and Tarandon. "And, too, he came to Castor to hunt Ned Crawford. I think he should have his chance to collect that reward. Kittery, will you accept?"

Kittery had a feeling this was Price's way of insuring he stayed with the vigilantes this time. The opinion of the crowd was revealed when the majority of them agreed with the appointment. They might harbor personal grudges against him, or think him a man of low morals because they believed him to be consorting with a married woman, but apparently they retained respect for his leadership. Or else they just didn't want the responsibility themselves. Most of them turned their attention to him now.

"I'll take the appointment."

Price let a half-smile part his lips. "All right. I want you to choose some of the men—six or seven, at least, to accompany you."

Kittery took the reins handed him and glanced over the gathering. He ticked off the ones he wanted with him and added them to a growing list in his head.

"All right," said Kittery at last. "I made my decision based on who has liberal workin' hours and not too many obligations. I also figured on six I know can handle themselves in a tight situation. So right off you qualify, Perry. And Bacon, you and Long work here, and you're handy. You have more of a stake in this than most, so you're both in. Adam Beck, of course. Then Bell and Ridgon, if you can ride along. I added a couple more in that I'd like to see ride with us if they can. Trull and Simpton, if it fits your schedule you're welcome."

Kittery turned to Price. "Now, beggin' your pardon if I seem to be takin' over—"

"Don't fret about that. As far as this operation is concerned, you are in charge."

"Good. Then I'd like to start tomorrow. I have some pressing business to tend to in the Baboquivaris when this is over. We may as well make this our meetin' place, Perry. I'll be here at dawn. All of you bring a stout rope and a good rifle. A couple pistols might come in handy, too, if you have them and can use them. And bring your best horses. If you don't have one I'm sure Perry would be more than happy to loan you one. And last, if you can't make it before the sun comes up, don't bother to come."

When Kittery was finished, the meeting broke apart. After Kittery helped McBride onto the wagon seat the banker turned and peered at him through the starlit darkness. "These men respect you, Tap, my boy—despite your little run-in with Perry and Tarandon. I think you'll do wonders for this outfit. Except I don't think Greene likes you much." When Kittery made no reply, he went on, "Listen, Tap. I trust your judgment, but it's all over town that you went and saw Maria Greene last night. That reached me before you got home. You'd do well to watch your step."

Kittery's stubborn streak told him to keep silent, so he just nudged the horses forward. The other vigilantes were riding away from the barn and corrals, too, filtering into the darkness. Hoof falls, the creak of saddle leather, and the jingle of bits and spurs were the only sounds that marked their passing.

And so the cleansing began.

The seven chosen ones met at dawn the next day, and with bedding and supplies tied behind their saddles they rode across the creosote flats toward the gray peaks of the Sierrita Mountains. The Sierritas were a small range running eight or ten miles from north to south and perhaps five from east to west. On the northeast flank, the range's highest point, Samaniego Peak, thrust up to nearly six thousand feet, its profile not so dominating as the bulk of Baboquivari yet still impressive in its insurmountable ruggedness. Like the Baboquivaris, the Sierritas consisted of granite and many layers of sandstone and limestone that appeared almost black from a distance when the sun was high. At the moment they were gilded by the early morning sunlight.

The vigilante band camped the first night at the base of sloping Horse Pasture Hill. In the dawn they moved out once more, eating jerky and cold biscuits as they rode. They crossed sign of ridden horses only once that day, and this disappeared in the gravel and sand of the lower, wind-swept slopes.

Three days later the vigilantes, moods sullen, returned to the Double F ranch. Frustrated by their lack of success, some talked of giving up the chase. But Kittery and Perry grew more determined than ever to see the hunt through to its end.

After supper, Kittery and Adam Beck rode back to Castor with Tick Bell and Fred Ridgon. For the first time in days they were able to relax and get a decent night's sleep.

Then things began to stir. Word came Saturday night that the rustlers were on the move again, and Sunday at dawn nine vigilantes hit the trail. North of Horse Pasture Hill, their search was over.

They rode in single file, almost as if in formation. Lafayette Bacon was in the lead, trying to decipher a trail they had cut an hour before. At eleven o'clock, Jake Long drew in his buckskin, jabbing a finger ahead of him.

"Take a look at that!"

Three horsemen had just boiled from a dusty draw two hundred yards distant, spurring away from the hunters. Topping out above the draw where the riders had been concealed, Kittery and his men looked down on a makeshift brush corral containing six head of horses.

Zeff Perry must have recognized the animals, for he spat a string of curses at the fleeing riders and laid spurs to his chestnut. The animal sailed out over the lip of the draw and charged down it in a roiling cloud of dust and flying rock. Somehow Perry managed to keep his seat, and the horse kept its footing to the bottom. They went up the other side of the draw and galloped along the dusty trail left by the three outlaws.

Kittery and Satan went over next, and the other seven streamed behind, faces grim, leaning back hard in their saddles. On the other side of the draw, in a long gallop, the slower animals dropped behind. Perry, Kittery, Bacon, Beck and Long surged ahead.

The bandits, realizing the desperation of their situation, pushed their mounts brutally to keep their lead. But the terrain wouldn't allow much speed. It was cut by brush- and cactus-choked draws and washouts, and one mistake could mean a man's life.

If the outlaw trio could reach the higher mountains and a place worthy of concealment, they might make their escape. They retained a good lead, although Perry and Beck and Bacon had closed the gap. Even Kittery had fallen back, his horse hampered by the big man's weight.

The rear outlaw, trying to negotiate a narrow corridor of cactus and rock, misjudged the distance between himself and a mighty saguaro. His horse made it, but an outstretched arm of the cactus seemed to reach out and sweep the man from his saddle into the brush.

Pushing to his feet, the outlaw sprinted for a brushy washout as Perry's horse bore down on him. As he reached the cut, Perry's horse flew by him, its shoulder slamming into the man's back. The outlaw sprawled in a cloud of dust to the bottom of the washout.

Perry's horse nimbly leaped the wash. On the other side they whirled. Perry, seeing the man lying stunned, turned and spurred the chestnut after the remaining two, and Beck followed suit.

Bacon stopped first where the outlaw had risen to his hands and knees, trying to regain his wind. Kittery galloped up and reined in, the dust swirling around him. Ridgon and Jake Long were the last to stop as the others leaped the wash and continued after Perry.

The downed outlaw managed to recover his senses. He struggled up and made a limping run down the draw. Kittery spurred the black down into the draw and cut off his escape. The outlaw went for his gun, but Kittery jerked his first.

"Drop it!"

The rustler looked around him at the bristling weapons and complied. He let the Smith and Wesson he carried fall into the sand. Without any order to do so, he climbed up the bank, digging his fingers in to get a purchase. Kittery guided the black to an avenue up the side, and they scrambled to the top.

When Kittery got close to the outlaw, he saw that the side of his face and neck and his right arm and side bristled with thorns. Blood oozed away from some of the wounds. The vigilantes remained mounted, their horses blowing and stomping their feet. Kittery's Remington remained in his hand, but it hung along his thigh.

"Where's Ned Crawford?" he barked, trying to catch his breath.

"Gone," rasped the shaken outlaw. His face was twisted with pain. "He rode out yestiday mornin'."

"Rode out where?"

"Don' know that."

"Tell me or die," warned Kittery, his voice soft, matter-of-fact. As he said the words, he knew he would have to back them up.

"I don't know!" the outlaw yelled with a grimace of pain.

Kittery nodded, his eyes hard. His glance flickered over at Long, Bacon and Ridgon. His chest had drawn up tight with what he knew was coming. He couldn't meet the others' eyes. He wanted to tell them to kill this man, but he couldn't. He had taken responsibility for this hunt. It was on his shoulders. He couldn't expect these men to do something he couldn't do himself. He had to see that justice was done. He knew Jake would never have tried to put such a horrible load on someone else's shoulders to bear.

Tappan Kittery had killed in the war. It wasn't so hard. He had killed many times. He had killed Rico Wells and Shorty Randall, too. Killing was in his blood, he told himself. But all of these thoughts rang empty in his head. He hated killing, hated it passionately. The war had numbed him to pulling the trigger on a man, but the thoughts of grief still snuck through from time to time. Why did it have to be this way? He wanted to scream out. He wanted to run. But he didn't. Jake was there steadying his hand.

Cocking the Remington, he raised it and shot the outlaw twice in the chest, grimacing like he was trying to swallow a bitter pill. The impact of the slugs threw the outlaw back into the draw, and Kittery watched him go down, the side of his mouth twitching. He spat, trying to rid himself of the taste in his mouth. Then, without looking at the others, he whipped the black about, crossed the wash, and rode on at a gallop after the distant riders.

The other men looked down at the dead man, their eyes full of shock. They glanced at each other, then in silence pushed on, no longer so full of bloodlust.

Ahead, Zeff Perry's gelding began to slow, pulling hard for air. Adam Beck gradually drew his own chestnut abreast of Perry's. He realized he was close enough to try a few shots at the outlaws. As he eased past Perry, whose curses colored the air, he drew his Army Colt and squeezed off three shots, his aim appearing almost casual. The gunfire seemed to excite his horse even more. The animal laid on speed enough to pull several lengths ahead of Perry.

A puff of dust lifted from the rear outlaw's coat at Beck's fourth shot. He left the saddle in a sidelong pitch, lighting in the brush and rocks with a crunching sound. He careened to a halt fifteen feet away in a nest of prickly pear.

Beck and Perry thundered by, and Bell, Simpton and Trull trailed fifty yards behind. They gave brief attention to the body in the brush, but there was no reason to look back.

Thirty yards ahead of Beck now, the final outlaw chose in desperation to turn and fight. Lather soaped the sides of his horse, and it heaved for air. Laying back on the reins and throwing his weight back in the saddle, he skidded the bay to a halt nearly on its haunches. He wheeled it around and pulled a Navy Colt pistol from his belt.

By the time Perry and Beck could bring their animals to a halt, they were almost on top of the outlaw, in the center of a cloud of swirling dust. Beck brought his horse under control, and as the outlaw's first shot whizzed past, missing him by inches, he pointed his Colt and fired. The bullet slammed into the spine of the outlaw's bay behind the saddle. Its hindquarters buckled and lurched sideways, throwing the outlaw.

Beck spurred in closer, yelling curses and orders at the outlaw. On all fours, the outlaw groped in the brush for his pistol. Perry yelled, a strange tone to his voice, "Leave it, Bandy! You don't stand a chance!" He became furious when the man continued his search for the pistol. "You gutless bastard!" he spat, looking over at the bay struggling to rise. He cocked his pistol.

"Perry!" Beck yelled at the same moment Perry's gun spat smoke and fire. The outlaw's horse lowered its head, shuddered and ceased to struggle. Beck gave a sigh when he saw Perry didn't mean to kill the outlaw. But he was surprised. There had been something deadly in the rancher's eyes.

"Help your friend up, Perry," Beck said. "Kittery's gonna want to talk to him."

Bell, Trull, and Simpton reached them and gathered their heaving horses around. "Looks like the captain's a-comin'," Trull said, pointing back down their trail.

"Yeah," Beck said with a nod. "Better let him handle this."

The outlaw, Bandy, had risen to his feet and stood haggard and spraddle-legged under the harsh gazes of the vigilantes. He looked from one to another, finding no friendliness. He looked at his dead horse and wiped blood from a cut on his jaw.

"How do you like workin' for a backshooter, Bandy?" Perry mocked. Bitter anger was undisguised in his words.

The husky outlaw's eyes narrowed. "You ain't fit to lick Crawford's boots. None of you. Him and Varandez'll see you all dead."

"They may, but for now . . ." Riding in close, Perry untied his reata from the saddle horn. Before the outlaw knew what he was about, he lashed him across the side of the head with it.

Dazed, the outlaw rammed his eyes shut and staggered to the side, clutching his head and almost going to his knees. He cursed the rancher. When he regained his senses, Kittery was still two hundred yards away.

Bandy wiped blood off his ear and spat, looking up at Perry. "You thievin' hypocrite. You're worse than Crawford. At least he ain't tryin' to fool his friends. You tell 'em 'bout Baraga?"

Without warning, Zeffaniah Perry raised his pistol and shot Bandy in the stomach. The outlaw sat down hard, eyes shocked, mouth hanging open. The others were as shocked as the outlaw, and no one made a move. Perry spurred close, cocking his pistol and firing in quick succession. One slug pierced Bandy's cheek while two others shattered some ribs and his collarbone. Bandy rolled over, kicking. He clutched at a clump of grass. In seconds he released his grip and was dead.

Kittery galloped up and sawed the black to a halt, covering the violent scene with a coat of yellow dust. His eyes jumped from one to the other of the vigilantes and to the dead outlaw. "What happened?"

No one spoke for a time. They all looked at Perry. Perry fumbled his pistol into its holster and swore. "That man used to work for me. Name's Pete Bandy. I killed him, Captain. He started spoutin' off at me. I lost my head and killed him."

Kittery looked around at the others, and none of them seemed willing to meet his gaze. The only sound was the labored breathing of the horses, and at last the growing rumble of the last vigilantes galloping in.

"We used to be friends," Perry managed to say, staring at Bandy's dead horse. "I thought so anyway. He took to stealin'. I fired him and told him to get out of the country. I never thought . . ." He shuddered. "We used to be friends."

Kittery let out his breath in a gusty sigh. He massaged his face with a gloved hand and looked around once more at the others. Beck was the only one watching him, and he looked away.

Zeff Perry climbed down and managed to unsaddle Bandy's animal by using his lasso and his own horse to jockey the dead horse around on the ground. He took the bridle and the outlaw's belt, boots and spurs while the others watched, bewildered. Perry found Bandy's pistol in a bush and stuck it in its holster. Finding nothing in the dead man's pockets but a locket, he gathered all of this plunder together and rolled it up into Bandy's blanket. Holding the bundle in one hand, he grasped his saddle horn and pulled himself with considerable effort into the saddle. Ignoring the looks of the others, he glanced over at Lafayette Bacon. "Do me a favor, Lafe, and hand Jake up the saddle and bridle. Pete's mama's bad off. She could use any money we c'n get from this stuff."

Kittery nodded his approval. He watched Bacon obey, and on the way back they did the same with the other outlaws and rounded up their loose horses, herding them along before them.

Halfway back to the ranch, Adam Beck dropped back to where Kittery rode at the rear of the group. For a long time he rode beside Kittery without saying a word, and Kittery could see the lines of concern etched in his face. He waited, confident Beck would work up to telling him what was on his mind.

"I reckon I oughtta tell you somethin', Tap. I debated about keepin' my mouth shut. But you oughtta know."

Glancing now and then at Zeff Perry, who rode at the head of the troop, he related what had passed between Perry and the dead outlaw just prior to Bandy's death. He remained silent for a long time, then— "Perry's a good man, most times. I trust him—or did. But . . ."

"Yeah, I know," said Kittery. "I'm glad you let me know."

Afterward, Kittery rode on behind the other vigilantes, who strung out in a line forty yards long. He pondered Beck's news. It appeared he had no other choice but to shift his suspicions back to Zeff Perry. Considering what Pete Bandy had said he had to suspect once more that Perry was the man they called the Mouse. In fact, it seemed a sure thing, except for the fact that the only proof they had now lay in the words of a known outlaw—and a dead one.

Kittery found it strange how fervent Perry had been in his desire to find these outlaws today. Perhaps that was because this operation itself had nothing to do with Baraga.

Kittery had done things he was ashamed of, things he didn't want known. But as for Perry, was it a shameful past he wanted to hide, or was he trying to mask his treacherous present?

The sickening thought didn't come to Kittery until they came within sight of Castor: considering the derringer and the comment made by Bandy, what if there were more than one "Mouse?" What if Perry and Rand McBride were in on it together?

Chapter Twenty-Four
The Broken Holster

The following morning, Tappan Kittery came in sight of the Perry ranch with a soft gray light in the east. He rode a gray gelding. After yesterday's heat he didn't want to be out there at all. Cactus wrens chattered raucously among the desert scrub. At the porch three horses stood saddled and ready, rifles in their scabbards.

Kittery, as always, was well armed. The Springfield carbine was a solid lump against his inner thigh, and the .44 Remington nestled in the black holster on his right hip. Today he also carried Cotton's Colt and the .44 Smith and Wesson American he had purchased at Hagar's store. The Colt was in his saddlebag, and the American sat in a belt and holster rig he had fashioned himself, except for its iron buckle, which Bison Sabala had forged for him. The belt and holster were plain, unembellished cowhide stitched with an awl and heavy thread. He wasn't satisfied with the stitching in the holster, but time had been short. He intended to amend the holster's faults later, but for this hunt it would have to do.

Lafayette Bacon stepped out the door wearing a fringed deer hide jacket that hung to mid-thigh, gloves and a large scarf drawn tight against the cool morning air. He toted a Remington rolling block rifle.

"Mornin', Captain," Bacon drawled. "Just finishin' breakfast. The others'll be out di-rectly."

After a few more moments Jake Long and Zeff Perry emerged from the house together, rounding out Kittery's lean force. They mounted and headed due west.

As they rode, Kittery studied Jake Long. He didn't know the man. Long wasn't inclined to talk much to anyone, and he had never exchanged a sentence of meaningful conversation with him. But Kittery's instincts told him the man would be a good hand, a man of strength and courage.

The puncher had long since passed his high-water mark of forty years, and he wore his shaggy dark hair well past his collar. Long's face was alley dog-thin, his neck wizened and scarred by brushes with too many thorns on too many cattle chases through the *brasada*—the southwest brush country. He wore his pistol high on his hip, but his veiny hand rested on its butt with telling familiarity.

Bacon, too, Kittery had a good feeling about. He was younger than Long by a few years, but old enough that he, too, had fought in the war—a rebel from out of deep Florida. Bacon was under average height, but compact and full of energy—always ready with a smile but quick to defend an ally.

Then there was Zeff Perry. He, too, had always seemed to Kittery like a capable, tough man. Yet a shadow loomed over the rancher that Kittery couldn't erase from his mind.

The morning wore on. No one had spoken much since the ride began. Kittery was the only man there who knew the day's destination. No one broached the subject, and Kittery didn't feel the need to reveal it yet.

Around ten o'clock they neared the makeshift brush corral where they had jumped the outlaws the day before. When Kittery pulled up at the draw, the others milled around him.

"This is where the work begins," he said. "You boys cast about for sign—tracks of other horses besides the three we jumped yesterday. I have a hunch Ned Crawford was a lot closer yesterday than we knew. That's the way I want to play the hand, anyway." He pointed toward the mountains. "If we can find a trail of one or two horses headed up there, I bet we find Crawford and the Mexican."

They searched the ground for half an hour, circling ever wider after tying their animals a good distance back. It was Lafayette Bacon who stopped and called the others to his position a hundred yards to the west of the makeshift brush corral. Up to that point the ground had been clear—obviously cleared by hand, for here a profusion of fresh tracks appeared, some of them unshod, two of the shod ones being ridden.

Kittery looked at the tracks for a few moments, then smiled grimly at Bacon. "Good work, Lafe." He threw his glance at the others. "I'd say there's at least ten head here, countin' Crawford and the Mexican's. That'll slow 'em some."

Perry sat in grim silence until a curse broke from his lips. "What're we waitin' for? Let's ride!"

They went to their horses. As the day before, Perry took the lead. He pushed his bally-faced sorrel without a glance back at his riding mates.

They made their dinner at the base of the Sierrita Mountains, nooning on sandwiches of beef and hard dry cornbread, then moved on. In the lower Sierritas, tons of loose gravel and soil buried the bedrock, hosting the usual melange of desert plant life. But as they climbed they watched the terrain grow ever more rocky and windswept, and sign became more scarce as they rode on. Progress slowed to a crawl as they studied the granite bedrock for ever-diminishing tracks.

They were approaching the foot of a giant monolith now. It towered one hundred feet or so above them, thrusting solid and bold into the skyline, a more-or-less sheer wall of salt and pepper granite.

Kittery's worst fears came to pass here. The tracks separated and continued on in two directions. One ridden horse went on with each half of the herd. Kittery pondered the trail ahead, and what this new development meant. There was no way they could tell which horse belonged to Ned Crawford. To Kittery, the paramount goal was to get that man. The Mexican was an added bonus. So the hunters would be forced to split up, too.

That would leave one man alone with Zeff Perry.

They climbed down to rest the horses, and everyone rolled cigarettes and squatted in the sand, pondering the turn of events. Kittery thought about Long and Bacon. What were those men like? Were they square? After all, the two of them worked with Zeff Perry—what if they knew everything about him? What if, after all, Kittery was the lone man of integrity here? That answer was simple: he would be dead.

But he must take his chances. He couldn't turn back when they were so close to Crawford and Varandez. The outlaws could be miles and miles away, in reality, but they sure weren't going to get any closer if Kittery got nervous and called off the hunt.

Kittery spat and looked at the others. "Well, we can split into pairs and keep following, or we could pick one path and stay together. On the other hand, we could call it quits and go home. Those horses could be seventy miles away by now."

Zeff Perry shook his head. "That last choice *ain't* a choice, for me."

The others didn't speak, but by their looks Kittery knew they were true riders for their brand. "All right, then," he said. "I hate to do it, but we can't afford to go after just one of them, in my opinion. So we'll take 'em both. Chances are they're tryin' to throw us and they'll meet back up somewhere ahead. Perry, I'll send you and Bacon on that fork." He pointed to the right one, which headed north. "Long and me will head south. And if it looks useless, don't hesitate to call it off, because I will. If you come back this way, either try to follow us up or just head on back to the ranch.

"And one more thing: if you've forgot what it's like to be in a war, this is it. Maybe you've forgot what it's like to have prisoners to drag around. I'm askin' you not to take any. Don't give these men a chance to surrender," said Kittery, forcing the feeling of guilt out of his mind. "Kill them when and however you can, because they will you."

Perry and Bacon rode north, and Bacon tugged off his gloves and laid them on the saddle to scratch at a trickle of sweat that ran down the back of his neck. He studied the rocks around them, feeling the heat of his saddle, wishing he had left his chaps at home. They hadn't been riding in the brush much, anyway, and this wasn't a day to be wearing any extra skins over his pants.

When sign became hard to find on the rocks, they could still see a light-colored hoof scar now and then. Fortunately, there was only one way the horses could go. Sharp limestone ridges lifted up on both sides of them, forming a narrow, quiet canyon. As they went they noticed the trail begin to curve to the west, through a maze of granite walls and eroded limestone upthrusts. They had gone no more than three miles when they saw the smoke.

It lifted up maybe two hundred yards ahead, one long, dense white cloud. And then nothing but a wisp. Bacon looked over at his boss.

"Could've been a smoke signal," Bacon suggested. "Some Apaches that snuck off the reservation."

Perry took a deep breath. "Naw. Too short. It looked more like somebody just dousin' a fire. Could be Crawford or the Mexican."

The rancher glanced to the left, where the limestone rocks leaned against one another in jagged disarray, rising up with oak and juniper to meld with the sky. "I bet we could climb up here. Maybe see down in where the smoke came from. We might even get a shot from up there."

Bacon studied the treacherous-looking rocks and made the imitation of a shrug with his eyebrows. "Yore the boss."

They had to tie their mounts to a juniper tree down below. Then Bacon pulled his Remington rifle from its scabbard, and Perry shucked his Winchester. Together they began the perilous ascent through the rocks and brush. Bacon was surprised to see the climb wasn't as bad as it had looked from the trail below. It took no more than twenty minutes to make the hundred feet to the top. He looked for rattlesnakes at every step, but gladly they encountered none. Once on top, they found a flat of barren rock, with one ratted juniper growing to the left and a little bunch of piñons on the far side. They stalked across the flat, crouched low.

The sudden, sharp crack of a pistol startled them. Its echo slammed into every slab of stone before rocking off into nowhere, and for several seconds neither man dared move. At last they started forward and reached the edge of the rocks on hands and knees.

A sheer cliff greeted them, a cliff that fell for a hundred feet, disappearing into a mass of shattered and wind-beaten boulders and shale. Dropping onto their stomachs, they peered over into this rock quarry. Every side was barricaded in by cliffs, with exits from the rocky prison but few.

In the center of the sun-bleached canyon, sprawled over the remains of a fire, lay a man wearing flare-legged *calzoneras*—Mexican style pants with conchos down the sides—and two gun belts. Another man stood over him and kicked him hard in the ribs, but the man over the fire didn't move.

"Crawford," whispered Bacon. "That's gotta be Ned Crawford."

Perry nodded agreement. "So the other one must be Mayo Varandez."

Bacon grinned to hide his nervousness. "Yeah. Looks like he's bein' roasted. Wonder if there's enough to go around."

Perry's face was stiff. There was no sign that Bacon's humor had registered.

Down below, Crawford opened the loading gate of a nickel-plated pistol. Bacon bored the man through with his eyes, knowing he should shoot him. There was no doubt the man was Ned Crawford, yet somehow he looked different than his descriptions. He was supposed to be a coward, a deserter, a backshooter . . . a woman killer. But he didn't look the parts. This man was tall and straight-backed, broad across the shoulders and narrow in the hips. His countenance appeared dignified, with black hair and beard both streaked with white and curried, giving him the look of a gentleman—a businessman. He wore a sky-blue shirt tucked into tight black pants, which in turn were tucked into beaded white deer hide leggings.

At the same instant that Crawford finished pushing the last of five cartridges into his cylinder and clicked the loading gate shut, Lafayette Bacon cocked his rifle and rested its barrel on the edge of the rock. He aimed at the outlaw's midsection. His voice rang out loud and clear in the rock quarry, and he was only faintly aware of Zeff Perry's curse.

"Crawford! Throw down your weapon! We have a bead on you!"

Cat-like, Crawford whirled. He drove in a running crouch for the base of the cliff where the two vigilantes were hidden. Bacon's bullet split the air where the outlaw's chest would have been.

Bacon cursed himself, unaware of the fact that Perry was doing the same. "I should have handled it, Lafe! Why'd you yell out? If you'd shot him . . ." Perry cursed, and Bacon just ground his teeth together and was silent.

"We gotta do like the captain said in the first place," said Perry. "No more quarter. I don't aim to miss, Lafe. Once is too much for that."

Bacon ignored the biting comment. Swallowing hard, he glanced over the edge of the cliff. "Maybe you oughtta wait. A shot'd be perty risky now. You'll have to lean over here. He'll have the same chance as you."

"Of course he won't," Perry gruffed. "He's only got a pistol. I have a rifle!" The rancher's voice softened a little when he spoke again. "Don't worry, Lafe. It ain't your fault. It's hard to kill a man with no warning. But now he's as good as dead."

Bacon gazed at him for several seconds, then sighed. "Be careful."

"Of course."

The rancher crawled farther along the edge of the cliff away from Bacon. When he had chosen a position, he looked over at Bacon and gave him a wink. Then he leaned over the edge.

The look on Perry's face was that of a confident man. But it disappeared the instant his eyes went from the sun-shattering rocks of the quarry to the shadows below the cliff. Bacon watched the transformation, and he thought Perry would pull back. Instead, he began to swivel the rifle barrel, searching below. He sucked in a breath and settled into his position. He had spotted his quarry.

The crack was loud in the confines of those towering rocks. It wasn't the crack of a Winchester rifle but of a pistol from below. Perry's head snapped up, and Bacon saw droplets of blood fly away from him. He let go of the rifle, and it sailed to the rocks below. Then Perry, on nerves alone, tried to rise from the edge of the cliff. He stared into space with the look of a man who no longer has control of his senses and pitched over the edge of the cliff.

Lafayette Bacon could only stare at the place where a matter of seconds earlier his friend and employer had lain in ambush, confident in his ability to bring the enemy down. Bacon's mouth hung open, and his rifle lay on the rocks where he had dropped it.

After fifteen seconds, Bacon struggled to a sitting position, still staring at the place where he had last seen Perry. Glory, how fast it had happened! A healthy man, an able man, dead in the blink of an eye!

Numbly, he looked down at his Remington rolling block. The urgency of the situation came home to him like a bullet, brought on by the rifle lying there. He picked it up as his face drew into tight lines, and pushed to his feet. His heart was pounding and his ears were ringing as he started back down through the rocks the way he and Perry had come. Crawford had no way of knowing if anyone else had been on that cliff with Perry, so he wouldn't dare leave his shelter yet. There Lafayette Bacon would find him.

Still pressed against the cool rock of the cliff shadows, Ned Crawford peered upward at the skyline. He searched the cliff's edge for sign of other hunters who might lie there in wait. He cursed and looked around him at the glaring rocks, then at his dead companion on the fire. Stupid Varandez! The greedy Mexican had wanted everything for himself. Crawford had sensed it. He had only done what the Mexican would have done had he been presented the chance.

A fly buzzed his face. A distant ring from the gunfire tormented his ears. He wiped his palms on his pants one after the other and looked up again at the edge of the cliff. "Hey! You up there! We're in a stand-off here! No need for anyone else to die!"

Silence answered him. Not even a breeze disturbed the day's utter quiet.

"Well? What you wanna do? It's nice and cool down here. You'll bake in the sun where you are."

Crawford's black eyes explored the cliff's ragged edge. Nothing. No one in sight. "There's no one up there," he growled at last. "He was by hisself." He glanced out into the sunshine again, then took a look toward the cliff. If anyone had been up there they would all have fired at him. But . . . what if? He wasn't a man known to take foolish chances. He had to make sure before he dared expose himself in the sunshine again. With his Remington pistol still in hand, he started toward the south end of the cliff, his eyes scanning above.

It was some time later before he discovered a place where he could climb to the top of the cliff. And as he rounded the rocks here he heard the jingle of a bit chain. Sneaking on, he found two mounts tied to a juniper tree. Both bore empty scabbards.

"Well, now." He smiled grimly, speaking to the horses as he walked to them. "So there is another one of them up there after all." He walked up and patted the two horses on the neck. Without another look, he started up through the rocks.

It took twenty minutes for him to reach the flat top of the rock, his breathing ragged. As he came up, he cocked the Remington and stole over the edge of the rim to lay eyes on the naked tabletop of rock. He looked around in amazement, giving the one juniper and some sparse piñons on the far side of the flat a moment's study before deciding they were too meager to provide a hideout for anyone.

His spine straightened like a rope drawing taut as he came out of his crouch. He grunted and swore, shoving the Remington back into its holster and walking to the edge of the cliff to look over. A moment's study revealed the dead Mexican and the would-be ambusher's body, dead in the rocks. Where was the man who rode the second horse? He grunted again in disgust and turned to make his way back down to his own horse.

Crawford's jaw slackened, and he felt like he had been punched in the stomach. His eyes widened. There, on the far edge of the plateau, like an omen of death, stood a big man in a blue army shirt and striped gray and white pants.

The big man caught Crawford's eyes flickering toward the revolver on his hip. "Crawford! Touch the gun and you'll die."

The outlaw licked his lips. They were so dry! He stared at the right hand of the big man, who had his left side quartered away. The man's right hand, like his own, hung loose and empty near his pistol.

A slight smile came to Crawford's wide mouth. Few men could beat Ned Crawford to the draw. Again he licked his lips. "Mister, you're a fool."

He sneered, and the fire came up from deep in his heart. He grasped the butt of his Remington and began to draw it free.

The big man turned in a fraction of a second, revealing his left hand, until then hidden. Just when Crawford knew he would soon feel the pound of his pistol in his fist, he saw the Smith and Wesson American already poised in the other man's left hand. It came up like the head of a snake and made smoke twice, and instead of feeling his own pistol pound against the palm of his hand, he felt two hundred grains of lead tear twice through his chest. He found himself lying on his back looking up at a vastness of blue sky he didn't remember ever looking so smoky before.

Tappan Kittery stood for a moment and watched the shuddering form of Ned Crawford. He lay on his back, staring upward, his hands flung out to the sides. His .44 had wobbled to a standstill three yards away.

Kittery moved forward as the last of the blue smoke swirled and faded away. He towered over the outlaw, who sent bright frothy bubbles out of the holes in his shirt with every breath. He was careful not to let his shadow fall across the man's face and give him any comfort.

"I need to know somethin'," he told the dying man, unsure if he could hear him. He knelt beside him and lifted up his head with one hand. "You know Zeff Perry?"

The outlaw's face bunched up, puzzled, but suddenly a light came into his eyes, and he seemed to look through Kittery. His voice was weak and low. "Yeah, sure."

"How?"

Crawford attempted to chuckle but stiffened with pain instead. "Used to work with me."

"You and Baraga?"

Crawford smiled, but his eyes were full of pain. "Me an' Ba—"

Crawford winced and raised up a little more. The light faded from his eyes as his head sank once more onto the palm of Kittery's outstretched hand.

Kittery let the dead man's head thump against the stone. He stood up, stuffing the Smith and Wesson behind his belt as he did so. At that moment, Lafayette Bacon charged breathlessly out onto the plateau, and the big man turned while drawing his Remington. He cocked it and aimed at the man's midriff before he realized who it was. "Where's Perry?" Kittery's voice was rock-hard.

Bacon shook his head and closed his eyes, waving toward the cliff's edge, behind Kittery. "He went over," he huffed. He looked down and gestured at the dead man. "He killed him."

Jake Long came up then, gasping like he was about to collapse. "Hard climbin', Captain." At sight of the dead man, he drew a deep breath. "Not fast enough."

"He was fast enough," Kittery countered. "But that holster I made fell apart—my lousy sewin'. I stuffed it in my saddlebag a while back and was packing the Smith and Wesson in my pants, but on the way up here I held onto it to keep it from fallin'. He was lookin' at the Remington in my holster, and he thought he could beat me. He never had a chance."

Long chuckled, sucking in and letting out a deep breath that this time was more like a sigh. "Any way that works. Just be thankful you're a lousy seamstress."

They went down below, and as they made the descent Kittery explained to Bacon why he and Long had returned. Varandez had led them in a circle. They turned the Mexican's body over out of the fire.

"Varandez," Kittery observed. In the man's holsters were two well-kept Colt Peacemakers. "Didn't get a chance to use 'em, did he?"

"Nope," Bacon agreed. "Shot in the back. Them posters didn't lie about Crawford."

Long nodded. "Yep. Backshooter."

"Sometimes a shot in the back is better, if you can overlook the bad reputation it might get you," said Kittery. "Safer, anyway. I admit I've made a couple myself, when the odds were down."

It was only then that Bacon admitted how his boss had died. Kittery could tell the brush popper blamed himself for it. He nodded when Bacon finished.

"Don't get yourself in a lather over it, Bacon. I know you and Perry were close, but I have to tell you somethin'. Crawford as much as told me he and Baraga were workin' together with your boss. Sounds like he was most likely our leak in the vigilantes. At least one of 'em."

Bacon and Long were shocked into silence. Kittery spoke no more either, and the three of them went about rounding up the horses and loaded the dead men on their mounts. At last, with the herd in front of them, they headed toward home. By then an hour had passed since any of them had heard another human voice.

Chapter Twenty-Five
A New Leader

For the next two months, Tappan Kittery was seldom seen in Castor. The vigilantes continued with each passing day to gain momentum, and with the share of the reward money Kittery had received by bringing in Crawford and Varandez, along with his money from Rico Wells and Tawn Wespin, he had no financial worries. He spent all his time with one vigilante force or another. The vigilantes had begun to draw recruits from as far away as Tucson, Tubac and Nogales, and on all the ranch lands surrounding Castor and Tucson, all the way to Mexico, they closed in on the outlaws like a pack of hounds on the wolf. The rats' nest had been set afire, and its occupants were dispatched one by one. No man with hard proof against him was spared. The vigilantes shot rustlers, killers and robbers or hanged them on the spot if they could find a tall enough tree. None lived to see jail bars.

Joe and Jed Reilly discovered a rustling operation on their R Slash R ranch, and, unable to fight the war alone, they were more than happy to accept Price's and Kittery's offer and join the crusade. Black Ramblene, powerful owner of the Empire Mountain Ranch, south of Tucson, also entered the ranks, bringing in a permanent crew of fifteen seasoned men. Jake Long and Lafayette Bacon, inheritors of Zeff Perry's Double F ranch, managed to keep riding with the vigilantes whenever the call came.

There were others who placed their names on the roster. Jed Polk, whose JP Ranch lay in the northern Santa Rita Mountains, brought with him the Beehan brothers: eighteen-year-old twins, Edward and Linus, and the oldest brother, Durnam, himself a reformed outlaw. An expert tracker and sometime Pima County deputy sheriff named Mort Dansing, and two gunmen staying in Tucson, John Laynee and Johnny Hedrick, completed the list of names Kittery knew well. But the vigilante numbers were well over one hundred now, and there were many he wouldn't have known if they stopped to shake hands with him.

Tappan Kittery was a leading factor in the vigilantes' success. The men came to see him as their leader, since he was the one who rode alongside them far more often than either Price or Tarandon. They looked to him to make decisions in the field. They saw in him something Beth Vancouver and others had seen—a gallantry, perhaps heroism. In Kittery, the men saw something they might have liked to be themselves. There was a certain hard-fisted drive behind that tall, broad-shouldered frame—a will to fight, to push success home at all costs. He wasn't rash and hot-headed, though he was human, as evidenced by his fight with Perry and Tarandon in the cantina. Since then he had given up drinking more than a shot of hard liquor a day. Usually it was a matter of weeks before he partook at all. The vigilantes could expect him to be a thinking man, brooding over each and every move he made. Then he never looked back, just forged ahead with all the deadly, controlled fury he possessed. He was, in every sense, a fugleman for the vigilantes, and many of them would have followed him through the gates of hell itself.

Yet Kittery was no tyrant. He rode with these men as a friend and partner. And anything he expected them to do was no more than he would do himself; he was a man who led from in front, not drove from behind. Jacob Kittery would have been proud.

To some of the younger members of the group, Kittery became something of an idol. Ed and Linus Beehan would have done almost anything he asked them. In a way, they had discovered someone to pattern their own lives after, someone who was free, who for years rode alone, unattached, bringing his own effective form of justice to the land. Something about that appealed to the idealistic, heroic image with which the boys had grown up, reading about Lancelot, Beowulf and Odysseus. Durnam was a decent brother, and a good enough man in his own right, but he lacked the charisma and the natural leadership of the bigger man. And his sidetrack into the realm of outlawry had left a negative impression on the boys only time might erase.

Of the Beehan twins, Linus was the more taken with Kittery, shadowing his every step. Someone told Kittery the boy wanted to become like him. He didn't say much about it and didn't try to stop it, though perhaps that was wrong. It made him feel good to have the boys look up to him the way they did, to think he was wanted and really needed by someone. Besides, they would change. He did not want them to become wanderers and gunmen like him, and he knew before long they would see the light. Perhaps by then he would be dead.

So through Kittery's powerful leadership, the inside-the-law tip-offs of Miles Tarandon, and the calm, sure planning of James Price, the Castor Vigilantes became the leading force of justice in the entire territory of Arizona, and before they realized it the rustling and robberies and killing had subsided. Even the mighty Desperados went into a forced—if temporary—dormancy. Any occasional outlaws who wandered in from surrounding borders, defying the power and name of the vigilantes, swiftly followed their fellows to the grave.

Kittery had never told anyone, not one single soul, about Rand McBride. Maybe he had a twisted sense of loyalty, maybe it was because of Tania. But he did not want to follow the trail of the derringer any farther, at least not yet. He didn't have the heart to challenge McBride. He couldn't help but wondering what Jake would have done. Jake was always so straightforward he would probably never have let a woman stand in the way of a question like this.

Kittery *knew* McBride though! He wouldn't be involved in something as low as treason for his own gain. Would he? But the more he thought about it the more the entire thing seemed to fit. Who would Shorty Randall have had better access to right after his escape than McBride? After all, the bank stood right next to the jail. And who would know more about financial affairs and the movement of money than a banker? It all fit so well Kittery could hardly stand to think about it. But all he had done was to tell Adam Beck to keep the banker in his sights, without telling him why. If McBride gave any overt hint of treason, Kittery would face him down. But not until then. He liked him too much, and he was too much a part of the woman Kittery still loved. If Price and the others didn't agree with his decision, which they surely wouldn't, that was too bad. Tappan Kittery had always gone his own way.

The first week in August, following an absence of two weeks, Tappan Kittery rode into town. He was hot, thirsty and tired, and gray dust hung from his three-week-old beard and rumpled, stale smelling clothes.

Without a glance around him, he rode up to Cardona's cantina and stepped down, tying the black. Besides a place to wash the trail dust from his throat, Cardona's was also a hotspot for recent news. He knew by now if any vigilantes needed him they would come here first, upon hearing of his arrival in town.

That evening, James Price made his appearance. Kittery lounged at his corner table with a warm glass of beer, crossed legs stretched full-length and feet resting on another chair.

Price sat down without comment. "I heard you just rode in." He lowered his voice so only Kittery would hear. "We're holding a meeting tonight, Kit, at two hours past midnight. There's a grove of sycamores on the right side of the road, two miles or so after the cemetery; we'll have a fire in there."

Kittery studied Price. "Sounds important. What's the deal?"

"Don't talk to anyone," Price warned. "Since your last visit to town, there's been some trouble. The Federal government has had some agent brought in—supposedly under cover. Word's out about us in a big way."

Kittery scowled but held the curse on his tongue. "Bound to happen, after the stir we've made."

"Right. We're lucky it hasn't happened sooner. One more thing: if a stranger approaches you, get out of town. I can guarantee you'll be one of their major targets. Vigilantes aren't looked upon in a kindly light by government officials—especially vigilante *leaders*."

At ten minutes before two in the morning, Kittery rode within sight of the fire, pausing under an ironwood tree whose branches still dripped now and then from the monsoon rains that had drenched the countryside all that evening. The sound of the Colorado River toads was almost over-powering, the croaking coming from every direction, and every time he moved his eyes he could see one or two more of them hopping around in the muddy sand. The summer monsoon season was the time for the little creatures to come to life, the time they lived, breathed, bred, and then returned to the mud, to be baked beneath the ground until the rainy season came again.

He reconnoitered the camp from beneath the ironwood, trying to distinguish individual faces. He couldn't hear any voices, for the toads were too loud. Satisfied that things were all right, he clucked his tongue, and Satan moved forward, halting beyond the circle of firelight.

More than fifty men surrounded the fire. They turned warily to see who had arrived. When they recognized the big man, there were smiles, respectful nods, and half salutes as he stopped at the edge of the group and stepped to the ground, lashing Satan's reins to a tree. Despite the warm welcome he received, Kittery sensed the tension in these men. It didn't go unnoticed to him that Rand McBride was not here.

Straggling groups arrived in twos and threes until a quarter past the hour. Then James Price rose, peering at each face across the flickering light, assuring himself of familiarity before beginning to speak.

His gaze fell last on Kittery, and they held each other's eyes for several seconds. With an almost imperceptible nod, Price slid his eyes away.

"I know you're all tired, and I know you're wondering what brought on this meeting in the middle of the night. Well, if we can get those toads to shut up we'll get it over with and send you along home." He gave a half smile. "I guess we'll have to go on without trying to quiet the toads. We'd be here two weeks.

"Our secrecy is for your own good tonight, believe me. For those of you who might not have heard yet, we have a leak somewhere among us. Consequently, the boys in Washington got together and had the governor send some agents into the area. They're somewhere near now, I'm sure—if they aren't among us tonight. Why they've not moved in I don't pretend to know.

"I called this meeting to tell you all that you are free to leave the vigilantes with no repercussions. In case the government hasn't yet discovered all of your names, there's no sense in waiting around until they do. After all, we've done most of what we set out to do.

"I, for one, am leaving Castor. It's not something I want to do so much as something I feel is a necessity, at least for a time. But of course, as always, there will be some of you who are more tenacious than the rest, who believe this work isn't completed and will opt to remain and face the music, so to speak. I am certain of one of these, and I will make it my last request that you accept him as your captain, as long as you continue to operate. Incidentally, most of you already know him as 'Captain'. He's Tappan Kittery."

There were nods and remarks of approval from those gathered, and Kittery glanced from them to the bearded hangman. He found it difficult to believe Price was running away out of fear, trying to escape punishment by the government—not with the undaunted courage he had so far demonstrated. There had to be something more.

"Why are you leavin'?" Kittery asked. "I think you owe us an explanation."

"So I won't go to jail," replied Price.

"You're not scared of jail. They can't lock us all up. What's the real reason?"

The corners of Price's mouth turned up in a faint smile, but his face went serious when he spoke. "I'm heading north, Kit, partly to keep out of trouble with the law. But you know I've been a bounty hunter from time to time—not unlike yourself."

The two stared each other down for several more seconds. The only sound was the grating rumble of the toads. Finally, Price dropped his gaze away. Then his eyes returned to Kittery's and he spoke barely loud enough to be heard over the croaking. "I'm going after Denton and Sloan."

"I guess not, Price." Kittery shook his head. "The Desperados are mine. You'd better stay."

"Sorry, Kit. Whether or not you take this job, I'm headed north. Don't become greedy on me."

Tappan Kittery pondered his situation. He wanted Denton and Sloan. But the men he really wanted were Baraga and Doolin. If he decided to go after Denton and Sloan, perhaps he would lose track of the rest of the band. Greedy? He had promised Joe Raines to see all the Desperados dead; well, Price was as capable of bringing that about as he. So then he would leave them to him, if that was the way it must be.

He chuckled without humor and drawled. "I guess I'm your man."

No one opposed him as leader, not even Miles Tarandon, Price's second in command. He was smart enough to realize who the real leader of the group was. Of the few men who quit the vigilantes that night, Deputy Miles Tarandon was one.

Little more was said. After Kittery appointed Adam Beck as his right hand man, the vigilantes, silent with their own thoughts and hidden fears, mounted and drifted toward their homes.

Chapter Twenty-Six
The Labyrinth

In mid-August Tappan Kittery rode into Castor from the north, flanked by the rancher-vigilantes, Jed Polk and Black Ramblene. Two wagons took up the rear of the procession, one driven by a Ramblene man, the other by young Linus Beehan.

Drawing past Albert Hagar's mercantile and hardware store, Kittery broke away from the others with a nod. Stepping down at the hotel, he tied the black in front and walked inside. Maria Greene's eyes brightened the instant she saw him. She got up slowly, hampered by the baby inside her, but when she was on her feet she came around the desk and greeted him with a bearhug, a custom between them by now. He returned it, pushing aside the twinge of guilt he had at being here with another man's wife. It was as innocent as a hug from his sister, anyway. They had always been nothing more than friends. At least that was what he tried to tell himself.

"I have not seen you in a long time," Maria said as she stepped back.

"I know—and I'm covered with dust and not even shaved." Kittery grinned. "But it sure is good to see you."

"You, also. And I do not mind a little dust."

Kittery didn't stay long. He gave himself only time enough to exchange news with Maria. And she insisted that he wait long enough to feel the baby kick inside her. That was all—that and a kiss on the cheek, after which he said so long.

He stepped next door to the cantina. Ramblene and Polk had left their wagons at the dry goods store and were now leaned up against the bar beside Miles Tarandon and Adam Beck.

Kittery spoke to Beck in a quiet voice. "I'm glad you're here. You been watchin' McBride?"

"I sure have," replied Beck. "I wish you'd tell me more about that. He hasn't done a thing I thought was suspicious."

That made Kittery happy inside, but he just nodded. Maybe if he could bring himself to tell his friend more detail he would be able to interpret some of McBride's actions.

"That's good," he said. "But I'm glad you're here for another reason, too. I wanted to ask you to take care of things for a while—rustlers have moved back onto Ramblene's place. I guess he prob'ly told you. But I have a little unfinished business that needs tendin'."

Miles Tarandon no longer took part in the vigilantes' affairs, but no one had any qualms about his overhearing their conversations concerning the work. He heard Kittery, and for the first time since the clash over Zeff Perry he spoke directly to him, leaning past Beck to say his piece.

"What's your business that's so important?"

"Doesn't concern you."

"Anything to do with Mrs. Greene?" Tarandon persisted. "I saw you go in the hotel. Her old man says you stop in there every time you come to town."

Kittery took a moment to calm himself while he stared at the cocky deputy. At last, he nodded. So Greene was staying on his fat old toes. He thought his wife was having an affair. He was wrong, but of course it wouldn't seem that way to him, nor perhaps to the rest of the town. But what difference did it make what men like Greene or Tarandon thought? Their opinions meant nothing to him except as to how it might effect Maria.

"You better watch your implications, Tarandon. We might have to go another bout. And I'm not drunk this time."

Tarandon's eyes took on a bitter memory, and he started to straighten up. Then his glance flickered and fell away from Kittery's calm, steady gaze. He settled back against the bar with a tight, humorless smile and sipped at his drink.

Kittery looked at him for a moment more and then placed a hand on Beck's shoulder as he turned and stepped from the room.

"Touched a nerve, I think," Beck mused after Kittery had gone. "I'd be careful around that man, or you're likely to regret it—again."

Rankled, Tarandon tossed off his whisky and stared at Beck. "His gun's no bigger than mine. The man's a louse. Anyone with half an eye can see it. Courtin' another man's wife in front of the entire town," he scoffed, refilling his glass.

Adam Beck's friendly expression had disappeared by the time Tarandon finished speaking. He let the weight of his torso rest on his left elbow as he turned to face the lawman more fully. "You might watch what you say about him when you're talkin' to me, too."

Tarandon shot the manhunter a glance, but his muscles relaxed some. "He's seein' a married woman. Anyone can tell you that."

"Seeing," Beck repeated with a nod. "Nothing more."

That evening, across a table from Adam Beck, Kittery sank down in his chair. After the news Beck had just broken to him, he could hardly respond. He let his friend's words sink into his mind. The vigilantes were finished. James Price had sent word to Beck from Tucson: the vigilantes would soon be broken up, its members prosecuted.

Francis H. Goodwin, in his last turbulent days as United States marshal of the Arizona Territory, had put forth his most valiant efforts to redeem himself in the eyes of Governor Safford. His had been a poor administration, and his publicity was terrible. So, knowing he couldn't hope to hold onto his office without some worthy deed behind his name, and having heard of the activities taking place around Castor, he had sent two special undercover men to investigate, secret appointees of Governor Safford. Now they supposedly held sufficient evidence to bring in the army and have each and every vigilante arrested and prosecuted to the fullest extent of the law. Goodwin would have done it, too, if he thought it would save him a dishonorable resignation of his office.

Yet despite his efforts, Francis Goodwin had been forced out of office three days earlier, on August fifteenth. The new chief marshal was Wiley W. Standefer, the deputy marshal who had wept at the graveside of Joe Raines. Now it was he who made plans to visit Castor in person to check on circumstances surrounding the town and to bring in the army, if need be.

In silence, Kittery pondered what Beck was telling him. He hadn't imagined the vigilantes would ever be brought down. And a list of all their names? The only way that could happen was through someone on the inside. It hadn't been bad enough to think someone in the vigilantes was working with the outlaws. Now they were being betrayed on both sides of the fence!

It was only moments before the realization struck him. McBride! Was it possible . . . Would he be so low as to be working with Baraga and at the same time betraying the vigilantes to the law? Well, why not? It was a perfect way to rid the Desperados of one of the biggest thorns in their side. In fact, it would be ridiculous to think he *wouldn't* work both sides, if he was working one.

Kittery was sick to his stomach with these thoughts that plagued him, but for a moment he forced himself to forget about McBride and to mull over the rest of the vigilantes. One by one, he ticked them off in his mind. But he could find no reason to suspect anyone else. All of them had been too much a part of the violence to risk being found out. But what about Miles Tarandon? That man hated him. What would stop him from turning in the vigilantes, knowing Kittery would be among those most harshly prosecuted, as their leader?

He mentioned his thoughts to Beck, but he put no stock in that notion. "He's been in the midst of it. He'd be prosecuted, too."

"Not if he worked a deal—amnesty?"

"I don't believe it, Tappan. He might hate you, but he wouldn't sacrifice the rest of us just for revenge on you."

Kittery raised an eyebrow. "Wouldn't he?" After a moment he said, "Maybe they won't have anything on the rest of us anyway."

"What do you mean?"

Kittery raised a forefinger in front of him as he sat upright in his chair. "Just this, Adam—I know the new marshal—Standefer. Not well, but I know he was a real good friend of Joe Raines. You know what that might mean to us? It might mean he won't prosecute any of us. He has to know what we're tryin' to do down here. What if he turns his head and looks the other way?"

Beck began to smile at the thought of it. At last he laughed outright. "Well, hallelujah for the demise of Francis Goodwin! And God bless Standefer!"

Tappan Kittery rode out of Castor before the sun topped the mountains the following day. He was clothed in his buckskins and wore the moccasins on his feet. His beard, along with his outfit, gave him the appearance of a mountain man, a trapper coming down out of the wilderness. He rode with his Springfield carbine across his lap, his eyes ever moving. He rode toward the Desperado Den.

He found the faint trail the Desperados used and put Satan onto it. For hours he rode with no sound but the clopping of the black's hooves. The sun's heat drove at him through the top of his black hat, and as if that wasn't enough it shattered across the sand and gravel on the trail and bounced back up at his face. Sweat made little streams down his face, making him use too much of his precious water. His seat was sore from the constant moisture there and from rubbing against the Denver saddle. How he wished he could hunt a hole. But he had to keep to this trail. Joe Raines and Cotton Baine were riding with him.

Twenty miles from Castor the trail ran through a mile-long stretch of gullies and broken rocks, where granite and limestone intermixed and the limestone had largely eroded away to create what appeared from the trail to be a series of deep, jagged mazes, snaking off to his left as far as he could see. To the right of the trail rose desolate clay hills with clusters of granite erupting from their battered flanks. Beyond that lay the best known trail to the Baboquivaris, the one where Joe Raines had met his end.

In the north and west, clouds with black intent were stacking up, some spilling their moisture on distant mountain ranges without any pattern. Now and then he could see the flicker of lightning on the boiling blue-black horizon.

A sound found his ears, a faint sound like wind whipping across the top of brush. He had heard that sound before, the sound of the wind in lonely country. But today there was not a breeze. And even if there had been, a man riding a strange, wild land would be a fool to pass off unidentified noises as simply as that. He had heard far-off shots make the same whisking sound.

Out across the heat-blasted rocks nothing moved. He heard no further sound. He urged Satan forward, now at a cautious walk. If that had been a gunshot, what might it mean to him?

As he came around a huge, round hill of mica-shot clay he cursed in surprise. There, no more than three hundred yards before him, five men sat their saddles. Five men, and one of them was missing his right arm at the elbow.

They sat studying the surrounding rocks. Three of their horses were looking toward him, but none of the riders had noticed that. He guessed they, too, had heard the distant sound and were trying to pinpoint it.

With a slight but steady pull of the reins, and moving his feet forward so they skimmed the points of the black's shoulders, he began to back up.

Then it was too late!

One of the riders chanced to look his way as he disappeared around the rock. The man issued a yell of surprise. With a curse, Kittery wheeled the horse. He jammed his heels into the animal's ribs, and Satan, sensing his friend's urgency, bunched his muscles, tearing back up the trail like a pronghorn.

From a stand-still to a full gallop in seconds, even with Kittery on his back the big stud was clearing ground at thirty-five miles an hour. But it was a speed they could not maintain.

Behind, the outlaws yelled and drove their horses to overtake him. Their curses rose to his ears, seeming to spur the black on faster. Kittery's head pounded with racing thoughts and fleeting images. He had to get a lead on these men, then find cover. He was too big to outrun them. Even the stallion couldn't carry him at great speed for very far. That wasn't an option. Fighting was the only option, and he had better fight well. He had to keep his promise to Joe Raines to avenge his death.

The brush and cactus flew by with incredible speed. But the sounds of the outlaws seemed to draw nearer by the moment. Maybe Kittery's senses just honed in more. Dust billowed up off the trail, rising in a long cloud behind him. The stud breathed in great chugging rhythm, straining against the reins. He reached out with his powerful front legs, pulling the ground back at them and shoving it off behind. Kittery wanted to leave the trail, but he saw no place the big horse could go. To the left, the slope was too steep. To the right was the maze of rock. The stud would die trying to cross either.

And then the black stumbled. He smashed through a three-foot high clump of cholla that ripped at his knees as it broke up and flew in every direction, scattering its evil thorns. The outlaws closed while the black tried to regain his momentum.

It was during this lull in Satan's speed that Tappan Kittery's instincts left him and his good sense took over. The outlaws were getting close enough now to try shooting at him, and he didn't want to lose this horse to a bullet. Here in this sun-bleached country of dust and rock and creosote, he would fight and he would die.

Heaving back on the reins, he leaped to the ground before they came to a complete stop, whipping the Springfield carbine from its scabbard as he went. He let the black go, and he charged away, leaving Kittery in the open.

At a crouch, Kittery turned and sprinted for the limestone and granite maze that began thirty yards from the trail and ran to the horizon. He came to a smooth rockslide where run-off had battered and worn down the stone over the eons. Launching himself into a sitting position at the top of the precipice, he went down, steadying himself with his free hand. He stumbled to a halt at the bottom, ten feet down. Twenty yards away was a patch of jumbled boulders where the main part of the stone labyrinth began. He could hold the outlaws from there for a while, he thought. He had once more loaded the Remington with an extra round that day, and he had enough cartridges in his pouch to last a long time. For the carbine, unfortunately, there was only a handful of shells in the pouch hanging from his belt.

Leaving the foot of the slide, he dashed for the boulders. He had almost made it when the crack of a shot jolted him. He felt a stabbing, thunderous pain in his side, as if someone had struck him with a bat. He turned to glimpse the outlaw who had fired leaping from his saddle. Then he felt himself go down.

His face bit into the searing sand, and the carbine left his hands. Rolling over, he clawed for his pistol and fired at the approaching outlaw. He didn't know if he scored, but the man disappeared behind the rocks.

A wave of dizziness passed over Kittery, and he felt his senses weaken, like he hadn't slept in a couple of days. He shook his head, but the feeling didn't leave him. The rocks spun around him.

Would this be the way it ended? Must he die right here where no one could ever find him? So Baraga would have him after all. All his work and trouble for nothing! No one could save him now, not even his own faithful Remington.

Suddenly, he heard another shot, but this one was farther away, back toward Castor. He heard cursing from the outlaws, and with elation he knew—somehow the people of Castor had found him. In one moment there were the frantic sounds of men regaining their saddles, fighting their horses to be still, then two pistol shots fired toward him in spite, then the thunder of hooves. In less than a minute, all was still.

He had started to come to his feet, to call out to his friends, when he heard a sound that made his blood chill.

The *ki-yi-ing* war cries of Apaches!

He realized it must have been one of them shooting which had cautioned him earlier. They might have been far enough away for a shot from them to sound like the wind. Now he could hear their screams as they came charging in on their horses and lit out after the Desperados.

The blood-curdling yells of the Apaches brought Kittery lunging to his feet on strength he hadn't possessed moments before. He fled farther down the nearest corridor of rock, trying desperately to plug the hole in his ribs with a finger as he ran. He knew if the Apaches returned and found one drop of blood they would know a man was near—and hurt. They would set upon his trail like a pack of timber wolves after a wounded elk.

There came another war cry, close again. It was an Apache, and his voice made it clear he had found something. Perhaps a track, perhaps blood, but by the nearness of the sound he sensed they would be after *him* now, not the Desperados.

With renewed urgency he struggled down the wash, clutching his wound. But still the blood wept between his fingers, so he stopped long enough to cut a strip of buckskin from the bottom of his shirt. He stuffed that with a grimace of pain into the bullet's exit hole. He did the same where the bullet had gone in, and this helped stanch the flow of blood for a moment. Steeling himself to the burning in his side, he moved on, careful not to touch any of the scorching rocks unless he had to catch his balance.

Sometime later the corridor he was following was divided by two huge granite backbones. The top edge of the rocks was two, maybe three feet above his head, enough to conceal him from view of the Apaches. He chose the cut that ran the narrowest and deepest and continued on. But as he went the channel began to constrict until it looked as if some distance ahead he could no longer squeeze through.

Would it end here, then—mere yards from where he had gone down the first time? Was he to find himself at an impasse, to sit and wait for his pursuers? It seemed that way, for suddenly the corridor abutted against almost solid rock. He was walking sideways, barely squeezing through, when he realized he could no longer do even that. He had been a fool to believe this passageway would continue forever. The opening, from ground level to about five feet up, widened to a little over shoulder-width a ways ahead. But beyond that wide spot it narrowed to a mere three-inch slit. A child could not have slipped through.

Despair swept through him. His hands shot out, pressing against the walls on either side of him, and he started to whirl back the way he had come. *Get a hold of yourself, Kittery.* The words seemed to come from somewhere else. But they were his. He remembered his mama saying that to herself when he was small. When she scolded herself mockingly she always called herself Kittery.

What would Jake do? It seemed, in Kittery's state of shock, that he heard the question out loud. What *would* Jake do? He would not give up before the game was over. Jake would never give up.

With a deep breath, Kittery approached and looked at the wall before him. It was eight feet tall. Not much taller than he was, a mere foot and a half. Sighting through the crack down its middle it looked as if the passage widened out again beyond. He didn't know how far that would go, but it was better than going back. Yet there was a problem: in his normal condition he could easily have scaled that rock, but with the bullet hole through his side and the blood he had lost . . .

The rock was straight up and down, smooth from millions of gallons of water and tons of debris that must have rushed against it through eons of desert flash floods. He didn't know if he could find a handhold. And he knew if he did, and if he got up there it would be at the price of great pain to himself. But he had to go. Otherwise, he had given up his life. Jake would have cursed him for that decision.

Stepping back, he surveyed the obstacle. He drew a deep breath into exhausted lungs. He bunched his leg muscles, and then, trying not to think about the impending torture, he threw himself up. His fingers caught the edge of the searing rock like claws, and like a crazed animal he pawed at the rock face with his feet. The pain in his side, though he tried not to let it through, cried for him to let go. But he had his foot over the top now, and he was grasping for another handhold.

Then his anchoring foot slipped, and he started to fall back down. He clawed with his fingers, landing across the edge of the rock on his ribs. A shot of pain like red-hot blades jagged through his side and chest. He wanted to scream away the pain. He wanted to die. But he didn't. He was alive. He had made it.

Reaching forward, he grabbed another hot, jagged edge of stone and pulled himself forward, gritting his teeth. With a motion like a slithering snake, he wiggled and wormed until his entire body lay across the rock bridge. Then he just lay there, lay until his eyes grew heavy and the pain became a dull throb. He had come to forget the heat in the rocks, for it was so much kinder than the pain in his side.

Forcing himself to his knees, then his feet, he limped across the rock, holding his ribs. When he reached the far side, he dropped back down to the gravelly wash. The pain of it jolted through his side, but he didn't flinch. It was nothing compared to what he had just been through.

The corridor widened here, as he had perceived through the crack in the wall. He started along it, clutching his side.

He knew that behind him he had left his blood all over the stone bridge. He glanced up at the glowering clouds that built overhead, and he wondered if the monsoon rains would come down today. Yet even though the rains would wash away the telltale blood, they would also create flashfloods in these corridors that he could not survive.

Kittery had not walked far before he came to a second fork in the rocks, and once more he took his chances, choosing the deepest, most difficult crevice. He would do his best to throw the Apaches off even while he put himself through hell to do it.

As certain as hands circling a clock, he grew weaker, seemingly with every step. The earlier surge of excitement had worn off, and his head began to grow light again. He felt like he would pass out three different times. He felt like he had to throw up. His body ached for rest. But if he chose now to lie down his muscles would stiffen. Could he rise again? If he did choose to rest he had to find a better place for it, a place where if he passed out he would not be so exposed.

He had a faint recollection of cutting into three different washes. He had walked over half a mile. He would never have seen the hole, hidden as it was by a slab of granite and some brush. But he was bent over, tightening the lacing on a moccasin. As he straightened back up, his eyes passed over the opening. It was in an old fault, and water had long since made it into a tiny, low-roofed cavern just wide enough, he saw, for him to squeeze into.

He thought about the danger of rolling onto a snake—or onto several of them—before he crept into the hole. But he was as safe with snakes as he was outside with Apaches. Anyway, he had just as well die of snakebite as of exhaustion or heat stroke. Besides, maybe the snakes would have sense enough not to make a den of a place prone to roaring flashfloods.

The weathered fault turned out to be a tunnel. By the air currents he knew it ran far back into the rock. Perhaps it had an opening somewhere ahead. The ceiling did not remain low hanging, as at the entrance. Instead of crawling, as he had been forced to do, he was soon able to walk crouched over.

He had gone what he estimated as thirty feet down the tunnel when his weary, fumbling hands found another opening. This one was along the wall to his right. As he crawled into it, the screech of a bat startled him. It dropped from the ceiling, followed by a second. His hand dropped to his pistol butt, and when he realized it, he gave a weak half-smile and lay down, curling into a ball.

Now he could rest. And since it was almost cold in the shelter of the rocks, he didn't thirst so mightily for water. Maybe he could . . . Maybe Castor wasn't . . . so far away . . . Maybe . . .

Kittery sat back up, shaking himself. He could not go to sleep. To sleep now meant to sleep through the night. That meant losing his best chance at escape. He needed the darkness.

Far down the tunnel the way he had come he could see dim light. What did he have—two, maybe three hours of light left? He would make his way out of here then. It wouldn't be hard finding his way along the maze in the dark. There were only so many ways a man could go. He just hoped he didn't stiffen up too much to make it across the stone bridge. If he did, no one would ever find him here. It was an ideal tomb.

Some time later, he jolted awake to the sound of thunder. It sounded oddly hollow this far under the ground. Looking toward the cave's mouth, he saw the day had grown very dim. He saw a flash of light, and again thunder rumbled fierce through his lair. He had to get out of this hole, out of the labyrinth entirely. He would be glad if a storm came to wash away his blood, but if he was down in here when it came it would also wash him away!

He rubbed his eyes and pushed at his side, testing it. As he got up and moved toward the dying light, he realized he was not as stiff as he had expected. Perhaps all the moving around to get to this point had been of benefit to him. And another thing, perhaps a gift from the cold, was that unless he thought about the wound in his side, there were moments when he didn't notice it.

Upon reaching the opening of the cave again, he saw the huge thunderclouds hanging low and black all around, and heat lightning danced behind them like fireflies behind a curtain. Electrical energy filled the air, tickling the hair on the backs of his hands.

It would probably be another hour before the sun went down, but if the clouds stayed then darkness would come much sooner. He began his painful journey, thankful for the light he had. In his physical state it was difficult enough scrambling over loose rock under these conditions; darkness would bring with it a waking nightmare.

There was a sudden yell from somewhere up ahead, only partially drowned by another burst of thunder. The Indians were too near, and by the tone of the voice a brave had found something. Drawing his pistol, Kittery stepped back toward the partial shelter at the edge of the rock. In his haste, he stumbled over loose stones and went over backward. He landed in a pile of rocks, and it hurt worse than being shot.

A dry buzz commenced to the side of his head, a mere three feet away. His blood ran cold, and the hairs lifted on the back of his neck. Moving only his eyes, he looked toward the sound. There, close enough to touch, was coiled a diamondback rattlesnake six or seven feet long and as big around as his arm.

Kittery could hear more voices now, in conversation where the first man had called out. It could not be far, it seemed, but the rocks and the corridors did strange things to sound, distorting and bending it to make it seem like it came from several sides at once. Or like it was nearer than it really was . . . or farther. The Apaches could have been around the corner.

The snake continued rattling, swaying its raised upper foot of length and its head back and forth as if trying to hypnotize him. It had not moved any closer to him, but it had not retreated, either. It was backed up as far as it could go against the rising rocks, and the only way out was through him.

A flash of light and the clack of thunder filled the labyrinth and made the snake flinch and drop its head. Kittery had thought for a moment it was making a strike.

He felt the cold sweat come to his face even with the heat of an oven still collected in the labyrinth. His neck was stiffening, but he didn't dare lower his head. The snake was primed. It waited only for a movement by him to strike. And the bite of a western diamondback was among the most deadly of all its ornament-tailed kind.

A rattler will not always inject its venom when it strikes. If its target is a man or something else it doesn't want to eat, the snake might save its venom for where it will do more good. Or on a whim it might not. It is not wise to test a diamondback.

The snake made a move toward him. He shoved himself with an elbow and rolled to the left. The snake feinted, as if to strike. Its narrow, slitted pupils gave it an evil aspect as it advanced, chasing him. Kittery could not kill it with his pistol. A shot would bring the Indians on the run. And he knew the pistol was too small to use as a club. The snake's lightning reflexes would sink its needle-sharp fangs deep into his flesh before he could withdraw his arm.

The snake came on until Kittery was backed up against the far side of the wash the way the snake had been. Here it stopped and coiled once more. It was within striking range of his legs. Kittery cursed the serpent vehemently in a whisper. He cocked the Remington, very much against his will, and aimed it at the flat, triangular head. The slitted eyes stared at his legs, tormenting him.

Again he heard the voices off in the rocks. The snake rattled non-stop. It reared back its head and stood the upper third of its thick body up on end. Kittery had the pistol's sights aligned, but he could not squeeze the trigger. With the snake's bite, he might live. There was no chance the Apaches would withhold their venom.

Chapter Twenty-Seven
Fight for Survival

With no warning, and for no apparent reason, the big diamondback curled back into itself and began to retreat sideways, keeping an eye on Kittery's legs as it went. Five feet away it turned and slithered off into the rocks and brush. When it was out of sight, Kittery could hear the rattle stir now and then, warning the man not to enter its domain again.

Kittery let the entrapped air escape his lungs. He lowered the Remington and closed his eyes, his entire body shaking. He couldn't hear the Apaches' voices any longer. Were they moving in on him, or away? He let the hammer of his pistol down but kept it in his hand as he crept forward.

When the clouds broke and passed overhead, the sun was gone behind the hills. The violet sky brandished long, milky streaks of vermilion tinged with yellow along its edges. Nighthawks circled and swung crazily above him, and somewhere a mourning dove made its song rise like a distant flute on the warm desert air. Coolness flooded over the rocks and sand, making them softly warm to the touch, in contrast to the searing sensation he had gotten from them so many times that day. It would have been a beautiful evening any other time, but death waited for Tappan Kittery to leave this sanctuary . . . or it would find him here.

Even in the growing gloom he found the sign. The Apaches had been not fifty yards from him during his encounter with the snake. They had turned back at the stone bridge. Yet surely they had seen his blood there. They knew it wouldn't take long to find him in the morning, not with the blood it was apparent he had lost. But the darkness had spooked them. If an Apache could help it, he would not fight at night, for darkness was sacred to his ancestral dead. Most Apaches, renegade or not, took their religion seriously. It would not do to displease their creator, Ysen. Of course, the fact that a wounded gunman awaited them in the rocks probably helped make their decision.

Kittery traveled the wash by feeling along its sides as it grew darker and darker. After an hour, he found his way out of the last of the narrow washes and stopped at the edge of the rocks. He searched the night for any brave who might have been left to watch for him. He could see no one.

Brilliant stars had replaced the wash of cloud in the sky, but their presence did little to brighten the labyrinth. It was enough, however, to see that his Springfield carbine was gone. He had expected no less. He found the slide down which he had come in fleeing the outlaws. His senses came to life with the pungent odor of mesquite wood smoke hanging in the air like incense, and he saw firelight's glow on the rocks across the trail. Managing to climb a ways up the slide, he could see the Apaches had their camp in the exact center of the trail. Each of them slept by his own small fire, in Apache fashion. Their only coverings were long shirts that reached to their thighs.

He could see six of them. Only six. That must have been why they chose not to chase the Desperados. No self-respecting Apache would ever fight a battle without the almost sure knowledge that he would come out the winner, and only outnumbering the Desperados by one man was not healthy odds.

Kittery played with the idea of opening up on them. There were six men . . . he had six shots before having to reload. It was tempting, but he had seldom seen a .44 bullet bring down a man instantly. And in the dark he doubted he could place all six shots to his advantage before one of them managed to fire back. Besides, there was probably at least one more man up in the rocks with their horses.

Tomorrow the Apaches would return to the stone bridge, ready to capture him—unless they saw his tracks leaving first. Either way, they would be on his trail by the time light was good for tracking. Perhaps with fifteen full hours of hunting light their luck would be better—and his, worse . . .

When he had slid into this rock pit earlier that day, he had run straight ahead, for the nearest point that would hide him from the outlaws' guns. But he could as easily have turned left, to the east, and it was that direction which he now chose. East, to him, meant Castor—and home.

He had no idea how far he had gone when he looked down and realized he was bleeding—had been for some time. With all his movement the buckskin plugs had worked loose. He stopped, painfully plugging the holes again, and then he cut a long strip from around his shirt, making a more permanent bandage.

He tried to step mostly on rock to hide his sign the best he could. But that had its disadvantages, too, since it would wear his moccasin soles out quicker than sand, leaving him barefoot. Soon his feet would bleed. Then nothing could hide the sign of his passage. He regretted not taking the care to put rawhide on the bottoms of the moccasins, but he had not planned on walking in them much.

He chuckled to himself, almost delirious. He was a fool to think he could hide his sign from Apaches. Any tiny speck of blood, any bit of leather worn from a tattered moccasin, any loosened prickly pear kicked by a careless foot in the dark would show them his trail. They would find his blood first, and the rest would be simple. Unless they were fools, or unless he made twenty miles that night and reached Castor—or unless something else drew their attention away from him—they would find him tomorrow. The Apache was the master of the desert. He was an expert at concealment, at stalking, at tracking . . . and at torture, if he so chose. They would find him, there was no doubt. He must go down shooting.

For all the good it would do him he would have to hole up when light came. Normally, in good shoes, he could have walked the entire twenty miles to Castor in four hours of easy travel. But in his present state he would need rest. He had no water supply, either, and wandering about under the desert sun would suck out what he had stored in his body.

On top of his lack of water, he had no food, and nothing to keep him warm. He had nothing at all that he really needed. Fortunately, his moving at night would keep his blood moving, even if it didn't keep him warm against the onslaught of the desert cold. As for his hunger, there were stories of people who had saved themselves from starvation by boiling buckskin to drink the thin soup it made. But normally they were people who had been snowed in, with nothing if not an endless supply of water at hand. Kittery was without water, or even a pan, had he been inclined to risk a smoky fire. Anyway, he didn't figure on getting hungry enough to drink the water from boiled buckskins in the day or two it might take to reach Castor. He would be dead long before he got that hungry.

One thing from his memory could help him, though. He was a son of Carolina, so the dry country was still foreign to him. But Cotton Baine had once related a story about a drought he had gone through in Texas. It was so dry Baine and the other hands had resorted to burning the spines from prickly pears so their cattle might survive on the moisture in the pads. They *had* survived, too, and fed on them until the dry spell passed. So he would do the same. He had no means of burning off the spines, but he could pluck them out or cut them off with his knife.

Toward two or three in the morning a warm breeze began to blow from the south, almost a hot breeze. Kittery turned and glanced that direction. The stars were blotted out by clouds. Or was it clouds? The dry smell of warm desert dust whipped past him on the wind. Then it came again, until soon it was regular, and he could no longer smell the cool night air and the vegetation.

The sky blackened. There was a cloud, sure enough. But it wasn't a storm cloud. It was dust.

Kittery whirled about, seeking for some type of shelter. He had survived Arizona dust storms before, knew the horror of suffocating hot winds and volcanic dust one would bring.

All around him the plain was flat, sprouting clusters of creosote. But in one spot off to his right heaped a jumble of boulders. He barely reached them when the bulk of the storm swept over him, blinding him, whipping him with sand and chunks of gravel. The wind battered his clothes, lashing around him in a frenzy and tearing at his face. He huddled down in the lee of a large boulder and buried his face in his arms, holding onto the dipping brim of his hat. On a sudden thought, he unholstered his Remington and placed it inside his shirt, against his chest. Then he resumed his position, protecting his face from the bitter wind.

The storm buffeted him, flinging the fringe on his buckskins around, puffing up the bottom of his shirt where he had cut it short and filling it with dust and sand. Gravel spat against the sides of the boulders, ricocheting off into the sand and tumbling away as if no more than feathers. The heat and intensity pressed down until Kittery was engulfed and it seemed like the storm would never pass. He could hardly breathe. He wanted to cry out in desperation, but he didn't want to open his mouth. Then, in less than fifteen minutes he felt the winds break, and slowly they began to subside.

When the storm was over and the winds gone, Kittery raised his head, and dust and sand tumbled from the brim of his hat and the disheveled hair beneath it. Dust covered him a quarter inch thick in places, and it sifted off as he rolled over and sat up, pulling the Remington out of his shirt. The desert was quiet and still, as if nothing had happened there, except that the usual night sounds were absent. And the northern sky was empty of stars, the path of the storm.

Kittery staggered to his feet, and with his bleary eyes to the east he pushed on.

In the early dawn, when the gray light rolled over the world like a fog and the stars began to blur, a Mojave rattlesnake in a jumble of rocks to his right coiled up and began to send its dry buzz. It was a clear warning to anyone coming within three feet that danger was imminent.

The warning whir continued with a frenzy when Kittery stepped closer. The snake slid into a better defensive position. Normally, Kittery would have continued on; this rattler had done him no harm. And his mother had taught him to respect life, killing only what was needed to eat—or in self-defense, in the case of ticks, chiggers and flies. That respect for life extended to the lowly rattlesnake, and the copperheads and water moccasins and coral snakes of Carolina. It took only one look into the late-night corn crib and wheat bins to see where the rat and mouse population would go if snakes weren't allowed to thrive.

But what he needed today was food, and preferably food that could be gained without wasting ammunition. So he would take what providence offered. Stooping, he selected a smooth, round stone . . .

As the sun lightened the eastern sky, Kittery searched for a cove in which to spend the day, the dead rattlesnake dangling across his fingers. He trudged up out of the gully he was in and gazed around. To his right rose a gentle, sandy hill covered with brush and rock. He made his way up it, searching for shelter as he went.

Several boulders littered the ground near the top. On a closer look he saw that two of them were covered by a third, flatter one, creating an overhang that offered shade. He chose to rest here, about twenty yards above the main trail. He flung the dead rattler in before him, then went about gathering brush, which he arranged in front of the shelter as camouflage. Next, he cut loose enough prickly pears to keep him busy plucking spines throughout the day and gathered them in his hat.

He spent the morning asleep. It felt good to let himself go, after the night's ordeal. Around noon he opened his eyes. His shirt and pants were soaked with sweat, and his hair was plastered to his forehead with it, despite his shady hideaway. His skin was feverish, and his throat parched. His feet were raw. It must have been over one hundred and fifteen degrees out in the direct sunlight as he began the tedious job of plucking the thorns from his intended source of moisture. There were patches of dark gray-blue clouds scattered across the massive expanse of sky, and some sent down tattered curtains of moisture, but they all seemed to ring the horizon on every side. Here where he lay the sun was undisputed master.

He had finished plucking several prickly pear pads and tried to shave away the smaller spines with his jackknife. But some of those managed to find their way into his lips and tongue anyway. What little flavor there was in the juice had a bitter edge to it. Throughout the pears ran small, stringy fibers. And for all his trouble they contained a very small amount of moisture. The summer had been an abnormally dry one, without July and August's usual onslaught of monsoon rains, and even cactus had lost much of its water supply. Still, he savored the moist flesh, and every bit was worthwhile. He knew it could be the closest he would come to water for a long while. Maybe the rest of his life . . .

Sometime in the afternoon he caught a bright flash from a metal object in the far rocks across the trail. Peering that way, he made out several Apaches, one toting the rifle responsible for the glare. They were hunting him. The dust storm had held them off for a while, hiding his sign. But they had guessed he would head for the closest civilization. If they found the sign he had begun leaving in the gully after the storm he would be theirs.

He lay there for several minutes, watching them crisscross the ridge. They stopped to stare across the trail in his direction a time or two, and soon he realized they might not be on the other side because they thought he was there but because it was higher and offered a better vantage point. If they spotted any sign that told them he was on the opposite side of the trail, it would only take them minutes to be here.

The big Indian who carried the rifle stopped. Catching the attention of the others, he pointed across the trail toward Kittery. Kittery caught his breath and clutched his pistol. His heart thudded in his throat like it would strangle him. The big brave started down the slope toward him. Kittery prepared to fight . . . and die.

The Apaches disappeared in the bottom of the gully, fanning out. Suddenly, a rifle shot broke the stillness. Kittery cringed, expecting a bullet to strike him. He heard a whoop, and then faint voices he didn't understand. After a few minutes, there were no more voices, and half an hour later he saw one brave struggle back up to the trail carrying a dead whitetail buck across his shoulders. He saw no more Indians.

Hours passed. Dusk was looming over the valley. Heat can stifle the odors of the desert, and now that it began to fade the smells of the land returned—creosote bushes, rock, dust. Smells that meant Arizona. A huge tarantula, his hind end looking like a miniature coconut, wandered past as if Kittery weren't there.

The devil's own palette exploded over the sky, searing it into bold streams of scarlet, gold, violet and pale yellow. The cactus and creosote and mesquite and paloverde trees took on the glow of the sky. These were nature's ways of reminding a man why he had chosen this land of heat and snakes and scorpions, where everything—alive or dead—wanted to hurt him in some vicious way.

The faint sound of voices reached Kittery's ears, and it took only moments to recognize the guttural sounds of the Apaches. They were much closer now than before. They were coming down right on top of him!

Kittery's eyes swept the area for a better place from which to defend himself. But it was too late. Here was where he must fight.

The voices ceased. Silence dropped like a stone in a still pond. Not a whisper shifted through the quiet. What had the Apaches seen? Would he die here now, or would they take him away to torture? Why the silence? Why didn't they move in?

It came to Kittery that moment how much the thought of death frightened him, though he had lived with it most of his life. He had never taken time to contemplate it much until now. What awaited him on the other side? Was it like his mother and the Bible said? Or was it nothingness, simply some endless black void where he would wander for eternity? And if he went there, wherever he went, would Tania grieve him? And would he meet Jacob there?

Twenty minutes of mental anguish dragged by. The grip of the Remington had long since begun to feel greasy in his hand. He waited, straining his ears, and then he smelled smoke. Smoke! Were they trying to burn him out?

Kittery whipped his head around. The odor went away but came again. He didn't dare look out beyond the rocks. He thought he heard a stone clatter against another. The Apaches had to be on top of him.

And then, for the first time, he saw the Apache standing there.

Through the veil of brush he had gathered he could see the Indian, standing as still as a saguaro. His long, black hair lifted with the breeze, and smoke drifted away from his lips. It must be the same smoke Kittery had been smelling—cigarette smoke.

The Apache looked like a spectre, lit up by the sun, his narrow eyes hiding back behind high, thick cheekbones. Kittery had seen mountain lions look like that, staring out over their domain. His thigh-length shirt fluttered. His hand raised to bring the cigarette to his mouth, and he drew deeply. Then other braves seemed to rise up as if out of the sand around him. He passed the cigarette to one of them, and the five of them started down the hill. No way could they know in twenty more paces to their right they would have been on top of their prey.

The air cooled swiftly as the next twenty minutes ticked by. There was a time when Kittery knew the Indians were gone, because he felt nothing. No threat of death. But why had they gone? There had been enough light to search for him and find him. Had they grown bored of the hunt? Likely they were hungry. Maybe they were headed back to camp to eat the deer the sixth Apache had packed off earlier in the day. Would the Apaches return tomorrow? Would he have to wait until then to wind up this awful trek through the desert?

Later, all was black. A huge centipede crawled across his hand. He saw it only very dimly in the starlight and felt its many legs glide coolly across his skin. An owl hooted back in the rocks, a lonesome sound, the only sound. The sky held no moon, and the stars seemed close enough to reach out and touch.

Kittery rolled over, groaning in pain. He tried to straighten his torso but couldn't, and he knew if he tried any harder he would reopen the wound. Still on his hands and knees, he dropped the naked prickly pears into his hat and draped the snake over his left arm. He crawled out of his hole and looked around him, struggling to his feet, half doubled over. With a deep breath, he started down through the thorny brush.

Rather than lessening, the pain in his side grew worse until almost unbearable. It felt like someone had heated a knife white-hot and then jabbed it through his ribs. But he pushed himself on through the night.

He knew he would not eat the snake until his hunger became extreme, for it must be downed raw. And it smelled none too healthy after its day in the heat. Even if he had wanted to risk letting the Indians see a fire, he had nothing with which to build one. He had seen Mort Dansing make one once out of a bow and drill set of yucca stalk and cord, but that trick wouldn't be easily mastered. Besides, he had no time to collect and shape the necessary components.

Finally, his hunger managed to overcome him. He had eaten more of the spiny pears while walking, which momentarily filled the gap. But now if he was to keep his strength he needed something more substantial. Right then, the danger of being poisoned by bad food didn't bother him as much as the thought of his body running out of fuel.

Removing the jackknife from his pocket, he began to cut strips of the tepid flesh away from the skin and force them down his throat. Twice, he nearly threw them back up, gagging and having to close his eyes and his mind to hold back the tide. When he had taken all he could stand of the smelly flesh, he tossed the remainder of the carcass away.

The going became extremely slow. His feet grew heavy and awkward, and it was hard to swing them forward for each new step. He began to stumble over the smallest of stones and clumps of brush. Behind him lay an intermittent trail of blood drops, blood from his feet, where the moccasins had given way and rocks and cactus now tore at the bared soles.

As the night dragged on, he was forced to crawl more than once just to keep moving. He would come out of a barely conscious state to find himself on hands and knees, inching along. When he did travel on his feet, he always did so doubled over, clutching his feverish side.

Once, he awoke in a kneeling position to find himself violently ill. He had thrown up, and his hat was lying in it. His stomach convulsed again, and his throat seized. But his stomach was empty. Nothing came up. He tried to focus his eyes. He knew then that the snake meat had been tainted. If he had been in his right mind he might never have eaten it all. But he had been so hungry. So hungry . . .

Picking up his hat, he tried to rub it off in the dirt, then started to crawl on. He could hardly see the trail before him. He fell over on his side like a dying animal, then pulled himself back up, trying to see what lay ahead.

At last came the inevitable final time he drew himself from semi-consciousness. He recalled some bizarre dream of Jake riding out of the desert to rescue him, smiling that old smile and with his blond hair whipping in the wind.

Kittery had lost far too much blood and too much water. It didn't surprise him to see that his wound had opened, and his shirt was plastered to it. The bandage was gone. He had clung tenaciously to his hat, but the prickly pears it had contained had been lost somewhere back along the way. It didn't matter. Food didn't matter. Water didn't matter either.

It took a moment, but at last he realized where he was—back on the trail! Somehow he had left the dry wash he had been traveling. Maybe it had come to an end. Whatever the cause, he now found himself out in the open, helpless. The Apaches would have no trouble finding him here. He must reach cover.

But he couldn't move. He couldn't rise. He felt almost paralyzed. With that realization, he slumped back to the earth. And he thought now that he didn't even want to move. Just as well die here as farther on. It was a good five miles to Castor, an easy jaunt on a normal day. In his condition it might as well have been a thousand miles.

Then a thought came to him, a thought that never would have come to a sound mind. The black stallion! He had forgotten him in all the struggle and pain of his journey. Where was the horse now? Had he made it back to Castor or been lost in the desert? Or could he still be somewhere near, searching for his lost master, his friend? A foolish thought, but it was the only hope he had. The horse might just be loyal enough . . .

It didn't hurt to try anything. A long time ago he had taught the stallion to come to his whistle. Would he hear if Kittery tried that now? Was there any reason in all the universe to hope he was that close? Even in Kittery's senseless state he told himself no. But he didn't care.

He tried a whistle, but his lips were too dry. He tried to moisten them with his club of a swollen tongue. At last, desperately, he touched his fingers to the sole of his foot. They came away wet with blood and sand, and he smeared it across his cracked lips and tried to whistle again. The second try succeeded, but it was feeble. He knew it would not work. His foolish hope was in vain.

Consciousness began to ebb. He was still in the center of the trail. He had the vague thought that he would rest here, then rise and move on. Rest. That was what he needed—a simple rest. And water. Food. He needed to rebuild his strength. Yes, when he awoke he would move on. Rest . . .

Then it came, touching the edges of his consciousness. It could not be, but it was. Hoofbeats! The black! He tried to wake up. No . . . Too many hooves. So it was the Apaches then. Or maybe the Desperados. It didn't make any difference. Either way, he was just as dead.

Part Three
Season's End

Chapter Twenty-Eight
The Winds of Change

Tappan Kittery lay still. An odd sensation of comfort and warmth surrounded him. Something here was out of place—not wrong, for it felt too good to be labeled thus—but strange. The ground beneath him should have been hard and rocky, and the air surrounding him hot, dry and permeated by the stench of his own stale sweat, blood and vomit. But neither was so.

Glancing about in bewilderment, his eyes tried to lay hold of something familiar. They came to focus on a hand-painted Indian vase sitting on the floor across from where he lay. It was the vase he had meant as a gift for Tania. The gift he had given to Rand McBride upon finding out his daughter was gone. This was Tania's room. Tania's bed.

To the right of the vase stood a tall figure, the silhouette of Rand McBride. He turned and slipped from the room, closing the door. Then once more the door eased open, admitting a flood of light against which he had to close his eyes. When the door shut, he opened them, and a slow, weak smile parted his lips. He knew the tall, straight, splendid form before him, the proud bearing and feminine scent. He closed his eyes as the woman moved to his bedside, and he felt her warm hand on his forehead. He looked up at her.

"Hello, Beth."

Her hand flinched, and she straightened up. "Playing possum, are you?" she asked in mock seriousness. "You've begun to look lazy, Tappan Kittery. Four days in bed. I declare."

"Four days," rejoined Kittery in a sharp voice, irritated by this revelation. "How'd I get here?"

"Adam Beck—and twenty others. They carried you in after your horse came back."

The vigilantes . . .

Kittery lay silent for a moment. His men—and his horse—had saved his life. *His men.* He had the vague recollection of hearing horses' hooves before blacking out. So what of the government agents? He put the question to Beth.

"They haven't moved yet, Tappan. But they're waiting at the hotel. I think today will be the day."

Beth started to quiz him. She wanted to know how deeply he was involved and who was with him. But he wouldn't tell her. Even when she tried to change the subject to talk of lighter things he remained quiet and brooding. She left him with her final bit of medical wisdom: "Why don't you just sleep, Tappan? It will do you good."

He wanted to laugh as he watched her disappear behind the closing door. Sleep? Unlikely! He'd been sleeping for four days.

Nonetheless, later he found himself waking from a sound sleep. Someone sat in silence on the bed beside him, probably the someone who had awakened him. He felt soft, gentle fingers brush the hair back from his forehead, and the hand rested against his cheek for a moment. He still had not opened his eyes. But every person has their own individual smell; Kittery had been close to few people in the past weeks, so he recognized this scent. He felt a strange chill. This was Maria Greene.

Suddenly, her soft lips touched his cheek, then her cheek pressed against his. She remained that way, her warm skin pressed to his sunburned, bearded face.

He whispered her name.

Maria sat up, touching her cheek in embarrassment and averting her eyes. "You are awake." She continued to look away from him, her face dim in the half-light. "They told me that perhaps you would die, so I came to see you. I have not missed one day since you' return. But I knew you would not die."

Kittery could only move his right arm, because the left was strapped to his chest to prevent the side wound reopening. With his right hand he tried to straighten his tousled hair.

"Perhaps it was by your faith I didn't die."

He took her fingers in his hand and squeezed, gazing up through the dim light. This woman had been kind to him. It meant a lot to have a friend like her.

"The baby kicks," she said, her voice excited. "Here." She took Kittery's hand and placed it on her abdomen. He could feel the weak struggling of the infant inside. The magic of new life.

"It is almost here. Any time. Now I will have a new little someone to love," she whispered. She leaned closer and kissed Kittery's cheek and put an arm across his body. "And still I have you, too."

"And Greene," Kittery added.

But Maria did not seem to hear that. And when Kittery dozed off once more she remained there beside him.

Outside, a considerable crowd had gathered around the arena and gallows. Dust rose in a thick cloud, stirred by the shuffling boots of nervous men.

Sheriff Luke Vancouver stood on the gallows facing the crowd. United States Marshal Wiley Standefer waited beside him, shifting his weight from one foot to the other in quick succession, his hands folded across the front of his trousers. He cleared his throat often, and his eyes flickered here and there, from the ground, to the crowd, to Vancouver and the other man with him. The latter was a government agent from Santa Fe, Nelson Chace.

When the crowd quieted down, Vancouver moved to the edge of the gallows steps. From his pursed lips he plucked a black cigar, which he studied with all intentness for several seconds before dropping it into the thick dust at the base of the gallows.

"Folks, I don't have to say why we're here. I'm not going to go into any lengthy speeches. There are others here who are far more qualified for that. I only want to introduce them and ask you to give them your full attention. This is Marshal Wiley Standefer." He indicated the man with a wave of his hand. "He has a few words to say to you."

Standefer stepped in front of Vancouver. He let his eyes scan the crowd before forming his words.

"I'm certain that by now you have all heard what took place two days ago up at the Empire Mountain Ranch of Black Ramblene. A man was killed." He removed a fat cigar from the inside pocket of his gray, pin-striped suit coat, but it remained unlit between his index and middle fingers as he continued to talk.

"For many years this section of our territory has had far more than its share of riffraff to contend with. I, as a marshal, know that better than anyone. It's understandable and an unfortunate fact of life when you're this near the border of a country as lawless as Mexico. And the marshal's department has never been granted sufficient funds to hire the staff it would need to handle the situation. It would take ten times the deputies we have.

"Before I became marshal, a good friend of mine, a man I worked with for many years, made a tragic mistake. The man was Joe Raines. He thought he could come down here all alone and change the way things have been for decades. He didn't show it on the outside, but during his time as an officer he became a very sad, frustrated man. He was sick of having the territory torn apart by the lawless faction. Deputy Raines could have made something of himself. He could have been a primary force in cleaning Arizona up if he had gone about it the way he was supposed to do, and through the correct channels of the law. But he didn't. He made a mistake, and now he's dead.

"I'm the first to admit that part of Joe's dream has come true. The country's clean for miles around your town. Clean—and scared plumb to death. Everybody's afraid that at any moment someone with a grudge against them, or someone who thinks they've done something suspicious, is going to come to their house and hang them from the nearest tree. That's the way it's been, isn't it? Don't deny it!" he burst out at the murmuring that began in the crowd. "That's why I'm here. I guess you know that. Because of you, the lawless vigilantes. You thought you could create a peaceful community by taking the law into your own hands and killing anyone you didn't like or trust."

Agitated talk arose amid the gathered. Seeming to realize the cigar was between his fingers, Standefer put it to his lips and lit it, drawing deeply as he waited for the noise to subside. He blew out a long cloud. "Every one of you knows I'm right. And you realize by now that we know every one of you.

"You're pretty popular with a lot of folks. You're a bunch of blasted heroes! You've done what the United States Government could not; you've cleaned up this land. But let me emphasize that you have done this outside the law—outside the courts." His voice had risen almost to a roar. "And who but God knows how many innocent men you have killed in the doing? You are outlaws yourselves, all of you who rode with the vigilantes!" Then his voice grew quiet. "The sad part is, we could have used you. Had we but known there were so many who were willing to work with us, for no wage at all, we could have deputized you and done this legally. Who knows how many lives we'd have saved?

"Now, as I said before, a man was killed up at the Empire Mountain Ranch two days ago. Just another outlaw, you assumed. You shot him without mercy and without a trial of any kind, ignoring his pleas of innocence. But you were wrong. Oh, yes, you were very wrong. Now you will see what the law has against vigilantes. The man you killed was innocent of the crimes you charged him with. And how do I know this? Because he, too, was a good friend of mine. He was an associate—a government agent under cover."

Gasps of disbelief came from the crowd. Shock in their eyes, the vigilantes gaped at one another.

"His name was Adam Rolph, and like this other gentleman up here he came to Tucson to investigate the rash of killings that were being reported."

Again, Standefer paused, eyes flicking over the crowd as he puffed his cigar in short, steady breaths. He turned to Nelson Chace and motioned him forward with a flick of his fingers.

"Mr. Nelson Chace would like to speak," he told the crowd.

Chace looked the crowd over, his pale blue eyes glaring from beneath shaggy gray brows. Defiant of the heat, he wore a long, gray linen duster. A mustache and ragged goatee hung from his face.

His opening words belied the anger in his eyes. "My friends, I am up here to beg only one thing of you: for God's sake and the country's, bring this carnage to a halt. I plead with you. Friends, you have done what you set out to do. You have cleaned up your community. The several of you we talked to I'm sure felt, by our harsh words, that we would prosecute you all to the fullest extent of the law. To be honest, we were considering that, and we could. But we don't intend to, if we can have your cooperation. Frankly, you have done Arizona too much of a favor, despite your mistakes and your evident disdain for the law." It seemed to cause Chace great pain to voice this dictum, as if he were speaking with the tip of a knife jammed against his spine. "For political reasons, because it's a political world, Governor Safford has offered amnesty to all of you—each and every one. But only if you will agree to put away your guns and halt the massacre. The Governor wouldn't make it long if he prosecuted you now; you're too popular. But look at what you've done. Adam Rolph was my best friend, and because of you he is dead. God only knows how many other innocents have died since you began your killing spree. It must stop, or we *will* prosecute, and to the fullest extent of the law. Governor Safford be hanged! If we must, we will call in the Army. Or you can make it easy on everyone. We know all of you—every last man. You can thank one of your own for that.

"Governor Safford wasn't able to come down and talk to you. He had obligations in Washington. But before he left he sat down and wrote up a statement which I have here with me." He took from his pocket a white paper and fluttered it for all to see. "It contains an oath. I'm sure you know all about those—particularly the secret kind. It reads as follows: 'I have been saddened almost beyond words to learn of the activities that have occurred in our glorious territory. As those of you involved most certainly realize, this kind of activity only damages our credibility when the time comes for us to make a bid for statehood. Our territory has become infamous—more so than even Mangas Coloradas and Cochise were able to manage—through these lawless activities. It is not the kind of reputation we need. And yet that is not the most important reason for refraining from this type of undertaking. After what you have learned today, you know that reason. I am grieved it has taken Adam Rolph's death for you to know why there are laws against vigilantism. I have no desire to see the people of my territory imprisoned for what they believed was the only course to see peace in their homes. It is because of this that I pen this oath, offering each and every one of you amnesty if you will but sign it and agree to live by it.'"

Nelson Chace looked around at the vigilantes and the other men and women gathered around him. There was no sound among them. An onlooker might nearly have heard them breathe.

"Here," went on Chace, "is the oath: 'We who have taken upon us the name of vigilantes, who have brought blood and death to the Arizona ground, do hereby solemnly swear to never again take up arms in force without first receiving the proper authority. We do this in the name of God and our country.' Now. You will sign at the bottom, and may God be your witness. And do not, my friends, think you can escape this oath by a mere failure to sign. As I said, we know each of you. The man who fails to sign will be placed under arrest immediately. He will be tried in Tucson for murder, a trial that will cause him to rest in jail for months, regardless of what the outcome might be. And then he will most likely be thrown into prison as the governor's scapegoat. That, my friends, is your alternative."

Upon finishing, Chace's features were flushed. It was not from the heat. He turned his back to the silent crowd.

After Chace had moved from the front of the platform, Luke Vancouver watched the faces in the crowd, taking mental note of their reactions. Most of those present were familiar to him. Some were his friends. One was his colleague, Deputy Miles Tarandon. Now the sheriff moved forward.

"I guess now you all know why the law can't tolerate vigilantes and mobs. You know why I hoped to stop you sooner. Yes, I knew about you—a lot sooner than you think. No one specific, but I knew about your organization. And I had hoped you would come to your senses. But it's taken the death of a good man to stop this.

"I want all of you who took that vigilante oath to come up now and sign this one, then return to your homes and hang up your guns before anyone else dies needlessly. One innocent man is far too many."

Behind him, Nelson Chace and Wiley Standefer were setting out chairs, each with a copy of the oath Governor Safford had penned. Black Ramblene climbed the steps first. His men followed him, and then, one by one, the others approached. When each man signed the oath that would insure his amnesty, he drifted away.

Tappan Kittery woke up to see Beth looking out the window down the slope toward town. "What do you see?"

Beth lowered her head, not looking at him to speak. "It's finished," she said. "Your men are signing up for amnesty."

Kittery drew a deep breath and closed his eyes. Amnesty! It was more than he could have hoped for. Vancouver must have taken part in pushing for that. It was just a guess, but it seemed like something he would do.

The door opened before he could say anything to Beth, and Kittery turned to see Rand McBride walking in. His heart sank. He couldn't avoid him now. He was bed-ridden. What would he say to Rand McBride? How could he act? He had no idea if he could trust him, and that knowledge tore him apart. He made up his mind right then to confront him. He was almost completely convinced that the law knew the names of each vigilante only because Rand McBride had given them the list. That shouldn't be hard to prove. And if it was true, he intended to also prove he was Baraga's informant, "the Mouse."

Beth turned to see McBride standing there watching her. Her gaze flickered from him to Kittery and back. "You two are like an open book," she said with a half smile. "I can see when I'm not wanted."

McBride laughed and started to protest, but she held up her hands to stop him. "No, I'm leaving, that's all." And she marched out and shut the door.

Kittery wanted to smile at Beth, but he couldn't. His heart was thudding in his chest.

He looked up at McBride as the banker was pulling a pistol out of his pocket. Kittery froze. He had doubted Rand McBride for so long now that he half-expected to be shot. But then he recognized the gun McBride had in his hand. It was the Stephens derringer! McBride was holding the weapon inert across the palm of his hand, scrutinizing Kittery's reaction.

"I could hardly wait till you woke up, Tappan," the banker said. "I'm filled with curiosity."

Kittery stared at him, waiting.

"My question is," McBride went on, a little red-faced, "I know you would never steal anything, my boy. I have no doubt about that in this world. But I have to ask you, where did you come by this pistol we found in your saddlebags?"

Kittery's heart hammered his ribs. He tried to read McBride's face, but the look there was inscrutable. There was no approach to this but to come right out with all of it. "That's the gun Shorty Randall tried to kill me with."

McBride's jaw went slack. "Shorty— My— Shorty Randall? I don't understand. How would he—" He clamped his mouth shut and stared at Kittery, who didn't know what to say. Finally, McBride sighed and dropped the gun to his thigh. "Tappan, this is my derringer! I swear to you. It's a very rare gun in the first place, and it has the initials 'C.B.' inside one of the grips. This is mine! I bought it at Hagar's store."

Kittery nodded. "I know that, Rand." He was too confused to go into more detail. This whole thing was too bewildering for him to know what to say.

"You *know?* How could you know, Tappan? And how could Shorty have—" The banker stopped, and his face paled as a memory burned into his eyes so bright Kittery could see it. "Oh, Tappan! For the love of— Tania told me there was a question of how Shorty got hold of a gun. Well, I know how! I rode my horse to the bank that day because I was running late, and I tied him at the rear of the bank until I came home at noon. I think that derringer was in my saddlebag! Oh, Tap! I— That was the gun that almost killed you?"

Kittery had sat up straight in the bed. He didn't know if he looked as white as Rand McBride, but he felt it. There was no doubt in his mind that McBride was telling the truth. No one could have convinced him otherwise. He cursed himself! He had doubted his friend for so long, and now he didn't know if he could ever tell him all he had been thinking. He felt like an utter fool.

McBride looked a little weak, and he fumbled for the chair beside Kittery's bed and sat down, clasping Kittery's forearm. "Tappan, how can you ever forgive me? I feel horrible. I was an imbecile to leave those saddlebags out there!"

Kittery smiled. His heart had never felt so light. "Rand, I forgive you. You don't know how much I forgive you, my friend. When I found out from Albert Hagar that was your gun . . . I can't tell you how I've fought with myself over this, Rand. I would never have told you, but you must have wondered why I avoided you so much." His head sank back into the pillow, and he let out a sigh. "I ain't been so scared in years, Rand. I was afraid you were Baraga's 'Mouse.'"

McBride began to laugh, and the hilarity spilled over into Kittery so that when Beth came to check on them she raised an eyebrow, shook her head, and stepped back out. They heard her in the hallway speaking to someone else. "I don't even want to know."

As the vigilantes finished wetting their throats at the cantina, some of them rode up to visit their bed-ridden leader.

Kittery listened with regret to the account of how Adam Rolph had died. The only good note was hearing that Standefer himself considered that part of Arizona a safer place to live.

Mort Dansing, vigilante and ex-lawman from Tucson, appeared in Kittery's room early on. A good-looking man of medium height, with dark hair, blue eyes and a groomed mustache, he carried himself with assurance and poise.

"Glad to see those federal boys haven't broke you, Mort." Kittery smiled and shook Dansing's hand.

"Not me, Cap. I'll just move on to other things."

"Like what?"

"I thought about giving the law another shot, if they'll have me. How about you?"

Kittery pondered that question for a moment. "Don't really know. My business with Baraga isn't finished quite yet." He avoided telling Dansing he had no intention of stopping the hunt, but the man caught his drift.

Dansing shook his head. "Bad idea, Cap. They'll nail your hide to the wall. That marshal has everyone's name."

"So I hear. But I was a bounty hunter before I joined the vigilantes. What's to stop that now?"

"The oath is worded to stop it. They'll clash with you."

"Maybe. But I'll wait it out and see. Maybe if they come for me I'll sign—maybe not."

"Good luck then, Cap. That's all I can say."

They shook hands once more, and Dansing left, allowing the line of callers to move ahead.

An hour had passed since the signing of the oath. The last man to stop in and visit was Linus Beehan. He moved toward the bed, eyes flickering to Kittery and away several times. He had always been closer to Kittery than his brothers, Durnam and Ed, who had already come and gone. He found it difficult to speak, now that all their rides together had ended.

"Looks like it's over, Line," Kittery broke the silence.

Linus inhaled deeply and held his breath for several seconds before letting it out. His sad eyes returned to Kittery. "It ain't near over, Cap. I ain't quittin'."

"You have to quit, Line. They tell me it's quit or be tried for murder."

Linus squinted his eyes in sudden distaste at those words, turning his head to stare without interest at the painting of an Arabian horse hanging on the wall. As his eyes came back, they passed over the Remington pistol and its belt and holster that hung from the bedpost.

"Cap, one time you told me you made a promise to Marshal Raines to see all the Desperados dead. You're gonna have to break one promise or the other. Are you quittin'?"

Now it was Kittery who looked away. "No."

"Then either will I."

"Did you sign your name with the others, Linus?" Kittery demanded softly, meeting the youth's eyes.

"Yeah, I did. But that don't mean nothin'. Just a piece of paper."

"More than that. That's an oath . . . and you signed it. Now, if you're the man I'd hoped, you'll live up to it. It was your own name you signed to that paper, Linus—a promise. Man that don't live up to his promises ain't worth the dirt they bury him with."

"I ain't quittin'!" Linus burst out, his face reddening. "Cap, you told me a man that quits somethin' he set out to do ain't no man at all. I don't see you quittin'. And you're the only one I follow."

Kittery smiled. "I'm honored, boy. But I didn't sign, and I don't intend to. You did."

Linus dropped his hands to his sides and bunched his jaw muscles, grinding his teeth. He turned to face the door. His brow bunched up with anguish, and he studied the crack along the top of the door, his watery eyes not seeing it clearly.

"I wanna be like you, Captain. I wanna be a bounty man like you—and free like you."

Kittery grimaced. He didn't want to ruin his young friend's dreams, but he had to. "Not the way to be, Line," he said gently. "You're only nineteen years old. You have your whole life ahead—don't throw it away on false freedom."

Linus turned to face him again, running a sleeve over his face. "Do you think you threw yours away when you brought peace to this place that didn't have none?"

Kittery sighed. "Maybe not. But the thing is times're changin'. Time'll come when you won't be seein' none of my breed. We'll be shot dead, fertilizin' the crops. Or old and arthritic and useless. It may be your generation will never see anything like the vigilantes again. Could be. Anyway, Arizona's not gonna be like it is for long, so you're gonna have to change with it. A man needs to change with the times. Even Arizona will be civilized someday. It needs builders, not gunmen. There's no glory in a gunman, Linus. No glory when your gun smokes and another life is gone, and you know it's because of you. And there sure ain't no glory when you're lyin' bleedin' in some dusty street with a bullet in your guts. There's no pretty girls come runnin', cryin'. No history books. Then you're just a man alone, watchin' your life pour out on the ground, cryin' because you're scared to death, and seein' that no one cares what you've ever done. Besides, I don't believe you were cut out to be a killer. Me, I might not have been either. But the war did its level best to make me one. I can shoot a bad man like you'd shoot a chicken-stealin' dog."

Linus didn't speak or move. Neither man spoke again until half a minute had passed. During this time, Kittery's gaze bored into Linus, whose eyes were focused on the holstered pistol near Kittery's head.

Linus looked at Kittery and tried to meet the smile that formed on the big man's face. "Maybe yer right, Cap. But as for bein' a builder, I don't know. I might stick to ranchin'."

Kittery's smile broadened. "A rancher *is* a builder, Line! One day you may be the biggest man in Arizona." And inside he knew this might not be far from true, for Linus was the kind who would go right to the top, no matter his course in life.

Linus stepped closer, his eyes misted over. "I'll miss bein' with you, Cap. You taught me a lot."

"I'm gonna miss you, too, Linus. I'll stop and see you, time to time." He reached up and clutched Linus's hand for several seconds. "Try for the stars. You can't miss by far."

Chapter Twenty-Nine
The New Marshal

Long after Linus and the last of the vigilantes had gone, Kittery heard three sharp taps at the door. He looked at it tiredly. "Come on in." Though the visitor couldn't see him, Kittery motioned with his arm. "Everyone else has," he added, in a lower voice.

The door edged open. Kittery looked over a short, compact man with a mustache and a tuft of dark beard on his chin. He held his hat in front of him like a shield, and a Schofield .45 butt forward on his left hip.

"Hello, Standefer."

Marshal Wiley Standefer nodded and stepped forward, thumping his hat back on unceremoniously. "Kittery."

"Long way from Tucson," Kittery said.

Standefer shrugged and dismissed that statement with a sideways wave of his hand. "I hear the Apaches almost got you. You weren't far from dead."

Kittery nodded. "Thought I was, for a while." He waved his hand toward the silver United States marshal's star pinned to the left lapel of Standefer's vest. "Guess what they tell me is true. You replaced Goodwin."

"Old Baldy?" Standefer chuckled. "Yeah, he was a little old for the job. Never had much backbone anyway, till near the end."

"Yeah, I know. And that was only to try and hold onto the job. Joe told me about his tricks."

At the mention of Joe Raines, Standefer's face grew sad. "It was Goodwin who drove Joe to that Desperado stunt. He wouldn't let us take a posse that far into the mountains."

"I know. Stupid on Joe's part, though. But after he lost Emily . . ."

Standefer clucked his tongue, and then there was a moment of silence. "So what of the Desperados, Kittery? Are they dead as well as the rest?"

Kittery shrugged. "How are things in Tucson?"

Perturbed, Standefer sighed. "About the same as you left 'em. You want to answer *my* question?"

"You know as well as I do they aren't dead. Tucson ain't *that* far off."

"Yes. Well . . . I heard about Rico Wells. This Cotton Baine he killed—wasn't he a friend of yours as well?"

"One of the best. And he stood no more chance than Joe, not against Wells."

"I guess the rest of them will ride free." Standefer studied Kittery's eyes, looking for a rebuttal. There was none.

"The rest of your group signed an oath promising to cease all vigilante activities. I guess you heard." When Kittery was again silent, he continued. "I brought a copy for you. Will you sign—if I find you a pen?"

Kittery stared at the marshal's badge. At last he looked up to meet his eyes. "No. I won't. Not in *your* lifetime."

It was Standefer's turn to stare. He glared into Kittery's eyes, his lips tight, and then he turned around. Clasping his hands behind his back, he made four slow steps toward the door, then pivoted once more on his heel.

"I had hoped you would feel that way."

When Kittery at length realized what Standefer had said, he cocked his head to the side. "Why?"

"Because. There's no need for you to sign."

Kittery drew his knees up under the sheets. "What about your list of vigilantes? What about the oath Safford wrote up? They tell me any man that doesn't sign will be tried for the murder of Adam Rolph."

"What about my list? Your name isn't on it."

"Why?"

"I took it off when your man first brought it to me. Call it inspiration. Besides, you weren't there when they killed Adam Rolph."

"Why would you take my name off?"

Standefer sighed. "Well, Kittery, I know Joe Raines was a good friend of yours. But he was the best friend *I* had in the world. We rode together for a lot of years. We saved each other's hides more than once. And now I want like hell to hunt down the men that killed him, and I can't. My hands were tied by Marshal Goodwin when I was under him. Now they're tied by Governor Safford. He won't allow funds to take a posse down in there."

"Sounds like the same fix Goodwin was in," Kittery mused.

Standefer chuckled without humor. "Yes. Embarrassingly so."

"You still haven't answered my question. I think I catch your drift, but I'd like to hear it straight from you. Why'd you take my name off the list?"

"Well, Kittery . . . We've never been friends. I don't know why that is. I suppose we're too different. But I've never disliked you. And I've never thrown you in that loose category so-called bounty hunters seem to fall so easily into for most lawmen. You're different. You don't like authority, I know. But you work in almost the same capacity as a deputized lawman. And you have a purpose in your head other than just making money. You're working for peace, too, it seems to me, and most of the bounties you collect are on live men—not dead. I have a feeling if you had been with the men who ran into Adam Rolph he would still be alive."

Kittery felt a twinge of guilt. He couldn't say anymore what would have happened had he been there. Hunting the Desperados had changed him.

Standefer noted Kittery's hesitation at his words, and he hurried on. "I don't know, perhaps things have changed a little since Joe and Baine died. But I know that was hard on you. It would have been on anyone. Your goal these days appears to be killing off the Desperados and their kind. Like me, you're tired of the slow process of the law."

"Everything you say sounds pretty reasonable," Kittery said. "But I hope you're not tryin' to get me to admit to anything. I'm an innocent man."

Wiley Standefer looked at Kittery long and hard. At last he shook his head, and his eyes seemed to hold a measure of disappointment. "If you think you are dealing with a fool then perhaps I have the wrong man."

Kittery held up his hand. "All right. All right. You know where I stand, and you can guess what I've done. Where does that leave us?"

Standefer withdrew a cigar from inside his coat. He lit it and took a puff. "I've been looking for a way to bring down the Desperados—dead or alive. I can't do it because of my responsibilities as marshal. Fortunately, you can."

"And that's why my name isn't on the list. You don't want me to stop."

"That's right. But there's more. If you agree, it won't really matter whether your name was on the list in the first place. It states specifically in the oath your group signed that you will not take up arms without the proper authority. You will have the authority. I didn't know how Governor Safford's requirements for amnesty would go when I removed your name.

"I guess you could say you'll be a volunteer lawman, Kittery. I can't pay you. But you want those men dead. I want to give you the authority and the badge. Takes care of legalities, you see. An appointed lawman, doing his duty. Unpaid."

Kittery raised his brows. "You want to *deputize* me?"

"Yes." Standefer made an arc in front of him with his cigar. "A temporary arrangement, say. Until the Desperados are captured, or . . ."

"Dead," said Kittery.

"Whichever. But it looks like you want it enough to use your own funds toward that end. Let's avoid legal complications, shall we? Take the badge." He produced a six-pointed silver star bearing the words, UNITED STATES DEPUTY MARSHAL from his vest pocket. "This makes it legal—if you'll take it."

"So my name's not on the vigilante list. Keeps Safford's hands clean, don't it? He takes no blame for having an ex-vigilante in his government." Kittery turned his eyes outside, musing. He looked back at Standefer. "I'll take your badge. On one condition."

"Yes?"

"I can deputize anyone I wish, at any time. No matter who, no matter when. Ex-vigilante or not."

Standefer bit his lower lip thoughtfully, then thrust the cigar between his teeth and jutted a forefinger into the air. "These deputies of yours, they mustn't be too many, nor too ostentatious. As with you, a low profile is of utmost importance. But, as my deputy, you will have full authority to temporarily deputize anyone you see fit."

"Well, I do like your offer. And badge or not, I won't stop chasin' Baraga. So let's do it."

"You know the legalities, I guess," said Standefer as he moved closer. "And you know your job. I won't swear you in. Here it is." He dropped the piece of shiny tin into Kittery's outstretched palm. "I trust you won't take overly long."

"Wait."

The word caught Standefer turning toward the door.

"One more thing: who gave you the list of vigilantes?"

Standefer shook his head. "Sorry, Kittery. My sources are confidential. He was promised anonymity."

Kittery almost smiled. He had taken the time to write down the names of every single man of the vigilantes who had come up to visit him after signing the oath. Only two of them were missing, and those two jumped out at him so forcefully he could find no reason to wonder if the informant was one of them.

"Was it Miles Tarandon?"

"Why, no," Standefer shot out in surprise.

"Then Greene. Thaddeus Greene."

Kittery caught the momentary surprise in Standefer's face before the marshal checked himself and cleared his throat. "As I said, my sources are my own."

Kittery smiled as Standefer opened the door. "I guess you said enough."

Almost the instant the door shut, it reopened, allowing Beth's entrance. She brushed it closed with her hip. Her face was flushed from cooking over the wood stove. The result of her efforts she carried in the form of a steaming bowl of chicken soup. She set a tray across Kittery's lap and deposited the bowl.

"Thanks, Beth," he said as he spooned down a couple of mouthfuls without waiting for it to cool. Within moments, he was sipping it from the bowl. His stomach felt as if he hadn't eaten in two weeks. He must have lost twenty pounds, and it was going to take a long time to gain it back.

Hands clasped before her, hair tied behind her head, Beth watched him a while with a smile tugging at the corners of her mouth. A yellow wisp of damp hair drooped down to her brow.

"What did the marshal have to say, Tappan?"

Kittery grinned lopsidedly. "Don't let it get around yet, but . . . he made me a deputy."

Beth's mouth dropped open. "You can't be serious."

His answer was to hold up the badge.

She looked at it, then back at him. "Because of Baraga, I presume."

"You presume correctly, my lady," he teased.

Beth rolled her eyes at the title. "So your trail won't stop then?"

"Was there any doubt of that? If you're worried about Luke, don't. No more vigilante troubles for him."

"He may not be around much longer *to* worry," she said.

"Why?"

"Prescott. He wants us to go, Tappan. If they offer him a job up where the children are we will go."

"Hmmm. Sorta thought that was a passin' fancy."

"No. No-o-o, much more than that. But it is his alone. I want to stay here. I'm tired of moving."

Kittery smiled his sympathy. "I know. I hope you get to stay. There's a lot of copper in these hills. I'm sure Castor will grow."

"It's not so much that I want to see it grow, really." Beth unclasped her hands and walked around to sit on the side of the bed. "In fact, I *don't* want it to grow. I like Castor small the way it is. I've seen cities. I've seen enough of them. No, it's not that. It's just that I have friends here—and so does Luke. We're too old to be starting over."

Kittery chuckled. "Sure. Old as Adam and Eve."

Beth laughed, then went serious once more. "No, Tappan—we're past our prime."

"Unless you want different, Beth, you'll always be in your prime. Just put it in your head."

"It's easy for you to say that. You still haven't begun to settle down. We have—three or four times. See, I lose track."

"Don't let me influence you, Beth. But if Luke wants to go, maybe it's best for now. I think he'll come back. This is your town. You started it. Just stay with him."

Beth rose and walked to the window. She gazed through the loose curtains out at the bright road below, the one running above town. She had been about to answer him but caught herself, putting her hand to her mouth.

At last, she turned from the window with a bored sigh. "I hate to be the one to tell you this, Tappan, but I think you have a visitor."

Kittery looked at her in disbelief. "Oh, please. Another one! I thought I'd seen them all."

Beth shrugged. "If you're too tired I could meet them halfway and tell them to come back later—when you've rested."

"I hate to ask you to, but please do. I'm tuckered."

"It's no problem," said Beth. "I'm sure they'll come back later. I hope you sleep well."

She placed the tray and the soup bowl he had emptied as they spoke on the chair beside the bed, then went to draw the curtains, then a set of heavy drapes, throwing the room into darkness. Picking up the tray, she left in silence.

An hour later, Beth opened the door and peered in, speaking Kittery's name. She caught a quiet snore in response, then stepped into the room. Taking a comb from the washstand, she brushed back his hair, combing his beard at the same time. "I told you you have a visitor," she whispered. "You must look your best."

As quickly as she had come, Beth left the room. She was replaced by the visitor, who eased onto the bedside chair. Already disturbed by Beth, Kittery stirred from his dreams of rocks and sand, cactus and Indians, blood and blazing sun. His eyes, when they opened, were on the person beside his bed. He remained immobile, staring. Light in the room was dim. But it could have never been dark enough for him not to recognize the shadowed face of the girl he had once begun to think of as his own.

Tania McBride had come home.

She rose, seeming not to see his eyes on her. She straightened the front of her light blue cotton dress self-consciously and walked around to the window. She opened the heavy drapes, letting sunlight wash through the window glass and the yellow curtains.

Kittery watched her back, her slow but nervous movements. Her hair seemed to have taken on inches since he had seen her last. It flowed black and shiny past her shoulder blades, hanging loose, the way she knew he liked it. When she turned toward him, the sides of it framed her olive skin, her soft, rosy cheeks. As she walked forward, she occupied her hands by pressing the cotton dress against her thighs. He thought then how beautiful she was, more than he had ever realized.

She came and sat down on the chair, and it was then she looked at his face and caught his eyes upon her. She drew a quick breath. She looked at him out of those big, dark eyes, still innocent, but now sad and wondering, the eyes of a troubled child.

He reached out and placed his hand over hers, which rested on her knee. Almost imperceptibly, she drew another deep breath. Their eyes held. Only when he saw the gentle sinking of her breast did he know she had exhaled.

His lips parted five seconds before he spoke. "Your father told me you moved to Las Cruces to stay."

"I did, when I thought . . . when you were gone." She brought her hand up, then dropped it again to her lap. She looked away and bit her trembling lip, then looked back while tears glistened in her eyes. "Father sent me a telegraph. He said they thought you might die. That was four days ago."

"You got here quick."

"I took the first stage."

When Tania tried to smile, the tears welled up and ran down her cheeks. She stood and threw herself across Kittery's chest, and now her tears kissed his cheeks, her lips, his. She rained his face with kisses, then buried her head against his neck, pulling herself over so that the entire length of her body lay against his.

"Hold me, Tappan, if you can. Please, just hold me."

Wrapping his arm around her back, the big man squeezed. Closing his eyes, he lay there and felt her pulsing warmth beside him, her young, strong body against his. And now he knew what the empty place inside him that summer had been—what he had tried to hide even from himself during the long, cold nights and the lonely days full of death. Now the empty space was filled, and he knew his own life would pass before he let this woman go again.

He mumbled something, and Tania's hair muffled the words. Raising her head, she looked into his eyes. "What did you say?"

He laughed nervously. "Nothing."

"Oh, please tell me."

"I guess I've waited long enough. I want you to know before anything happens. I said . . . I love you, Tania."

"And I love you, Tappan Kittery," she beamed, hugging him once more.

"Then will you marry me?"

Tania pushed away from him with her hands, her eyes searching his. But she never spoke. She only leaned forward and let her tender lips meet his for a long moment, and that was answer enough.

Inside his dry goods store, the obese Thaddeus Greene clutched a broom handle in his beefy hands until his knuckles turned white. His bleary eyes jumped about the room, looking at nothing. Spinning, he stalked to the other side of the room, paused, then returned.

"Damn that woman!" His bare hand lashed out toward the wall, and Maria Greene's photograph crashed to the floor in a shower of glass and broken frame. "Twice up to see him in one day." He burned his own eyes with his whisky breath. He flung the broom away, watching it collide with a stack of tomato cans, toppling them to the floor. One can rolled toward him, resting against his shoe. He gave it a kick into the others, and like billiard balls they flew apart and rolled to a stop about the dusty floor.

"She wants him." He glared at his blurry image in the mirror hanging behind the counter among the cans of peaches and sacks of coffee beans. "And he wants her. So they can have each other. But when I'm good and ready."

As if willed there by him, shoe heels clicked along the porch, and the door swung wide. Maria Greene looked about the cluttered room with confused eyes, bringing her hand halfway to her mouth. The confusion turned to terror as Greene crossed the room, jerked the door from her hands, and slammed it shut. Reaching over, he placed a CLOSED sign in the window, then proceeded to close the shutters and turn the key in the lock.

At last, he turned, and Maria stood in the middle of the room, facing him. Her hands held her abdomen, as if hoping somehow to protect the small life within from what she saw would now come. Her eyes pleaded with his, watching his bulk move in. But her lips were silent.

Chapter Thirty
Good News ... and Bad

Thaddeus Greene's first blow smashed into Maria's cheek like a brick, and she folded to the floor with a soft whimper. Powerful hands ripped her upright, tearing the top of her dress half off. His left hand held her by the neck, while the right slapped her back and forth across the face. He cursed, then spat into her face and threw her to the floor, kicking her in the ribs as she went down.

Somehow, that kick seemed to bring Maria back to consciousness, and she screamed as loudly as she could before the toe of his shoe smashed her in the mouth, turning the scream to a whimper. She only had wits left to curl up into a ball and fold her arms around her knees as her husband began to kick her, trying to get at her abdomen and the life he knew meant more to her than his. He kicked at her legs, tearing the skin of her shins, then tried to pull them down, away from the baby.

Now Maria was strong, and Greene could not get her legs to move. Cursing, he reached out a bloody hand and grabbed hold of what was left of the top of the dress that clung to her shoulder. He ripped as much as he could of that from her, then reached for a coiled rope, unaware of the thudding sounds against the front door.

The door flew open with a crash as the first rope lash drew angry red welts across Maria's back. So intent had Greene become on punishing his wife's supposed infidelity, he didn't turn to see Randall McBride standing in the doorway. The rope lashed again, and Maria whimpered and whispered, "My baby, my baby, my baby."

For a moment, Rand McBride's shocked eyes stared down at Maria Greene. Then, with a furious curse, he lunged across the room. His body collided into Greene's with a terrible impact.

Greene tried to whirl toward the banker, but both of them stumbled across the floor and went down. The glass gun case seemed about to catch them but had no prayer against Greene's two hundred and ninety pounds combined with McBride's weight. The glass buckled and flew into a thousand sparkling pieces, and Greene fell there amid the confusion of pistols, knives and broken glass, McBride on top of him.

Using Greene's chest, Rand McBride shoved himself up and staggered back, looking for a weapon to use against the much bigger man. His eyes fell on the twelve-inch blade of a Bowie knife, and as he crouched and reached for it he glanced at Greene. He was not rising.

Greene lay there in the glass, his flabby fists opening and closing convulsively as a foamy red bubble formed at the corner of his mouth.

Rand McBride paced the floor in Doctor Hale's office. Now and then he glanced over at Tania, seated on the sofa against the wall. If the situation hadn't been so serious she would have laughed at him.

The wailing of a baby broke the silence. McBride's pacing ceased, and he turned in mid-stride to stare in amazement at his daughter. Then his face turned serious again, and he whirled back toward the closed door.

It opened, and out hurried Doctor Hale. Tugging on his beard, he looked from McBride to his daughter, seeming uncertain to which he would reveal his news. He turned to McBride, dropping his hand from his beard. "It's a boy! A boy, and darn healthy!"

McBride smiled with relief. "Good! And the woman?"

Doc Hale dropped his chin. "That scum hurt her something terrible, Rand." He glanced at Tania, then back. "But she's tough. She's going to live, I think. And you know why?"

Tania rose and walked to her father as both awaited Hale's words. "Why, she'll live for the baby—and only for the baby. I think that's the one reason she's alive now."

"What about him?" McBride gestured toward a second closed door.

"I'm afraid, no thanks to you, he'll live too."

"Well, you can't say I didn't do my best."

"Uh-huh. Come with me—I'll show you what you did."

Doctor Hale walked to his desk and picked up his black bag. Working it open, he peered inside, then rearranged his tools to draw forth a cloth parcel. He unwrapped it carefully to reveal a five-inch shard of glass covered in blood. "This," he said, "is what I pulled out of Greene. It went through his back, between two ribs and into a lung. You near did a job, Randall."

McBride studied the glass, the side of his mouth giving a little twitch. Then he scowled. "Sometimes I think fate doesn't know what's best for this old world."

"Right as rain, Randall. If you had left the man five minutes longer he'd have died—bled out like a butchered hog. And maybe the woman and her child would be dead, too."

Inside the closed room where Maria lay, the baby continued to cry. Hale looked toward the sound, then back at Tania and Rand. "Beth's in there alone, folks. Guess I'd better go. Don't wait on me."

"Good, Doc," McBride said with a nod. Turning on his heel, he offered an elbow to Tania and moved toward the door. "Don't care too well for that miserable wretch, Doc," he spoke over his shoulder. "And thanks for coming so quick."

The last days of August swept by. Peace reigned in the flatlands, and the outlaws stayed silent in their mountain lairs, waiting for the smoke of the vigilantes to clear. Thaddeus and Maria Greene's recovery dragged, but Tappan Kittery's was swifter, and by the end of August he was moving around freely once again. He still favored his side, but the bandages had come off.

The second day of September fell on a Saturday, and early in the morning ex-vigilantes began drifting into town in twos and threes. Until midmorning, they hung about the cantina and café, but when someone began to bang on a triangle out behind the sheriff's office, they headed that way, beating the trail dust from their dress pants and Sunday shirts. The crowd gathered around the wooden arena. One hundred and fifty or so were men, dressed in the best they had. Perhaps forty were women in lacy dresses that no one ever saw except on special occasions. Several of them carried parasols.

A hush fell on the crowd, as if someone had fired a gun. All eyes turned toward Vancouver's office, at the corner of which stood Tappan Kittery, dressed in black broadcloth, with a white shirt and string tie. Beside him stood Tania McBride, blushing cheeks obscured by a veil, her slender form covered in a flowing royal blue gown. In that day it had not been accepted as strict decorum for the bride to clothe herself in white for her wedding; her mother had been married in red.

Luke Vancouver, dressed in a light gray suit, broke from the crowd and walked toward the striking couple. Beth followed behind, and Adam Beck, the best man, took up the rear. Kittery and the woman stood waiting, smiling, and the threesome stopped before them, all eyes watching.

"I'm sure sorry this town has no church." Vancouver smiled. "But I hope this day will be special for you. We'll do our best to make it so."

Kittery glanced at Tania, then back at Luke. "You already have, my friend."

Beth stood beaming in her dark blue, lace-trimmed dress, holding a bright bouquet of daisies. Beside her, Adam Beck shifted from foot to foot, shining the tops of his boots against the backs of his pant legs and glancing nervously at the crowd behind him.

Luke Vancouver opened a small, official-looking black book to a page he had previously marked. He began to read the words printed there, emphasizing those he found most important. He spoke of life-long love, of loyalty, of family. At last, he looked up at Tappan Kittery, and his next words were drawn from memory.

"Do you, Tappan Kittery, take this woman to be your lawfully wedded bride, to have and to hold, in sickness and in health, until death do you part?"

Kittery nodded. "I sure do."

The sheriff smiled. "And do you, Tania McBride, take this man to be your lawfully wedded husband, to have and to hold, in sickness and in health, until death do you part?"

"I do." Tania blushed beneath the veil, to some thought all her own.

"The ring, Adam," Vancouver said to Beck.

Adam Beck stepped forward and handed the small gold band to Kittery, who slipped it on Tania's finger with a smile.

"I now pronounce you man and wife," said the sheriff. "You may kiss."

Tania was looking up at her husband as he took her veil and let it whisper down behind her head. He bent and took her in his arms. As their lips met, the crowd cheered.

The celebration went on for hours, taking up the entire town. The orange sun began to melt into the hills, and dust hanging loose in the air gave up its orange cast and turned purple in the shadows. It mixed with the pungent scent of tobacco and gunsmoke. Broiling beef and pots of simmering beans lent their aroma, too, along with the smell of whisky, wood smoke and sweaty bodies. A dog barked persistently from its hiding spot beneath the gallows, answered by others at various points around town. Children ran to and fro, taking turns chasing one another and playing hide-and-seek. Boot heels echoed down the wooden walks, a small caliber weapon went off, and a horse standing before the dry goods store shied and rolled its eyes.

Tappan Kittery stood and puffed his black pipe and watched the celebration. The tobacco smoke curled up under his hat brim, then found the easiest passage into the evening air. The pipe he had taken to smoking since the death of Joe Raines put him in mind of his friend, and he felt the bite of regret that the marshal and Cotton Baine could not have been here to share the day with him. Near a straw-filled arena, where a few die-hard couples still danced, rancher Joe Reilly was turning a side of beef over a long bed of coals that now glowed intense in the deepening shadows. His inebriated brother, Jed, hoorahed him with every turn and gulped from a whisky bottle. Tania was seated at a long table, laughing and talking with Beth Vancouver, Priscilla Hale, Sarah Hagar and several of the other town wives.

A game of horseshoes was going between the gallows and the arena. A shoe clinked against a metal post, and someone swore good-naturedly, then guffawed.

Kittery walked toward a pile of loose crates near the corner of the dry goods store, and as he was about to sit down on one of them he glanced up to see a lone horseman turn off the street and ride toward him. The last brilliant sliver of sun had oozed behind the Sierritas, but plenty of light remained to recognize Mort Dansing. Dansing stepped off the chestnut gelding and ground-reined him beside the store, then, tipping his black hat back a little on his head with a knuckle, he stepped toward Kittery.

Kittery walked to meet him smiling, but Dansing's face was grim. They paused four feet apart, and Dansing rubbed his whiskered jaw.

Kittery's smile faded. "What's wrong, Mort?"

"Bad news, Cap. Any whisky around?"

"Come on over."

Dansing walked behind Kittery to the table closest to which Joe Reilly was performing his meat roasting. Kittery's heart was thudding dully.

"Good to see you, Danse." Joe smiled. When the smile was met only by a nod, he cleared his throat and looked at Kittery, then back. "Somethin' wrong?"

"Yeah, Joe," replied Dansing. He turned to the younger Reilly. "Jed, any of that *aguardiente* left?" He indicated the bottle with his finger.

"Sure, fella," Jed spoke thickly. "Plenty."

He reached into a crate beside the table, and fumbling about for a second he came up with another long-necked bottle. "There you go. Drink hearty." He hiccuped and covered his mouth in shame.

With a motion of his chin, Dansing invited Kittery to follow and walked back over to the pile of crates. They sat down, and Dansing took several swallows of the liquor. Then he handed the bottle to Kittery and swatted some of the dust from his vest while Kittery drank.

Kittery lowered the bottle and passed it back. "So what is it?" He had a grip on his heart now, but he could feel the tension in the air.

"You noticed Jed Polk and the Beehans didn't make it today."

"I thought about 'em. And you."

"Yeah. I don't like being the one to tell you this, but here it is: Ed and Linus had a run-in with Silverbeard Sloan and Crow Denton. It came to shooting, and they killed Ed."

Kittery's face darkened with the tightening of his guts. "And Linus?"

"He's shot up some. But not dead. Polk and Durnam Beehan went after Sloan."

Kittery swore. "Who do they think they are? They can't take Sloan. Ah, that Ed was a good boy."

"That's not all," Dansing went on. "They also got John Laynee and Johnny Hedrick." Kittery recalled the two men as gunfighters from Tucson who had joined the vigilante crusade during its last days of glory.

A huge bonfire springing up in the arena made Kittery glance over. They were burning the straw they had used for the dance. Everyone watched in the half-light, their faces glowing orange and their eyes flickering as the sparks flew high into the air and the heat increased. Priscilla Hale was talking to Tania, but Tania's eyes were intent on Kittery.

With a heavy heart, Kittery turned again to Dansing. "They dead, too?"

"Hedrick is. He died on the spot. As far as I know, Laynee's still alive. He was when I left six hours ago, but he's crippled up. He won't use a gun again. Sorry to ruin your celebration, Cap. I figured you'd want to know."

"You did good, Mort. But why you? How'd you end up as messenger?"

Dansing's smile was grim. "It goes with the job." He reached into his left vest pocket and withdrew a Pima County deputy sheriff's badge that glimmered in the firelight.

Kittery nodded. "You're a hero now, so they take you back. Funny how life goes. Where are Sloan and Denton now?"

"Somewhere in the Catalinas. I trailed them after the shooting in Tucson. That's where I ran into Polk and Beehan and found out the twins had been shot four days back. I also met an old hunter up there named Creech."

"Judd Creech."

"Yes. He said he knew you. He wanted me to tell you he found out what you do for a living, and he found out about Joe Raines and your cowboy friend and the vigilantes. He'd like to help you out now. Then he said something I wasn't supposed to understand, something to do with revenge and justice. Said for you to remember the grizzly, and you'd know why he wanted to help you."

"I guess I know," Kittery drawled. "Does he think he can take me to the outlaws?"

Dansing shook his head. "James Price is up there after them, too, and he's probably closer to finding them than anyone. There's a whole passel of fellows combing the mountains to get that reward. Creech said he'd get you to Price's cabin, but from there no one can pinpoint Sloan. He and Denton are running a horse rustling operation and keeping out of sight."

"Creech could find them," Kittery said. "All he'd have to do is try."

Dansing shrugged. "I was raised an Apache, Kittery. But I don't know if I could find them myself." He stood up and pulled his hat brim down. "I know you'll go up there, Cap. I'd like to accompany you if I could."

"Mort, you're always welcome."

The deputy nodded. "All I can say is be careful. Laynee and Hedrick were fast men."

"Well, you know I've never been fast, Mort. Just wily. Could be speed isn't what I need to beat Sloan."

When Dansing headed for the hotel, Kittery looked over to see Tania still watching him. The other women had moved away from her now, even Beth. She made her way over as if she didn't want to confront him but knew she must. She stopped in front of him and took his hands in hers. "I saw that man show you a badge, Tappan. Something's wrong."

Kittery cleared his throat. "That was Mort Dansing. You've heard me talk about him. He's a deputy in Tucson now, and he brought bad news. Tania . . . I'm gonna have to leave you."

"Tappan," she pleaded, dropping his hands and gripping his arms at the elbow. "You can't go. Not now. Not on our wedding night." Her eyes beseeched him.

"No. Not tonight. You know I wouldn't do that to you. But somethin' bad has happened, Tania, and I'm a deputy marshal. I have to go."

"The Desperados," she said, barely above a whisper.

Kittery nodded. "They killed Ed Beehan and Johnny Hedrick and shot up Linus and John Laynee. I'll have to chase 'em. But I wouldn't leave you alone on your weddin' night." He forced a smile. "Nor tomorrow. Nor the day following. But when a week's gone by, I'll be fixin' to ride."

Tania Kittery looked down and swallowed hard a couple of times. When she raised her eyes, tears glistened in them and on her cheeks.

"Tappan, you're a man with drive. And a vengeance in your heart. I still remember the day you buried Joe—that look on your face. I married you because you are what you are. A loving man, fiercely loyal to his friends. And I'll never regret that, no matter how short our time together is. If this is something you must do, for peace within, then go. But I'll be waiting for you every night with a lighted candle in the window. And when you fulfill your quest, you'll come back to me."

Chapter Thirty-One
Beginning the Hunt

Cold brushed the dawn air, swirled thick about the darkness. The coyotes made their last rounds in silence and an owl sent its final cry from the top of the house. A bright star careened past the bedroom window, drilling a hole through the jagged peaks of the Santa Ritas with such force that its headlong flight never slowed.

Lying immobile, Tappan Kittery held the blankets under his chin and stared unblinking at the world before dawn. The pendulum of the big clock in the corner was still. Dim light reflected off its smooth golden edge. Beside him, Tania's breaths came slow and steady, warm against his ear. She moaned and whispered an unintelligible word. The muscles of the arm she had draped across his chest tensed.

Easing her arm off him, he rolled onto his side, inviting the cold rush of outside air that flooded down between their bodies. Tania whimpered again, and he leaned over and kissed her soft, warm cheek. He rested his arm across her side, feeling her firm body beneath the loose cotton gown, and sadness came over his face. The last several nights were still vivid in his mind, especially last night, when they had lain in each other's arms and made love until far past midnight. The time hadn't seemed to matter. What a shame to leave her now, with so little time between them. But outside it would soon be growing light, and go he must. Mort Dansing would be waiting in the street.

Slipping from the covers, he eased them back to minimize the flow of cold air. They had said goodbye before falling asleep. He didn't want to wake her now.

He pulled on his striped denim jeans, his high-topped boots over them. He opened a drawer, groped for his spurs and buckled them on. He donned a long-handled undershirt against the cold, then his blue wool shirt as he opened the door and stepped down the hall to the main room.

He didn't need to light a lantern, for his gear and duffel bag with spare clothes were already prepared. Almost without thought, he picked his gun belt from the back of a chair, buckled it on and snugged it down against his hips. He took Cotton Baine's long-barreled Colt .45 and the American .44 from the tabletop and let them drop into his clothes bag with a couple boxes of shells. Then he took his coat, a battered Springfield carbine he had purchased at Hagar's store and a sack of biscuits and beef and stepped out onto the porch.

The air slapped him in the face. The Sonoran desert, broiling hot in sunlight, could in comparison seem bitterly cold on these fall nights. He shrugged into the sheepskin-lined green corduroy coat, then shivered in the dark while he set down the rifle and grub sack and rummaged through his clothes to find his pipe and matches. He tamped the pipe and lit it, and then, clamping it between his teeth, he crossed to the stable, tugging on a pair of elk hide gloves.

The black stallion whinnied as he smelled his master and leaned his head over the top bar of the corral. "Mornin', Satan." He scratched the stud's nose and between his ears, straightening the forelock that hung almost to his eyes. Doubling over to step between the corral poles, Kittery moved to the sacks of oats in the shed. The black tagged along but knew better than to nuzzle; he had learned that lesson long ago.

Kittery poured a bucket half full of oats and corn, and while he threw on a doubled up blanket and the Denver saddle, the stallion, concentrating on his meal, never twitched. Only when the cinch was drawn up and the cold latigo touched his hide did he shiver and glance at his master almost plaintively. Then he went on inhaling the last specks of oats and waited while Kittery replenished them and moved off to the adjacent corral to feed the mares.

In minutes, Kittery worked the pump handle and water splashed into a bucket, and by the time Kittery left the mares' corral with a full trough, the stallion had sucked the bucket nearly dry. The man walked to him, holding a bridle, and as the horse lifted his head Kittery slipped it on over his ears.

They walked out, and Kittery left the gate hanging open to the now empty corral and picked up his belongings from the porch, then climbed into the cold saddle and moved down the road to town. A soft yellow light seeped into the sky, chasing a light blue that moved into violet, bumping out the stars as it advanced.

The glowing red stub of a cigarette lit the whiskered visage of Mort Dansing where he stood beside his ground-reined chestnut near the hotel. Reminded, Kittery glanced down at the now cold bowl of the pipe between his teeth. He took it and knocked out crumbs of unburned tobacco on the saddle horn, then let it fall into his coat pocket.

Dansing mounted as he saw Kittery coming, meeting him in midstreet. "Morning," Dansing said quietly, in reverence to the dark.

"Hope you haven't waited long." Kittery reached and shook the lawman's hand.

"No. Just woke up."

They trotted north up the ghost-silent street. Neither man spoke. Hooffalls and creaky saddle leather alone told of their passage. Kittery's eyes and mind were on the house up the ridge, where his little love slept. He burned with the desire to return to her, and he cursed the devil who had sent him on this crusade. Savage Diablo Baraga.

A couple miles out of town it was still not light, but the world showed signs of life. A coyote began to yap, saying farewell to the stars, and the fuzz-tipped wings of a home-traveling owl made no sound as it sailed past them. Those wings had been created silent to sneak up on little creatures in the dark. It was the same silent way Kittery would have to dispatch the Desperados.

Sheep bleated a hundred yards from the road, and though Efraín Valesquez was not in sight a dim light glowed from his wagon, diffused by the white canvas covering. A shadow moved across the front of the light.

The sun took over the world, and at ten o'clock, when it was well up in the sky, the steep southern face of Black Mountain came into view on the left. Kittery reached into his food sack while he rode and ate a couple of biscuits and a chunk of beef, offering Dansing the same. Tucson was less than ten miles away.

The flat-looking, square adobe buildings of Tucson greeted Kittery's and Dansing's eyes at eleven o'clock. They could be seen shining among the stands of mesquite and paloverde and the saguaros that stood sentinel over everything. Mixed among the native vegetation grew vineyards and cotton fields and orchards of peach and walnut, orange and pecan trees, irrigated by springs and ditches and by the wide shallow Santa Cruz. The road ran straight through a foot-deep stretch of the river, and the horses splashed through and up the opposite bank.

Tucson was alive with activity. A train of freight wagons stretched along half the length of Main Street, with thirty mules hitched to a string of six huge, high-boarded wagons with wheels seven feet across. A couple of swampers walked the line, checking every hoof. A heavy-jowled man with a flat-brimmed black hat, holding a coiled bullwhip, turned from supervising the swampers to watch Kittery and Dansing plod by. Kittery was used to his uncommon size drawing attention, and he nodded at the freighter. The man nodded a greeting back to them, then, for their benefit, yelled an obscenity at his closest mule and its immediate ancestors and cracked his whip smartly in the air.

A Texas and California stagecoach rocked in from the northwest, scattering litter and debris from the filthy street with the wind from its wheels. It nearly ran over a loose hog as it drew up in front of the stage station, and the driver wasted no time in climbing up to throw luggage down to his shotgun guard, whose feet barely touched the dust before the first bag lit.

Loafers roamed the streets, now and then kicking aimlessly at garbage piles deposited without care on nearly every street corner. The refuse would in time be carried away by wheels, hooves and scavenging animals, but more of the same would replace it. Soldiers from Camp Lowell mixed in with the others who lazed in front of the scattering of cantinas and brothels. The latter were especially prevalent after Kittery and Dansing turned onto Congress Street, the *Calle de la Alegria*, and then onto Maiden Lane, the *Calle de la India Triste* (which in Spanish meant "Street of the Sad Indian Girl"). Indians and Mexican peons tugged wooden carts here and there on Maiden Lane, some of them with dogs walking along in the shade underneath. They touted their many and varied wares—chickens and turkeys, pigs, curly-horned goats, sheep, grain, wild hay, melons, mesquite firewood and drinking water. But as for business competition, they fought a losing battle against the Mexican, Chinese and white "maidens" who frequented street corners and advertised their own wares, not so many nor varied, but more tempting to the male passersby.

Inside a cantina dubbed La Ventana, a woman was screaming in Spanish, and someone smashed a bottle. Kittery and Dansing picked their way through the midday crowd of this ever-narrowing area between Maiden Lane and Congress Street known as "the Wedge" and pulled in at a dusty adobe sporting the faded words, THE BORDER HOUSE.

Kittery looked a question at Dansing, who frowned. "This is where Hedrick bought it."

They tied their horses and, considering the street's common loafers, slid the rifles from their boots so someone else would not do it for them. They pushed through a sheet of colored beads into the inner coolness of the establishment. Like most of Tucson's buildings, this one was built with walls over a foot and a half thick, and with only a few windows set high in the wall. That kept it cool enough to stand in the summer and maintained its heat in the winter, but it also kept it dark as the inside of a kettle. The sickly glow through the oily chimneys of a dozen lamps couldn't stir up enough light to shave by.

Eight bare, round-topped tables were scattered at hazard across six hundred square feet of hard-packed dirt floor littered with coarse sawdust. Four hand-fashioned low-backed chairs sat at each table. All were unused at the moment, seats shining in the half-light, polished by the pant seats of the thousands of customers who had slid across them in the years since the place had opened. To the left, three half-empty bottles of dark and amber liquor and a scattering of empty glasses littered a plank bar. A short Mexican with one spot of hair in the middle of his dome of baldness and a patch over each ear stood behind it, wiping glasses on his stained apron and offering a smile with only half the teeth God had intended for it.

"Buenas dias, Deputy Dansing. What can I do for you, *mis amigos*?" The man stood straighter, setting down the glass on which he had been working and raising his eyes farther and farther until they settled on Kittery's face. He gave Dansing a wide-eyed look of amazement at the big man's size.

Kittery let a dollar clink on top of the bar and teeter to a stop. "Two beers and some talk."

The Mexican complied with the beer, filling two clay mugs to an invisible mark indelibly branded into his brain. Shutting off the valve, he set the second mug in front of Dansing, smiling all the while.

"Now you weech to talk?"

"Uh-huh. A while back you had some trouble in here—a shooting."

"Oh, jess!" The Mexican nodded vigorously, looking from one to the other with widening eyes. "Eet ees night I 'member much. One man, hees name ees Sloan. He come een tha' night weeth hees frien' . . ."

It had been sometime after one o'clock in the morning when Silverbeard Sloan and Crow Denton entered the Border House. Sloan must have felt cocky because he had gunned down Ed and Linus Beehan without so much as a scratch to himself. When he came to the bar he began to shove the other patrons out of his way and challenged several of them to fight. The bartender sent for his two friends and adopted guardians, ex-vigilantes John Laynee and Johnny Hedrick. That was what Sloan wanted.

Johnny Hedrick was first through the door, and when one of the patrons, in relief, spoke his name, Sloan turned with a pistol already drawn and shot the surprised gunman twice in the chest. His next shot shattered John Laynee's collarbone, thrusting him back out on the porch, where he collapsed. All this time, Crow Denton remained placid beside the door, cradling his rifle. Sloan sipped his whisky down in peace and then stepped outside.

There, Sloan shot Laynee through his good shoulder and broke his arm at the elbow with a final bullet. The pair rode out without Pima County Sheriff William Barden lifting a hand to apprehend them.

The bartender finished his story with a sigh as a longing look seeped into his eyes. "John an' Johnny. They were ver' good men, I think. An' ver' good frien's. But Silverbeard, he ees evil. I weesh I ha' never call them here tha' night." He shook his head, his eyes misty. "You know, before tha' night there were many who come to fin' thees bad man who steal horses. Now all are gone. He ees no one to face with the gun."

Kittery and Dansing remained at the Border House for a dinner of beans, eggs and tortillas, washed down with water drawn out of the Santa Cruz. Next, they went to the livery stable, where Kittery bought two geldings, a bally-faced bay and a short-backed sorrel. He knew the Santa Catalinas enough to realize he couldn't ride Satan every day while hunting Sloan and Denton. He was a tough horse, but not tough enough to make a go of that country day after day. Then they moved on, despite the heat of the day, toward the Santa Catalina Mountains. On the way, they passed the gray adobes of Camp Lowell, its grounds silent now while the troopers waited for the heat to subside.

The sun sank, and the day's heat died with it before the pair pulled into the rugged mouth of Molino Canyon, in the Santa Catalinas. As they rode in sight of Creech's cabin, his big black dog, Griz, began to growl, and the hair stood up on his back. He walked stiff-leggedly toward them, eyes ferocious, but he never barked.

"Hello, Griz. Hello, the cabin," Kittery called.

A mule brayed annoyingly and thrust its head over the corral, twitching its long ears. The door of the cabin opened to reveal Judd Creech, dressed in a filthy long john shirt, his crooked legs clad in greasy buckskin, his feet bare and callused. He gazed toward them, trying to decide if he would need the Sharps rifle hanging from the fingers of his left hand.

Kittery nudged the black in closer, tugging the geldings behind, as Creech stepped away from the cabin and walked forward, pace quickening as he drew near.

"Pilgrim!" he beamed. "Hey, Pilgrim. Shut up, Griz."

Kittery stepped from the saddle, followed by the deputy. "Howdy, old man." He thrust out a big hand as Creech came up, and they shook.

"Ye ain't dead yet, Pilgrim. Injuns ain't got yuh."

"Nope, but they tried." Kittery touched the scar on his ribs without thinking about it. "And the outlaws did, too."

"So I heard. Wal, come on in! Nice t' see yuh ag'in there, Dansing." He nodded at the other man then herded them both toward the *ramada* outside the front door of the cabin. It was built of slender tree trunks laid by side and overlaid with leafy branches. "It's still a mite hot in the cabin," he said. "But it's cool out here—the water is, too."

After the horses were hobbled and left to graze, Creech forced Kittery and Dansing to be seated on the two hide-covered chairs while he went to an olla that hung against the wall. He ladled two clay bowls full of water and handed one to each man.

"How was yer trip?"

"Fine, under the circumstances," said Kittery with a grim nod.

"Yeah. Yes sir. I hear yuh lost some friends. On top o' them yuh lost afore. I found out all about yer misfortunes."

"That's what Mort tells me. I'm here for Sloan and Denton. You think you can find 'em—for me?"

"Yeah, Pilgrim, I figgered that's why you was here. But I'm afeered the answer's no. I cain't he'p yuh."

"Why?"

"To be right truthful, they're too good. That Denton fella has a heck of a instinct fer hidin' a trail."

"And you have an instinct for findin' one."

Creech chuckled. "Yeah, I knowed yuh'd say that. I thought so too, but I reckon it ain't true. Like I told Dansing, here, I wanted t' he'p yuh, so after he left I went out t' look fer where they was. Couldn't find 'em. Nothin'. This Denton's a breed, they say. Could be that gives 'im an edge. Yeah, I even worked with yer friend Price fer a while—till we had us a fallin' out."

Kittery raised his eyebrows. "Fallin' out? What happened?"

"That fella has the head of a mule. Won't listen t' nobody but hisself. I couldn't take it. So I left."

"Sounds like Price. You can at least get me to him, then."

"No problem a-tall—in the mornin'. But listen now. You two c'n lay up here t'night. I'll fix us up some sourdough cakes 'n' sugar syrup, an' I got a painter yestiday, too. We'll have us some steaks—have a real wingding!"

Kittery nodded, reminiscing on the loneliness of this place. He glanced sideways at his partner. "Sounds fine to me."

"And me," Dansing put in. "Never tasted painter before." He smiled at his usage of the mountain man's term for a cougar.

Creech clucked his tongue and winked. "Best meat there is, mister. Pilgrim, why don't yuh start a fire in the hole?"

Kittery nodded and pushed up from his chair. A small rawhide box beside the door held wooden matches, and he picked it up as he stepped to the fire hole. He began to gather small twigs and dry grass, which he laid together in the blackened pit about a foot deep and two wide that sat off to one side of the doorway and four feet out. Dansing collected firewood and arranged it atop the tent of kindling Kittery had already started. It was tinder-dry. One match and the entire pile soon crackled away.

Behind the cabin beneath some cypress trees, Judd Creech had fashioned a box with numerous holes in it and covered it with burlap. Above it he had suspended a basket woven of grass nearly waterproof with pine pitch. Periodically, he would fill the basket with water, which found its way, drop by drop, through the few tiny openings. This continuous dripping on the burlap, which was in turn kept cool by the breeze, served to keep meat fresh for much longer than otherwise would have been possible. It kept Creech from having to store all his meat supply in brine or jerk it to keep it from rotting.

Creech sawed off some steaks and carried them back around the house to where the fire popped and crackled.

"Put that on there when the fire dies down—but don't let 'em get too done, now. Takes the flavor away. I'll fix them cakes."

Within the half hour, the trio was forking down the sugared cakes and mouthfuls of the mild cougar meat that Dansing commented was reminiscent of bighorn sheep.

Griz lay a few yards from the fire, watching it flicker, shifting his stony expression from Creech, to Kittery, to Dansing. Whenever one of them glanced his way, his tail thumped the ground once, then again went still.

Finished with their meal, the three remained seated, talking the evening by while a cool wind ran off the mountainsides, flickering the glowing coals to life.

"I hear Geronimo an' his bunch jus' kilt a couple o' settler families down by Patagonia," Creech said. "Come out o' the rocks at dawn and wiped 'em out in bed. Killed 'em all, even the little ones. If y' ask me, the only thing good about an Apache is they don't gen'rally scalp. Any other tribe would've."

Kittery's glance shot to Mort Dansing, who had lived with the Chiricahua Apaches from the age of nine to his early twenties. The Apaches had been more influence on him than any white man ever had. But Dansing's face was placid, his eyes dropping to the ground as he pretended to concentrate on a piece of meat in his teeth.

"Heck of a nuisance," Kittery said. "Cochise made peace an' then died, an' we got Geronimo and Juh and Nolgee to replace him. Some bargain."

Dansing changed the subject, making Kittery have to hold back a smile. "You heard about Custer and the Sioux, up north, didn't you?" He directed the words to Creech, who shook his head. "You know who Custer is, of course."

"I hear tell of him, off and on," Creech admitted. "Nothin' much."

"Well, he's dead now. He managed to get himself and about two hundred and fifty troopers killed up on the Little Bighorn River, in Montana."

Creech had come erect. "When did that happen, by tar?"

"This has been a good two months, back toward the end of June," Dansing recalled. "Made all the papers. The Arizona Citizen made a big hullabaloo about it, comparing the Sioux and Cheyenne to the Apaches."

Creech clucked his tongue. "Two hunnerd fifty men. Huh. Don't imagine that many 'Paches ever looked at each other at one time."

"They say there were a good couple of thousand warriors in the bunch that did in Custer. If the Apaches could get together like that they'd *own* Arizona by now," Dansing said.

"Reckon that was a heck of a note fer the big Centennial celebration," Creech mused.

Dansing shrugged. "The news didn't reach a lot of places until after the celebration."

Talk of Custer's death reminded Kittery of Alan Peck, Tania's one-time beau, and that in turn reminded him of Tania. He couldn't stand to think of her now, when she was so far away, so he interrupted the conversation. "They say Sloan and Denton are stealin' horses. Where from?"

"Ever'where." Creech threw his hand sideways. "There's five or six ranches east o' here that raise stock for Lowell and Camp Bowie, over in the Chiricahuas. Good animals."

"And you can't follow their trail?"

"You'll see, Pilgrim. These boys're good at what they do. Even if I thought I c'd find 'em, in time, I don't *have* the time. Hides're too good right now an' gettin' better. I cain't lose a minute's huntin'."

Kittery glanced over at Dansing. "What about you? We haven't talked about tomorrow. Which way'll you be headed?"

"Tucson. I'd like to stay and help you, pard, but I should've been back before now. Sheriff Barden hates my guts anyway, and he'd like nothing better than a reason to fire me."

"Then I'll need to find James Price for sure." Kittery turned back to Creech.

"Can you take me to him in the morning?"

Creech stood up, using only his wiry legs, and looked down at Kittery. He pointed at the fire. "Better put that out before you bed down." Then he nodded. "Sure, I'll take yuh t' Price's cabin, an' then I come back here." He jabbed a finger at the ground. "But if yer goin' after 'em, you keep a gun in yer hand, Pilgrim. Word is, no one alive can outpull Silverbeard. Not even Colt Bishop hisself."

Chapter Thirty-Two
Price's Cabin

The next morning the three men saddled their mounts in the dim light while the sun was a pale blue wish behind Piety Hill. Twenty miles to the northeast, the jagged ridges of the Galiuros looked foggy in the half-light. Several wisps of cloud were tinted pink on the horizon with the coming sunlight.

Dansing drew his cinch tight and shoved his Winchester into the scabbard, then swung aboard. He reached down to shake Kittery's hand and dipped his hat brim to Creech.

"See you boys in Tucson. Cap? Good luck, and take care of yourself."

Kittery put a hand on Dansing's knee. "Dansing, I wish you could stay, but I know you have a job to do. There's one thing I need to ask you."

Dansing waited, face noncommittal.

"If I don't ride out of here, I want you to look in on Tania."

Dansing sighed and nodded as his face went a little sad. "Thanks, Cap. Thanks for trusting me." The mask of an Apache settled over his features, and he whirled his horse away and put heels to its ribs.

Kittery watched Dansing's retreating back. Hearing no sound from Creech, he looked over to find he was watching him.

"You prob'ly should be headed that way with yer friend, yuh know. Yer out o' yer element here."

Kittery's mouth formed a half smile. "I'll make do. I always have." He put a foot into the stirrup and swung onto the stallion's broad back. "You ready?"

Creech shrugged and swung onto his mule in one easy, flowing motion while it was already walking. Without looking back, he angled toward the rim of the canyon.

For four hours, Kittery and his horses followed Creech and his mule over the steep, rocky southeastern flank of the mountain, leaving behind the oaks and juniper. They rode in the timber now, ponderosa pine forest with white pine scattered in now and again as they ranged around six thousand feet.

A full six hours found them on Guthrie Mountain's northern slope, and there, where the mountainside began to rise once more to become Evans Mountain, Kittery had his first glimpse of the cabin. It was hugged by ponderosa pine, some of their limbs lying over the roof. Kittery guessed being hidden meant more to Price than any worry about lighting the place on fire with sparks from his chimney. The cabin would have been easy to miss if a man hadn't been looking for it. Its front wall and one window looked down-canyon toward the desert and the southern edge of the Galiuros, in the distance.

Here Creech drew rein, and after studying the cabin for a moment he turned to Kittery. "Wal, that's the place."

"You comin' in?"

"Naw, I've gotta head on back, Pilgrim. I got my own business needin' tended to."

Kittery nodded. "All right, then. I reckon I'll be seein' you. Thanks for showin' the way."

Creech spat. "Yeah, an' now you jes' be keerful an holler out afore yuh go up t' that front door. That feller's crazy, an' he'll shoot yuh."

Kittery laughed. "You have a good trip home, pardner."

When the hunter had ridden away, Kittery let the black have his head and pick his way across the timbered side of the mountain, towing the geldings behind. Soon the ground leveled out, forming the bench where the cabin sat.

Thirty yards from the cabin door, Kittery drew in and hollered out. No answer, so he let the black walk on in. The dusty, pine needle-strewn yard was quiet. The corral to the left of the cabin, woven among the trees themselves, stood empty, its long pole gate hanging open. The cabin was a simple structure of rough-hewn logs, the spaces between chinked with white clay. Deerskins sewn together hid the single window, keeping out the heat of the day. An olla half full of water hung from a corner eave, dead flies littering the dusty scum of its surface. Kittery dismounted here.

Taking care to brush away the flies and dust, he sipped of the brackish liquid. He let the horses drink from a nearby trough in front of the corral, and the first time they stopped sucking and raised their heads he pulled them away and tied them to the corral's top rail.

With his back to the cabin, Kittery removed Satan's saddle and the packsaddles from the others. A sweaty horse should not be curried, but it didn't take long for the sun to dry the horses' hair, and then Kittery proceeded to rub them down thoroughly. All the while, his eyes scanned the valley and the surrounding trees, keen for movement or hint of anything out of place. But all seemed right; nothing except a squirrel and a handful of gray jays weaving through the tree branches stirred to disturb the afternoon.

Taking off the stud's bridle, he turned the three of them through the corral gate, then walked to the cabin and eased open the door. He half expected to find a dead man in there, but nothing so dramatic greeted him. Only more silence and a packrat scurrying across the floor to dive under the lone bunk.

Newspaper pages and wanted posters layered the four walls, along with advertisements for horse salve, soap, coffee and a host of other commodities. One poster advertised the big 1876 Winchester, the latest in powerful small arms, and one showed a layout of the Smith and Wesson "Russian" revolver and its barrel lengths and variations. A wool blanket covered the bed that rested against the far wall; another was folded at its foot. One blackened cast iron pot sat on the cook stove near the head of the bed, and against the left wall stood a handmade table and two chairs. The several items of gear in sight around the room were arranged in corners or hanging from nails on the wall, nothing out of place. Kittery had expected as much of Price.

The hangman had nailed a wooden crate to the wall for a cupboard. After checking the empty pot on the stove with mild disappointment, Kittery turned his attention to the crate. It contained four cans of tomatoes, one tomato can full of coffee grounds, a cotton sack stuffed with biscuits, and several unidentified cans and items in bags, along with a paper bag full of rye crackers.

Removing the can of coffee grounds while he chewed on one of the stale biscuits, Kittery set the can on the floor, then went about putting a fire in the stove. The day had some heat yet to give, but temperatures plummeted this high in the Catalinas, so the wood's heat posed no discomfort. With the fire going, he set a pot of coffee on to brew, then occupied his hunger by chewing on a handful of crackers and another of the day-old biscuits. He thumbed through the pages of an English grammar book and an old copy of *David Copperfield* retrieved from under the bed. Price had all the entertainment he needed here, thought Kittery with humor. He even had an edition of Mary Shelley's *Frankenstein* and Shakespeare's *Hamlet* and the big black tome of Blackstone, *Commentaries on the Laws of England*.

Growing bored, he rummaged once more through the cupboard and found a damp gunnysack containing a slab of bacon, and another with potatoes, carrots and onions jumbled all together. Nodding with satisfaction, he withdrew both sacks, picked up a jackknife and began to prepare supper for two.

Later, as deep afternoon shadows slunk across the little room and the slanting yellow rays of sun coming from the cracks above and below the door died, he lit a lamp and sat at the table perusing a copy of Montgomery Ward's new mail order catalog. That was a handy little item like nothing anyone had had the use of until recent times—1872, to be exact. A man could order clothes, household goods, guns. Someone said there was even talk of selling houses out of them. He would have to see that.

Now and then he tested his bacon soup. As yet, there was no sign of Price. During a pause in his reading, his thoughts wandered to Castor and his lonesome bride, sleeping all alone in their bed the night before and tonight. She must miss him now, as he did her. Her eyes appeared and reappeared in his mind, gazing up at him full of love. One of the hardest things he had ever done was ride away from her.

Thinking of Tania in Castor, his mind turned to Maria Greene. She had seemed in good spirits the last time they talked, nursing her new baby boy, Miguel. During his last days in town she had been healed up enough to come and eat with him and Tania and Rand, and they had talked a good deal. She told him in private how happy she was that he and Tania could find each other—someone to really love. He felt sorry for the Mexican woman. Her life had been filled with nothing but pain and loss. She had lost her family, then Emilita, then nearly her baby. Now she was in the process of waiting for a formal divorce to be finalized. What a sad life. He was sorry he could do nothing to help, but only time would heal her physical and emotional wounds.

As for Thaddeus Greene, he had returned to working his store, but his business was decimated. Castor had taken out its own brand of justice against him, in spite of the amnesty oath they had signed, and the first day Luke Vancouver went out of town several masked men had taken the merchant out to the arena and flogged him with a cat-o'-nine-tails. It was a not-so-uncommon punishment in the West for a man who beat his wife.

Outside, stars filled the sky. There was a sliver of a moon. Stepping to the window, he replaced the deer hides he had removed earlier, then returned to the table and blew the lamp. He put the catalog he had been reading in its place, then sat on the bed and leaned against the wall. Only the orange, flickering flames in the stove dimly lit the room. The coffee simmered in its pot on the corner of the stove, growing stronger—and better—with each passing minute and filling the room with aroma. He sat in the dark, lost in his thoughts of Castor.

He was holding Tania in his arms, whispering in her ear. He ran his fingers through her long, full hair and kissed her bare white shoulders. She ran her hands over the curled black hair on his chest and down his back, pulled him down and kissed him and led him to their bed.

The shrill neigh of the black stallion rent the night. In an instant, Kittery was crouched beside the foot of the bed. Pistol in hand, he waited for the call of another horse from the darkness. It could have been some wild animal the stallion had detected, but not likely. He was used to scents and sounds of the night, and prowling wildlife didn't bother him. He seldom sounded a false alarm. Quieter than the stud, but there just the same, he could hear the nickering of the other horses.

Without warning, the door swung open. No one was there. In another instant, a man stood silhouetted, silent. He struck a match, and in that fleeting second Kittery glimpsed a bearded, harsh-looking face and a man in a knee-length brown corduroy poncho. His rifle pointed at Kittery's head.

"Light the lamp, and take it slow and easy," ordered the man at the door. "I don't enjoy the prospect of having to shoot a man on an empty stomach."

Kittery obeyed, smiling to himself. When the light flooded the room, he grinned. "Hello . . . partner."

James Price said, "Kittery, don't you know that's how angels are born? In the home of another man, you should leave a light burning."

"Not necessarily," Kittery replied. "If I had, and you were Silverbeard, I'd be dead now. Anyway, you saw the black outside. If you hadn't known it was me you wouldn't've exposed yourself in the doorway."

Price grunted, leaning his rifle behind the door. "You always have everything figured out, don't you? What brings you up here?" As he spoke, he slipped the poncho off over his head.

"I had to come. For Ed Beehan and Johnny Hedrick."

"Ed Beehan?" Price's brows lowered. "I only heard about Laynee and Hedrick. What happened to Beehan?"

"Silverbeard killed Ed and shot Linus up."

Price swore. "That's a shame. Ed was a good boy. How's Linus?"

"He's alive. I haven't seen him. I see you haven't done too well; Denton and Sloan are still walkin' the earth."

Price looked irritated. "That they are. And so are Baraga, Dixon, Bishop—"

"All right, all right." Kittery held up his hands. "I figured we vigilantes could work better together."

"*We* vigilantes?" Price repeated. "I heard they made the whole lot of you sign an agreement to lay down arms, in exchange for a pardon."

"No, not me." Kittery shook his head.

"Why not you? As captain you should have been their biggest target."

Kittery reached into his breast pocket, then held up his deputy marshal's badge between thumb and forefinger. "This is why."

Price stared at the piece of tin. "They hired a vigilante as one of their deputies?"

"What vigilante? Wiley Standefer's the new marshal, and he took my name off the list."

As Price dished a bowl of soup up for himself and sat down at the table, he listened to Kittery explain the friendship between Raines and Standefer and the latter's consequential reasons for giving him the badge.

When Kittery finished, he added, "Standefer also gave me authority to appoint any deputy I please. How about you?"

Price shook his head as if he had anticipated that request. "No. I'm a free man—a public-minded citizen doing my duty as I see it. I don't need a badge."

"Suit yourself," Kittery said with a shrug. "It all comes out the same."

Later, James Price sat on the bed's edge, sipping coffee. "What about Durnam Beehan? He used to be a bad man—a hired gunman. Seems he'd be up here to help Silverbeard chew daisy roots."

Kittery chuckled at this last, then— "He is. He and Polk both. I guess under the circumstances you can't expect them to remember they signed their names on that oath."

The next day, Kittery rose when the lightening sky turned the eastern horizon into a broken-toothed saw blade. Carrying Price's double-bit axe, he stepped out into the dim light, moving around the left side of the cabin where Price had drawn up a loose pile of branches and wind-toppled logs. The sweet scent of pine made perfume of the wind, not just from the logs but from the live trees whose needles shuffled against the shingles of the cabin when the breeze nudged them, and from the rotting needles that carpeted the ground. Kittery filled his lungs over and over with the primitive, luring odors. He was once more out where he loved to be, away from the hustle of town.

He set to work with exhilaration, liking the solid feel of the axe handle in his hands, the whack and subsequent echo of metal biting into wood. He reveled in the cool breeze on his cheeks, the perfumed pungency of the fresh-cut pine that came apart looking unnaturally white in the dawn.

When he finally stepped back into the cabin with an armload of wood, he had cut and stacked enough of it for a dozen meals to come. Price, meanwhile, had coffee on the stove and cornbread baking. Steaks from the backstrap of a doe he had brought in the day before sizzled in bacon grease, filling every corner of the cabin with their aroma.

Kittery nodded approvingly at the advancement of the meal. Then he stepped outside once more, moving to the corral. Rationing out some oats, he brushed both horses down and threw their saddles on as they munched away.

The man's enthusiasm and the brisk morning air and creaking saddle leather all had their effect on the two animals. They champed at the bit and stomped their feet, anxious to be on the move.

Kittery shoved the Springfield into its scabbard and Price's .45-75 Winchester into its own, thinking as he did so that it was about time he gave in to progress and bought one of the big Winchesters for himself. The Springfields were accurate and generally reliable, yet it was hard to laud their single shot capability against the rapid firepower of the repeaters. But he was a man living in the past in that way, he guessed. He couldn't quite relinquish the single-shot Springfields, which had taken him so far and which he knew so well. It had been hard enough converting the Remington revolver to fire cartridges.

He returned to the cabin for his meal. Both men wolfed their breakfast without a word, and when they finished Price threw the crumbs into the yard for the birds, then straightened the room. Checking their pistol loads, they walked outside to the horses.

Saddle leather was like ice when they swung up. Both men wore their coats against the autumn night's remaining chill. But the sun soon tipped over the eastern hills and warmed them.

As they rode, Price couldn't help going over the situation he had been up against since coming to the Catalinas. It was one of frustration, like tracking two wisps of smoke.

"It's the half-breed," he said. "The sixth sense is remarkably strong in him. And he can hide his trail like a ten-year-old lobo. I also tend to believe he's lived in these mountains before. He must know them intimately."

"But he'll slip," said Kittery. "And we'll be there to clean up."

Unfortunately, this was not to be the day. Their search was a long and fruitless one. They rode trails where Price had seen tracks on other days. They left no rock unturned, no valley unridden, putting in over thirty miles, which in that upturned country was a solid day's work. Still they saw nothing. By all appearances, the two outlaws had vanished.

Chapter Thirty-Three
The Letter

Months slogged by. Snow fell infrequently in the valleys, once as low as Tucson. Up high in the Catalinas, meanwhile, the wind seemed always to blow chill, and soft rich white blanketed Mt. Lemmon, Mt. Bigelow, Rattlesnake and Rose Peaks. Kittery and Price were thankful they had taken a couple of weeks to lay in a supply of pine wood, stacked high and comforting beside the cabin. Temperatures dropped to below freezing at night, forcing them to keep a fire in the stove. They ran trapping operations along the creeks, eking out a living on the profits from selling beaver, mink, otter, coyote, fox and raccoon pelts. And they hunted deer, elk, wolf, cougar and sometimes bear along the timbered slopes and in the sheltered valleys.

The days were short and miserable except when the sun came out to lift their spirits. Otherwise, snow pattered like bird feet against the cedar shakes and against the door, or the skies hung low, socking in their valley with a sea of gloomy cotton mist. Discouragement plagued the cabin. No matter how far and wide the search took them, the Desperados could not be found. In fact, it was only assumption that they remained in the Catalinas. If they were there, Denton was too wily, like an old mountain lion living for years in the same mountain range yet seen by man for no more than a chance glimpse. He and Sloan were like ghosts on the land. By all signs, they had holed up for the winter, as the only mark in the snow was from the manhunters themselves or from some roving wild beast. But then, it was an enormous land with secluded, hanging valleys, cliffed-in canyons and vast tracts of rock and timber—places the manhunters had yet to discover.

Kittery and Price spent most of their time working separate trap lines. When they were together at the cabin they sat reading or brooding in silence. Most of all, they enjoyed the trips into Tucson, selling their green hides and restocking supplies. A black bear head brought two dollars, its pelt eight, and a grizzly brought twice that amount. Other game was less valuable but more plentiful, especially now that the bears had become largely inactive to wait until spring thaw.

Besides food supplies, by far the most important purchase that accompanied them back to the cabin was reading material, books by Dickens and Cooper, Melville, Shakespeare, Jane Austen, even Plato and Socrates. There were days when snowfall buried the valley and deadened their appetites and for hours on end the only sound to be heard was the occasional turning of a page. Once in a while Kittery would laugh, especially at Charles Dickens, and Price would just look up at him over his reading spectacles, study him for a moment, then return to his own reading.

On one of their trips they stopped at the Border House, where John Laynee spent most of his time since his run-in with Sloan. Once a friendly, if not carefree man, the ex-gunman now took up his hours drinking and fashioning gun belts and rawhide whips for anyone who could spare a dollar. He was a shell of his former self, a morose drunk.

But they chanced to catch him in one of his few sober moments. While working with an awl, punching holes in the edge of a holster, he spoke to them of Silverbeard Sloan.

"He's poison sure." Laynee stopped and rested his awl on the table. "He's like the devil hisself. So's the breed. Both would rather kill you than look at you, and they know they can do it. You boys better get out of it while you can—heed my words. You'll end up dead if you stay."

But neither Kittery nor Price was intimidated by the gunman's warning. Neither could give in, and neither considered it.

The hunt wore on.

For the first few months, Kittery made the fifty-mile ride to Castor every three weeks or so. He spent four or five days each time, hoping to make up for the days and weeks he and Tania were losing.

The first time, he had ridden into a quiet yard. October evening chill touched the air. Many birds absent since spring were moving down from the north, settling in for the winter. They added beautiful little splashes of red, yellow, blue and orange to the grayed vegetation that dotted the yard and the lane leading up to it. A young gila monster whisking across the walk seemed to be clearing a way for him as he came to the door, and there were fresh javelina tracks and droppings in the yard.

Tania greeted him with all the joy of a new wife, hugging him and showering him with kisses, thinking he had returned to stay. They spent the night in each other's arms, their love hot and young and strong.

Not until the following evening at supper could he find the courage to tell her he must return north. She took the news with grace—much more so than her father—then left the table and cried away her grief.

Each time it was the same. Tania wept, but she never asked him not to go back. And his love and respect for her only increased.

One thing had changed since Kittery's departure in September, a change that pleased him deeply. Maria Greene and Miguel were spending a lot of their time at McBride's house. In fact, during some of his visits the child had been with Tania every daytime hour, and Maria went directly to the house in the evenings. McBride's attempts to conceal his interest were in vain; everyone could see he was falling in love.

An army lieutenant from Camp Lowell had come to Castor in Kittery's absence. The man's name was Eddison Babar Peppridge, but his friends called him Bab, abridging the most unusual middle name his parents had given him in respect for their relation to a famous Turkish prince and descendant of the mighty Ghengis Khan.

Bab Peppridge was a short but sturdy-looking man with thick, flaming red hair and a mustache that grew into his heavy sideburns. His presence was obvious wherever he went due to this fact and the well-kept uniform.

Peppridge, like Kittery, had an assignment to bring the Desperados Eight to justice. Word had it that it was Baraga's men who had captured an army payroll two weeks before, wiping out a military guard of ten men in the process and escaping into the mountains with ten thousand dollars in silver and gold. It was the army's responsibility, but due to a severe shortage of seasoned troopers, Peppridge was passing the buck once again to the civilian law to find the outlaws' hideout and lead the army to it. Yet Vancouver had his own troubles to deal with. The disbandment of the vigilantes by the government had given banditry free reign, and the Desperados and others of their ilk made the most of it, plundering and murdering with a vengeance across the land.

Early in December, Jed Polk gave up his search for Sloan and Denton and returned home to his JP Ranch, leaving his foreman, Durnam Beehan, hunting alone in the Catalinas, a man possessed. Kittery and Price knew the cabin where Beehan stayed, but they never visited him. He was a man whose only goal in life was the death of Silverbeard Sloan. He lived on the desire for that death, and it consumed his soul with hate.

Five months had passed since Kittery's initial departure from Castor. On his most recent return he went to a wedding—a wedding every able-bodied citizen in Castor attended. All, that is, except Thaddeus Greene. His one-time wife gave her hand to the distinguished, beloved town banker, Randall Maximilian McBride.

Kittery glimpsed Greene the following morning on the street. He had heard little of the man in the past few months, other than the fact that his business was slowly dying. After the brutality perpetrated on Maria, most of the people around had changed their patronage to the mercantile across the street, which had to drastically upgrade its line of stock to please its customers. Its business had boomed, and Solomon Hart, the one-handed black man, had become its manager. He was then supervising the addition of another section at the back.

Greene's business downfall flooded over into his very appearance. He hadn't shaved in a week or more, nor combed his hair nor washed his face in as long a time. He wore a shirt stained with sweat, grease and other substances, threadbare pants and scuffed brogan shoes.

Greene stared, bleary-eyed, as Kittery passed him by one day. His dull eyes revealed no expression, portraying the soul of a beaten man, torn apart inside by his own barbaric deed and his failures as a man. He was loved by no one, tolerated only as a necessary evil because he had money when many had little. He lived to work, to eat and sleep. Kittery couldn't help but pity the man.

The final night of his stay, Kittery sat down across the table from Tania. She smiled, but it was a sad, forced smile, and she glanced worriedly at her father.

McBride had grown aloof with Kittery during his last few visits. They hardly ever talked, and when they did they never laughed, nor even smiled. Kittery understood the man's feelings. McBride believed his son-in-law was no good for his daughter, a fact he must have been sorry not to have seen sooner. He thought Tappan would break her heart. He believed him obsessed, forsaking any real occupation to pursue two will-o'-the-wisps, paid only by his trapping operations and an occasional bounty he might collect. Kittery hated having McBride continue to provide for Tania, and he would have loved to quit and come home, but nothing would end until the Desperados died, and he knew it. He was no longer avenging the dead, as people believed, although he still wanted to honor his promise to Joe Raines. Now his true mission was to protect the living.

That night, they ate their meal without talking. Kittery was conscious of the looks McBride threw Tania throughout the meal. Each time she reacted with a slight shake of her head and looked down at her plate, concentrating on her smallest bite.

Finally, Tania clicked her fork down on her plate, and after handing her father a stern glare, turned her eyes to Tappan.

"Tappan, there is something I must say. Will you come outside with me?"

Kittery glanced at Maria McBride, who sat at his left, then with some slight irritation at Rand, on the right end of the table. He looked back at Tania. "Why leave? Say it here."

"I—" She looked helplessly at Rand, then dropped her head and shut her eyes. "I can't say it," she whispered.

"Then by the gods, I can!" McBride thundered, shoving off his chair. He glared at Kittery, his knuckles pressed hard upon the tabletop, turning white with the pressure. "Tappan, this is something you should have known long before now, except that for some unfathomable reason this naïve wife of yours has wanted to hide it. But nothing will keep me from telling you now."

McBride had set his mind, but he made the dire mistake of glancing over at his daughter, of looking into her dark, pleading eyes, begging him to keep the secret she could not tell.

She started to stand. "Father, please . . ."

"Ah, girl," he said, his shoulders sinking. Then he raised his hands in despair and plopped back down into his seat. He glared at his plate for half a minute, then rose and stalked from the room.

Baby Miguel began to cry, and when Tappan and Tania looked his way, Maria smiled apologetically. Taking the baby, she followed her husband into the bedroom.

Kittery turned his eyes back to Tania. Hers were lowered, and tears glistened on her cheeks. She made no sound.

"Tell me, Tania. What is it? It's something pretty important."

"Tappan . . ." She looked up at him, tipping her head to one side, holding her napkin in both hands. "I can't tell you. I really can't."

"Why?" He stood and leaned halfway over the table, supporting himself on the tips of his fingers.

"If I told you now our lives would never be right. They would never be good for us together." She bit off her next words, stood awkwardly and dropped her napkin, hurrying down the hall to their bedroom.

Kittery stood alone, and feeling defeated he let his weight drop back into the chair. He toyed with a venison rib, gnawed on it briefly, then returned it to his plate with a look of disgust. He sipped his water.

He had heard Miguel crying in the bedroom for some time, and only now became conscious that the cries had ceased. Feeling someone's presence, he turned to see Maria watching him. She leaned against the doorframe, holding her small, blanketed bundle to her shoulder. She smiled her understanding, and Kittery tried to smile back.

He stepped toward her. "Maria, you have to help me. I need to know what's happenin' here—before I go away again."

No response came for several moments while she searched his eyes in silent decision. She looked away, then back, and stepped away from the wall. "Tappan, I once said to you that you mus' follow you' own mind. This is still true, you know. You mus' do what you feel to be right."

"No, Maria. No. That's not the way it is. Not this time, anyway. How can I make the right decision if I don't know all sides of the question? Those two are keepin' somethin' important from me."

Maria's eyes held his, feeling for him, hurting with him. She seemed about to speak, and he waited. But she just smiled, bowed her head, and returned to her room.

So Tappan Kittery pulled out his carbine that night, cleaned, oiled and returned it to its scabbard. He polished the worn Remington, the Colt and the American, stuffed a sack full of biscuits and another full of jerky. He threw some clean clothes into his saddlebag with cartridges and powder and ball. And nine hours later he left Tania weeping in the cold, quiet dawn as he saddled the black stallion and headed for Tucson. He wouldn't come home again until his hunt was done.

The following two months crept by like an eternity for Kittery. He heard from Tania weekly and answered her letters the best he could. But it was not until the middle of March that Tania revealed her secret.

He had purchased supplies at Warner's store, in Tucson, and on his way out of town he pulled over at the post office, the same as every week before. A letter awaited him, and he paused on the porch to slit it open and read. He expected some news of Castor, of baby Miguel, or of Randall McBride and his business. He anticipated anything, in fact, but what he read.

Dearest Tappan,

I miss you and hope you are well. I know how you must feel, that I have kept my secret for so long from you. Believe me, it has hurt to keep this to myself. I wanted so badly to tell you, and most of all for you to be happy about it. I have wondered and desired and prayed to know if I should tell you, and now I know that the time has come for you to know.

You have been gone from me for seven long months now—longer than any couple should be apart. The visits have been nice, but never enough. I knew you had a job to do, and hoped you could do it and hurry home. Now I don't know how long this will take you, so here is the secret that I dared not tell you in January.

You are going to be a father. I am with child and have been since one of our first few nights together. I am surprised you never saw it. I have gained so much weight since our wedding! You must have thought I was just eating Father out of house and home.

The reason I could not tell you this when you were here is because I knew you needed to finish your job. You have this to do for Marshal Raines and Cotton Baine, and for all of your other friends, and for those who may meet Silverbeard in the future. I just pray you can find these men soon, because my time will soon be near, and this is not a good time for a man to be away from his new wife.

Father is still angry that you left. He can't understand your drive. But he will forgive you, I know, when he sees the life we will have together.

As for me, I would never hold this time against you, and I love you now more than ever. Please do what you must, and then hurry home to me and our approaching child.

With all my love,
Your wife,
Tania

Seven months of fruitless hunting for Sloan and Denton seemed like it had taken a lifetime now. He and James Price, together with Durnam Beehan and Jed Polk, had enjoyed no luck in their quest, seen neither hide nor hair of the two Desperados. It was almost as if they no longer existed. But now they *had* to be found. He must find them or return home with his task unaccomplished. He had to hurry home to Tania, to be with her when she needed him most, and to try to regain some of the respect Rand McBride had once had for him. He *must* find the Desperados, now, at all costs, or that part of his oath to Joe Raines must go unfulfilled.

And somewhere in the mountain fastnesses, death was waiting for Tappan Kittery.

Chapter Thirty-Four
The Manhunters

Tappan Kittery swayed in the saddle, but his wary eyes and the way he clutched the Springfield across his saddle horn belied the danger he expected at any moment. He scanned the rocks and trees, searching for sign of Silverbeard Sloan and Crow Denton, who had recently passed this way.

Judd Creech had ridden up to the cabin two evenings before with a letter Mort Dansing had brought him from Tucson. The letter had arrived there one week earlier from the settlement of Clayton, forty miles to the north of the Santa Catalinas. Its simple message was that a recent report concerning Silverbeard Sloan and Denton had been substantiated, a report that the two outlaws had spent their winter in New Mexico territory. But now that most of the snows were gone, several people had seen them passing through Clayton, headed toward the Santa Catalinas. Price and Kittery had thanked the old hunter, but it was a bitter pill to swallow knowing they had wasted their entire winter hunting two men who were not within hundreds of miles from them.

Then, the following morning, a rancher had stumbled on Price's cabin. Upon learning that Price and Kittery hunted the two outlaws, the rancher revealed his reason for being there. Two rustlers—he supposed them to be Sloan and Denton—had ridden across a corner of his ranch holdings the previous day, pushing ten head of his horses before them, riding hard for the Catalinas. One of them rode a sleek, long-legged chestnut. He had given pursuit, but he was a lone man against two and dared not push a fight. Then daylight had passed, along with hope of seeing the outlaws again.

Morning came, and the trail had vanished. But yes, he said, he could show Price and Kittery which way the outlaws had ridden, and he would offer a reward of five dollars a head for the return of his animals, to be claimed at the stockman's building in Tucson.

Now here rode Kittery, alone. A mile or two separated him from the trail of James Price. They had discovered the tracks of the horse herd during the mid hours of morning, after spending the night on the ground. The trail led them several miles to a narrow canyon that took them up and over Oracle Hill, on the Catalinas' extreme northern tip, and onto its west slope. Here, they lost the trail in heavy talus but knew there were only two directions the outlaws could ride.

Now Price and Kittery rode alone.

While Price's trail led him toward the western slope of Samaniego Ridge, Kittery's took him to the rugged valley between Oracle and Samaniego Ridges, past Cañada del Oro and Pig Spring. He took to Samaniego Ridge, which offered a better view of the country, and here he trod rocky outcrops he would have trusted to few animals besides the black. The going was harsh and slow, skirting clumps of shin-dagger agave that tried to jab their knife-like points into the black's pasterns, scrambling over broken shale, swinging wide of the clutching arms of ocotillo, then riding through occasional stretches of oak and scrub brush as he rose higher.

It was by chance he saw the hand.

Where the ridge fell off toward the canyon below, the frozen gray fingers protruded above the rocks. Kittery dismounted and moved closer, eyes not only on the ghastly hand but scanning around him.

He jolted to a stop. The body lay hung up in the jagged clutches of the rocks. The eyes stared upward, black pools in a mass of white clay, and in the center of the forehead was a small round hole. The Beehan family had shrunk by one more, and Durnam Beehan, one-time hired gunman, had failed in his quest. His career ended here on this wind-swept ridge.

The black horse walked forward and nuzzled his friend. Kittery felt the warm muzzle against his cheek.

"I'm all right, boy."

He looked down at his former riding mate. The black started to approach the body, then caught wind of it and shied away. Beehan had been dead for some time; his pale, bearded face was stiff and cold when Kittery stooped and touched it. He thrust his hands under Beehan's arms and tugged him out on flat ground and found a second bullet hole surrounded by the splinters of the man's breastbone.

Looking over the rim, Kittery saw the gunman's rifle forty feet below, lying near his hat amid the rocks. If not for the jagged lip of the rock catching him, Beehan might have been down there too, probably never to be found.

Kittery set to work carrying stones, with which he half-covered the body. He hoped to return for Beehan before scavengers found him, but for now the outlaws awaited him. Beehan's body was enough proof they had passed this way.

Stepping back, he searched the ground and found the print of a boot heel five yards from the body. The outlaw had moved in close to view his handiwork.

He followed the killer's tracks to a spot along the ridge where the descent was not so steep, and the trail of the now-mounted Desperado led down into the narrow valley. At the bottom, the single set of shod tracks converged with a dozen others, some shod, some not—a typical range herd. But only one appeared to have been ridden, now making it two.

The looming bulk of Mt. Lemmon jutted up four thousand feet above. The well-worn trail rose swiftly to meet it, and the air began to thin as Kittery made his way into the lower timber, the aspen and pine. Up here it frosted heavily at night, and though much of the snow had melted, leaving crusted patches in the shaded valleys, he was forced once more to don his corduroy coat. He had reached a point in the herd's path where it leveled out, around seven thousand feet. He moved around the mountain, rather than up its steep, snow-laden side.

Here, due east of Mt. Lemmon's highest peak, the horses had stopped and milled for a while, leaving a profusion of tracks. One man had dismounted and smoked a while; Kittery found the stub of a hand-rolled cigarette and a burned match. But when he tried to follow the trail away, all sign had vanished. He circled the tracks on foot, peering at the rocky ground. Nothing.

He stopped to scratch his head, glancing over at the black as if beseeching his help. He thought of his own trick of masking his horse's hooves with rawhide. It would keep down the sounds of the horses moving and prevent the sharp hooves from cutting deep into the soil, simplifying the task of covering the trail. This was an Indian trick Denton would surely know.

Yet even though the rawhide might help hide the horses' tracks, in this narrow valley there was only one way for a horse to go. Kittery headed the black that way—straight forward—and his grip on the carbine tightened. Unless there was another escape from the valley, he could follow the trail without wasting time looking at the ground.

After a few hundred yards, he stopped to climb down from Satan and peer at a tuft of grass a hoof had ripped loose from its dubious hold on the rocky ground. Someone had tried to hide it by straightening it when they would have been better served plucking it and throwing it far away. It stood out like a black eye. *You're pretty thorough, Denton,* thought Kittery. *But you ain't close to thorough enough this time.*

To his right, the granite rocks rose up on two sides of a narrow crevice, creating a sharp V. This drew his attention. The path that ran on ahead was fairly level, easy going, the natural way for a horse to go, and especially a large herd. But this narrow V that left the valley and cut up into those rocks intrigued Kittery. It was large enough, it appeared, that a good horseman might coax a horse through. The horse would not like its narrowness, but its body would fit, and there were ways of persuading animals to go into unpleasant places.

Any ordinary man might have continued on the easy trail. But an ordinary man would have been captured by now. This was Crow Denton.

Kittery could not explain his interest in this V in the rocks, so forbidding to drag an unwilling horse into. All he knew at this point was that Denton and Sloan were somehow escaping the grasp of their pursuers. They had to be using extraordinary measures to do that. It wouldn't do to pass up any possibility. And Kittery had the eerie feeling this narrow passage might be the hidden piece to the puzzle of the outlaws' hideaway. What an ideal way to shake an already discouraged tracker!

He turned to the black. "Well, Satan, you're not gonna like this. But if I'm wrong, the trail will always be here. Let's go climbin'."

Turning, he took the black's reins in his left hand and made the ascent into the crevice, tugging the horse behind. Once within the passageway, he found its narrowness to be illusory. It was wider than it appeared, and Satan passed through it not only without hesitation, but almost gladly, as if something up ahead interested him. Did the black know other horses had passed this way? What did his nose tell him that Kittery had no way of knowing?

Wind—or man—had swept the rock in the passageway almost free of dust. Whatever had done it, there were no tracks, nor any white scars that a shod horse would have left.

Kittery stopped cold, staring. A chill went up his spine, and his heart began to pound until he could hear it in his ears. The short hairs on back of his neck stood on end, his scalp growing tight. It didn't matter now that he didn't have the black's sense of smell, for there to his right a tuft of chestnut winter horsehair hung from a protruding chunk of granite. Farther along, he found more and thanked God for these rocks that leaned out into the passage. He knew at last his quarry was near.

As a precaution, Kittery stepped around to the right side of the black and shoved the carbine back into its scabbard. In its place he drew the Remington. Though not as powerful nor far-reaching, its extra loads might be to his advantage.

He liked the assurance of the .44 in his fist. The words of John Laynee still rang in his ears. *He's poison sure. So's the breed. Both would rather kill you than look at you, and they know they can do it."* He didn't intend to give them that chance—only a fool would go into a confrontation with his weapons sheathed.

The crevice continued on for another fifty yards, now not so deep and dark but much wider, perhaps thirty feet across. Boulders and gravel cluttered its floor, and sign was much more abundant here. Careless hooves had overturned a hundred stones, and no longer did anyone attempt to hide that fact.

At last, the crevice disappeared altogether, where in the back side of the granite mountain was a hollowed-out area, a flat clearing one hundred yards wide all the way around. And in the center of this clearing stood a pine pole corral, ten head of horses inside. By the scant amount of manure there and its freshness, he guessed this was the stock of the rancher they had met.

Kittery skirted the corral with hardly a look while his heart beat like a drum. The black cast a disdainful glance at his caged fellows and made no show of fraternity. He followed close on the man's heels.

On the opposite side of the enclosure a trail broke away, running in plain sight for fifty yards, then dropping off at the far edge of the clearing, where the crowns of a dozen ponderosa pines protruded. Kittery halted and draped the black's reins over the saddle horn, allowing it to continue on behind at its own pace. They moved slowly, wary for any sign of danger. The only sound was that of the curious nickering of the horses behind them.

They had not quite reached the timber when Kittery caught the faint waft of wood smoke drafting up through the forest. It was there for only a second, but it was enough to let him know on the other side of the trees waited death . . .

Passing into the heavy timber, Kittery picked his way among the fallen trunks and rust-colored needles, the cones and broken branches. A whisky jack began to cry annoyingly back in the trees and was soon joined by a chorus of them more distant. Kittery cursed them and their warning cries and held the Remington out ahead of his body, ready to swing into action.

They had traveled two hundred yards through the trees, moving along fairly level ground, when the earth swept away before them and plunged into a grassy bowl a quarter of a mile around. Brush and grass studded the clearing, snow shone in patches all across it and among the stunted evergreens along its border. All around the bowl towered pine and Douglas fir and a scattering of budding aspen. There, on the far edge of the clearing, smoke curled from the stone chimney of a tiny log cabin. The structure stood alone, away from the trees close to the far side of the clearing. In a corral along its right side grazed six head of horses.

Tappan Kittery smiled grimly and started down the needle- and deadfallstrewn slope, the strong scent of wet, rotting pine in his nostrils, and a cold breeze coming up from the shadowy snow banks and brushing his cheeks. The going was treacherous here, slippery in places where the snow had melted. Satan slipped along behind him, skidding down several times through the slick carpet of pine needles and mud. Halfway down, Kittery's feet went out from under him, and he slid four feet before he could catch himself. Standing up, he dusted the needles and what mud he could off and holstered the Remington; he wanted both hands in case of another fall.

At the timber's edge, he crouched down and waited for the black to join him. In front of him was brush and stunted trees, new grass sprouting up among the old. A few yards farther on they would leave the forest completely and enter a long, flat stretch of grass and low brush that undulated off toward the cabin. He took the black's reins when he came up and lashed them to a branch of the nearest ponderosa.

"Stay here, boy. They'd see you a mile off if they're lookin' out that window." He patted the black's neck, then started away from the aspens and big timber, crossing to the left through the young pines, away from open view of the cabin. If Denton had had anything to do with building this cabin, there would probably be another window, or at least a set of loopholes, on its backside. But with luck the sides would be blind.

At last, he left the trees and started into the open, crossing the meadow in long strides. He had made no more than twenty yards when a voice stopped him cold.

"Don't even think about touchin' that iron, friend. Turn around—real slow."

With a silent curse, he obeyed and looked beyond the stunted trees, back toward the edge of the forest, thirty yards away. A man stood at the timber's edge. He wore a wide-brimmed, flat-crowned hat over dark hair, and the long, dark poncho hanging over him, that would have blended with the dark trunks higher in the forest, stood out stark against the aspen boles.

The man's face showed no expression. Kittery made no move for his revolver as the Winchester-toting stranger stepped out from the trees, drawing up fifteen feet away. His was a lean, strong face, hawk-nosed and deep brown, with narrowed eyes, a thick, wide mouth and a long, dark mustache.

"You Crow Denton?" Kittery asked conversationally.

"I am."

"I'm in luck then. I came to pay you a visit—you and Silverbeard."

As Kittery talked, his mind thrashed around for a way out of the predicament his carelessness had gotten him into. Denton's rifle pointed to one side, rather than directly at him. That would have placed his odds about even, but he had holstered his pistol. He was going to need a distraction—some kind of edge—to come out of this unscathed.

"I figured it was you shot the fellow up on the ridge last night. You wasted a good rifle."

"I saw it fall."

"Those your horses yonder?" Kittery jerked his head in the direction of the cabin.

"Mine—and Sloan's." Denton's voice was soft.

"And the ones up above—you just brought those in?"

Denton nodded. "Last night."

"I hear you're pretty good with that rifle." He motioned toward the Winchester. "The dead fellow on the hill agrees."

"This rifle—any rifle," said Denton. He sized up the big man through slow-moving eyes. "You're the captain, the one Baraga hates so much."

"That's right. How'd you know?"

"Word travels. They call you a giant among men. Say you're as big as Rico Wells. But not quite."

"Bigger, now. I killed him like swattin' a fly."

Denton emitted a grunt of half-hidden respect. "You don't give up a trail easy. They said in Tucson you'd be comin' for us when Sloan shot them two. Been six months."

"Seven, to be exact. We didn't know you left the territory."

"Didn't advertise it. But—why're we so important?"

"Several reasons, besides the two men in Tucson."

The Desperado cocked his head to one side.

"You and Sloan killed other friends of mine. Mostly Sloan, but you're with him, and you killed the man on the ridge last night. He used to ride with me."

Denton lifted his head in understanding. "He was crazy."

"Sloan killed one of his brothers. Crippled the other. How would you take it?"

Denton shrugged. "I hated my brothers."

"I want Doolin most of all," Kittery volunteered, ignoring the comment. "For the U.S. marshal."

"Good friend?"

Kittery nodded.

"You're the best I've come across, Kittery. Only one that found this place. Musta been fifty that tried. Your friend last night's the only other man ever come close."

"Zeff Perry's dead," said Kittery, on impulse. His mind still sought a way to beat Denton to the gun. He knew the killer would be growing bored with this talk.

"Zeff Perry? The horse rancher?" Denton nodded with satisfaction. "Who killed him?"

"I did."

The outlaw's eyes narrowed. "Killin' one of yer own?"

"One of my own? Ned Crawford told me he was workin' with you."

"No—hell, no. He worked with Crawford quite a few years ago, not us. He's been honest Injun of late. Last summer Sloan offered him a chance to work with us. He turned it down. Sloan woulda killed him, but I stopped him. I admired his spirit."

"That's it? It wasn't him sellin' information to Baraga?"

"No, not Perry." A satisfied smirk crossed the outlaw's face. "Looks like you killed the wrong man."

Kittery was silent for a time, weighing Denton's words. He guessed the outlaw had no reason to lie about it to a man he thought he would soon be killing. "I'm glad about him, because I told you a lie. It was Ned Crawford killed Perry. Then I killed Crawford. Who's this 'mouse'?"

"Can't say. We thought it was Perry, too. Me and Sloan tried to find out, but we never could. We just found out it wasn't Perry. I don't know for sure if Baraga even knows. Somebody Shorty Randall knew."

"Well, we'll find him one day. Can't hide forever."

Denton nodded, a bored look on his face.

"You gonna kill me?" Kittery indicated Denton's rifle.

Denton shot him a surprised look. "You don't seem like a man afraid to die."

Kittery moved to the plan he had made while talking. His right hand hung loose along the holster on his hip. He had merely to bring it up to grasp the gun butt. As he did, he kicked his feet out from under him. When he hit the moist soil on his elbows, with pistol cocked, he didn't fire. Instead, he rolled once to the left.

Denton did not expect the roll. And he hadn't already jacked a shell into the chamber of his rifle, perhaps fearing it would warn Kittery he was behind him. Those were his mistakes.

He jacked the first round now and fired, but he aimed where Kittery had dropped before the roll. The bullet plowed up dirt where Kittery's torso would have been—not three feet to the right, where he now lay.

Kittery squeezed the Remington's trigger. It exploded. It erupted with smoke and fire. Its grip slammed into the palm of his hand, and the .44 bullet sank in under Denton's breastbone. The outlaw's reflexes alone fired the second round, a wild shot high over the trees. He slumped forward on his face, the rifle pinned beneath his knees.

Kittery drew the Remington's hammer back again as he pushed to his feet, lining the barrel on the back of Denton's head. Behind him, he heard rapid hoofbeats and swore as he whirled about to see a rider on a long-legged chestnut galloping at top speed for the trees on the other side of the cabin. Even if he could have reached his Springfield on the black's saddle he couldn't have brought down Silverbeard Sloan on a galloping horse at that range. He sighed and watched him disappear into the timber.

He turned his attention once more to the fallen man, holding the Remington cocked and poised. With great effort, he shoved Denton's body over onto its back with a boot toe and was surprised to see the dark eyes looking up at him. As before, they held no expression, but Crow Denton was alive.

Kittery picked up the Winchester rather than kicking it aside, and Denton gave a grudging nod of gratitude. He didn't appear able to move any other part of his body.

"I always wondered somethin', Denton. Legend says you've never killed a white man. Any truth to that? Before last night, that is."

Denton surprised him by showing he knew how to smile. "So . . . that's my legend. Huh." Blood trickled from the corner of his mouth. "No. Man last night . . . he was third. You were gonna be four."

"Ten odd years a bandit, three men dead. You don't sound like such a murderer. I'm surprised Baraga kept you. You could've killed me with one shot to the back ten minutes ago."

"Can't feel nothin' but my tongue," said Denton, in a casual voice as if they were sitting on a porch watching the sun go down. He laughed without humor. "Reckon you done killed me."

Kittery nodded. "I can't say I'm sorry."

"Hell, no," Denton replied weakly, a trickle of blood running out the side of his mouth into his hair. "I wouldn't either." His words sounded very wet. He closed his eyes for a moment, and the chill wind stirred his hair on his forehead.

"Sure is cold today."

"Yeah, it sure is."

When Kittery holstered his pistol and removed his coat to stoop and cover Denton's torso, the outlaw squinted his eyes quizzically. Then he shivered. His brow knitted, and his mind seemed to mull over something he couldn't quite understand. He glanced down at the coat's sheepskin collar brushing his jaw. He shivered again and looked up into Kittery's merciful eyes, and his own clouded over.

Kittery moved swiftly. He took up his coat and ran to the black, untying him and throwing himself into the saddle. Drumming his heels against the animal's ribs, he put him to full flight along the trail of Silverbeard Sloan.

A hawk cried overhead, making giant circles against the sky, fleeting circles leaving no sign of its passage. Back in the forest a squirrel chattered, cursing the intrusion of the silver-haired man on the chestnut gelding.

Silverbeard Sloan had pushed Denton's fleet-footed horse as hard as he dared after leaving the cabin. They had to pass through a mile and a half of heavy timber, where the going was achingly slow. But as they worked along the side of a canyon where Sloan would have been fearful on foot, the horse, picking his way, never gave him cause for worry.

They began to descend, and as the miles and the trees fell behind them the landscape changed. The pine and fir, stubborn as it was, receded, replaced by oak, red-limbed madroño and juniper. The granite of the higher mountains changed to dark sandstone, with quartzite outcrops here and there. If he continued on this path, Sloan would enter a flat, rocky plain ten miles wide that separated the Santa Catalinas from the rugged Tortolitas.

Silverbeard rode unwary. He had left pursuit behind. He was a light-bodied man on a great horse, famous for its speed and endurance. No one could catch Silverbeard Sloan. Whoever was back there had found the breed. That was regrettable, for he had been a worthy partner. But he was just a breed. At least his death had given Sloan the time to make his escape. Now he rode at a jog, down into the lowlands, feeling the weight of the pistols strapped around his hips, knowing he was invincible.

Where he rode now, rocks rose up on both sides, dusty sandstone supporting the miserable lives of a dozen varieties of desert shrub. The trail sloped upward for fifty yards before him.

He thought of the deserted cabin behind him as his horse's hooves clattered over the surface of the trail. He cursed himself for not bringing a canteen. Or a bite to eat. What would he do now? Where should he go? Could he find another partner? Another hideout? Probably neither so perfect. Another state, perhaps. He had dreamed of California, of going back to see if his father was still around. If so, he would be in his sixties, but no matter. He would have loved to make him eat some of the dirt he had fed to his only son. If his father was not around, maybe he could go north. There still might be gold to be stolen up there somewhere. Montana was good for it. Idaho, too. It was time for a move, anyway. Hah! Had the men who killed Crow Denton—if he was dead—believed they could bring down Silverbeard Sloan? The man so lucky had not been born.

"Sloan! You're a dead man!"

The words, barked from the throat of a deep-voiced man, snapped Sloan's head up. He saw the hollow bore of a big Winchester lined on his chest. A man clad in a long, brown poncho and black pants and boots faced him fifty feet away, in the brush at the edge of the trail. Narrowed eyes stared from beneath the wide gray brim of the man's hat, and a hard-set mouth showed through a brown, gray-streaked beard.

No time to think, to consider death. Sloan was the fastest man alive. He had heard it said. A rifle against Silverbeard's Colt—at this close distance? The man was a fool to try it. His right hand flashed to his holster, and in less than a second the silver-plated gun reared, aimed at his foe. The rifleman winced and pulled the trigger. He jacked in another shell. Once, twice, three times, swift but methodical, he fired. All three slugs battered and ripped through Sloan's narrow chest. He heard them strike him and felt their power, but no pain. The whining ricochet of his own bullet added a violent exclamation point to the Winchester's powerful statement.

Sloan felt the chestnut horse disappear from under him. He landed among the clutter of rocks and brush on the trail. He felt his pistol fly from his hand. It struck a rock yards away. His mind reeled. He swore. What scratches was that going to leave on the gun? Every time he took it in for repair it cost him dearly. Another trip to the gunsmith! He wasn't looking forward to that. He hoped at least the grips weren't hurt.

Sloan's eyes rolled back in his head. The world went gray. The veins of his left hand swelled when he gripped the second Colt. He thought of Colt Bishop's lightning hand and cursed and cried out. Far away, he heard a Winchester's lever being worked, and an explosion smashed his ears. His head slammed to the side, and he felt warm blood ooze down the side of his head into his hair.

Tappan Kittery rode up with his carbine across his leg twenty minutes later. He had recognized the report of James Price's big Winchester—or at least what he assumed to be that rifle—from afar. He guessed the outcome of whatever fight had occurred when he heard the final, belated rifle shot.

James Price sat on a sandstone boulder cleaning his Winchester, his hat on a rock beside him. Price looked up without expression, then down at Sloan, who lay on the trail at Satan's feet. The latter lay still with four little holes in his chest. A fifth slug had left a hole in his temple and torn away a good portion of the other side of his head. His pistol lay in the sand where he had flung it, as if he had gained a sudden distaste for it at the end.

"I saw him half a mile away," Price remarked, looking up as his cleaning rod stilled in the rifle barrel. "He still almost beat me to the trigger."

"But not quite," observed Kittery. "Denton's dead, too."

On hearing this Price closed his eyes, running a bandanna over the barrel of his rifle, which he now lowered to point at the ground. He opened his eyes and looked at the palm of his right hand and absently wiped a streak of oil on his poncho.

"Then after all those months, we go home." He ran his fingers back through his hair, letting it wisp about his sun-darkened temples. "I'll never have to smell your dirty feet again."

Chapter Thirty-Five
Into the Fire

They buried Durnam Beehan where he had fallen, on top of Samaniego Ridge, in the mountains where he had gone to fulfill his destiny. His prayer was the whisper of the wind, his dirge the cry of a red-tail hawk that circled above. The Desperados' cabin they left as they had found it, loading Denton and Sloan on two of the horses. With the rest of the herd, they headed for Price's cabin on Guthrie Mountain.

At noon the following day they arrived in Tucson with the herd of horses in front of them and Sloan and Denton tied in blankets over their mounts. Sheriff William Barden greeted them happily when he discovered their success, content to have two of his greatest headaches gone. They left the bodies with him after filling out the reward claims. Their next stop was the stockman's building, where they collected the fifty dollar reward their rancher informant had promised them.

After going to the tannery, where they made seventy dollars off their hides, they stepped into the Border House for a drink.

John Laynee smiled up from his table when he saw them walk in. "Howdy, boys!"

"How are you, John?" Kittery walked over and shook his hand, followed by Price.

"Real fine, Captain. I hear you brought down Sloan and the breed."

"Well, they're dead. I got Denton. Price took out Silverbeard."

Laynee nodded with approval at the bearded hangman. "You did the world a good service, Price. You have my thanks. I c'n rest easy now."

"I'm glad to be of service, Laynee. It's good to see you in such fine spirits."

"Spirits? That reminds me." Laynee turned toward the bar and raised a finger. "Juan! Bring these gents a drink."

The short bald Mexican Kittery had spoken to before came waddling over carrying a bottle and two glasses. "With pleasure, John. With pleasure." He beamed up at Kittery and Price. "So, you have won. The *banditos* are dead. What will you do now?"

"There's a little business to tend to," said Kittery. "And then we're headed home."

"Ah, yes. To Castor. Home ees good. You have done well."

"And we're tired," added Price with a smile.

The trio drank and talked lightly, and for the first time in months they relaxed and let themselves laugh. And that day James Price and Tappan Kittery were toasted by the town.

That evening, as dusk fell thick and cool, Kittery rode into the ranch yard of Jed Polk, leading his two geldings and Denton's chestnut, which he had kept for himself. He had left Price in Tucson sometime after four o'clock and headed for the chore he had dreaded since burying Durnam Beehan up on the windy mountain.

"Hello the house!" he called in.

In a moment, he saw Jed Polk standing in the doorway with a rifle. "Captain? Is that you?"

"It's me, Jed. I was headed by on my way home."

"Well, by George, come on in. Supper's just ready."

Kittery rode closer and dismounted, tying the horses to a hitching rack. He heard boot heels clicking inside the house, and within seconds Linus Beehan appeared in the doorway.

"Cap! Good to see you back."

The young man had recovered well from his wounds. Other than a limp, he looked the same as always. He came off the porch, holding out his hand to Kittery, and they shook. Kittery threw an arm around Linus's shoulders and squeezed.

"I knew you'd pull through, Line. You're a tough number."

Polk met them on the porch and pumped Kittery's hand a couple times, slapping him on the back. "It's good to have you back. I saw your little wife in town one day. Told me to get you home."

Kittery laughed, following Linus into the dining area. The three sat down to beans and barbequed pork, tortillas and stewed hot peppers.

Jed Polk's wife was a young Mexican woman possessing a bare knowledge of the English language. She came to sit at the table, bowing her head shyly before the stranger. "How are you?" she asked.

Kittery smiled. *"Muy bien."* He shrugged. Her English was better than his Spanish.

After blessing the food, the four ate without more than the usual conversation. Despite the initial happy greetings, the Polks and Linus seemed aware of a tragedy in the air. They knew the Desperados must be dead, or Kittery would not have come back. That was good. But another man had gone to hunt Sloan and Denton, and he should have come home, too.

The meal continued, and conversation lagged. Tappan Kittery finished his last morsel and scooted his plate aside with the back of his hand. He looked about the table, swallowed and absently picked up a cracker. He began to break off its corners.

"Jed, Linus. I'm glad to see you, but I'm not bringin' happy news."

Polk shook his head and closed his eyes, holding up his hands to halt Kittery's words. He glanced sorrowfully at Linus, then at his wife, and looked back at Kittery.

"I once had a good friend who went mad with a driving hate. I used to have that same hate, but with time I realized how useless it was. My friend didn't. You don't need to tell us—Durnam is dead."

"He came close to winnin'." Kittery's eyes met Linus's. He went on and told how the older Beehan had met death at the hands of Denton, and how each of the outlaws had tasted death, in their turn. But those two deaths were meaningless, and the victory was empty after the loss of Ed and Durnam.

In the morning, Kittery saddled the black, loaded the packhorses and the chestnut, and with his final farewell rode toward Castor. The ten-mile ride seemed like an eternity before he broke off the ranch road onto the Camino Real. Five miles farther along, not far out of Castor, he saw a wagon rocking toward him. He recognized the driver as ex-vigilante Tule Simpton. Simpton sawed the rig to a halt in a cloud of dust.

He said, "Cap'n! Wish you'd been here earlier. Walt Doolin and Slicker Sam Malone came through town an hour ago, thereabouts. They shot up Dabney Trull perty bad and killed Albert Hagar outright. Nobody knows why."

Kittery swore bitterly at this Desperado killing that seemed to have no end. "Which way'd they go?"

"Back towards the Baboquivaris, couple hours back. They's a posse after 'em right now. You'll never catch 'em, Cap'n, if that's what you're thinkin'. Best let Vancouver handle it."

Kittery nodded, his mind elsewhere. He thanked Simpton and traveled on at a lope.

Kittery rode into Castor without looking around. He went straight to the livery, where he swapped the black for a stout bay. Without stopping to see Tania, he lit out on the posse's trail. He rode hard, keeping the bay to a steady lope. Past the brushy Twin Buttes they went, past the southern edge of the Double F Ranch and Horse Pasture Hill, and the southern foot of the Sierrita Mountains.

At times, the going was rough and slow, and he made poor time. But he knew this trail well and had some shortcuts of his own. It was growing dark when he reached the Black Hills, southwest of the Sierritas. Though he hadn't been able to catch the posse, he made camp here in the shadows of a butte that rose five hundred feet above the desert floor. It would be useless and dangerous to follow farther in the dark. With a sigh, he rolled into his blankets, knowing dawn would come too soon.

With first light, he was in the saddle, sticking to the posse's tracks. The twelve-odd horses made an easy path to follow. He loped across the creosote flats, aimed for the distant bulk of Baboquivari Peak. It was yet early in the day when he spotted the imprints of two horses leaving the trail, headed northeast. Sitting his saddle, he stared at the tracks and scratched his head. Two horses. Two riders. It could have been Doolin and Malone, but why had they left the trail here, when it was likely they would have escaped to the Den had they continued on? Yet it was more unlikely, he mused, that these tracks belonged to members of the posse, who would have traveled back along the main trail. If this weren't Doolin and Malone, it would cost nothing to find out. If it was, and he didn't follow them, it might cost everything . . .

Kittery broke off the main trail, following the new path in the sandy soil. As he rode, it became obvious that the trail, unless its course changed drastically, would lead him back to Castor.

He found the remnants of a fire near the base of Gunsight Mountain, where rainwater had collected in a natural dish in the rocks. The coals were still warm, and he pushed the bay hard, knowing he could not be far behind those who had made the camp.

As he rode, it came to him how the posse must have missed seeing this trail. Knowing they were after the Desperados, Vancouver must have assumed they would ride straight for the Den. To make time, he had pushed the posse into the night. It would have probably been sometime after dark that the outlaws veered off.

As Kittery loped along, gray clouds bunched along the horizon. Wind gusted up, bringing the scent of moisture and dampened dust. Then a fine rain began to mist the land, each tiny drop soaking into the soil. Not until half an hour later did the steady drizzle begin to leave its mark, sending tiny rivulets of brown rushing along the trail.

This time, he passed Twin Buttes on their northern side, and an hour later, with rain driving furiously at him now, he followed the nearly obliterated trail into Castor. It descended from the ridge east of town, taking him past Vancouver's place and down onto Main Street, where he lost the trail. The odor of mesquite wood smoke from the houses of town permeated the air like perfume.

Kittery let the horse amble up the thoroughfare as he watched both sides of town, scanning each quiet, gray, rain-washed building in turn. The water ran off the ends of his long hair and into his three-inch-long beard, trickling inside his coat. But with his heart pounding as it was the cold water went unfelt.

At the end of the street, he turned around and rode back again. He dismounted and lashed the bay at the hitching rail of Greene's store. He climbed the porch, stood in the awning's shelter and watched the meager activities of the town.

Down the street, before the warehouse, a freighter clad in an ankle-length buffalo coat was tying a tarp over his high-boarded wagon. He couldn't hear the man, but he could tell by his lips he was cursing the tarp and the weather; he batted the former, then glanced up at the glowering clouds as if daring them to continue raining on him. The rains seemed to take that dare and come down harder.

A pair of women walked shoulder to shoulder, huddled beneath a large black umbrella. They stepped tentatively into the muddy street and hurried to the café, disappearing inside.

Three children took turns running through a mud puddle between the hotel and the doctor's office. Their dog followed, watching over them and getting soaked in the process. A good dog, thought Kittery; he never barked.

Seeing nothing amiss on the street, yet knowing the outlaws had been here, he decided to search from building to building. He would work first one side of the street, then the other. He couldn't help but smile at the authority the badge gave him.

Turning, he stepped inside the dry goods store. Greene, standing bleary-eyed behind the counter, gave him a hard look. "What do you want?"

Kittery's eyes narrowed. "Is that any way to treat a customer? You seen any strangers in town?"

"Not a one. Lookin' for anyone in particular?"

"Yeah. Walt Doolin and Sam Malone."

"Then you're a little late. They left yesterday with a posse after 'em."

The sarcastic comment riled Kittery more. "Yeah, and now they're back. I just followed 'em into town."

Greene clutched the edge of the counter and stood up straighter. "That can't be! What would they want here?"

Kittery shrugged, glancing past Greene at the door back of the counter.

"You think maybe they're here?" asked Greene acidly. "They're not. Do you believe me, or would you like to search the place?"

"Maybe later. You seem guilty of somethin', Greene. I only asked if you'd seen them. Simmer down."

"Why don't you get out of my store, Mr. John Law? This is one place in the world you don't have the run of. And I'm one man you don't own."

Kittery chuckled without humor, trying on purpose to vex the man. Instead, he allowed Greene to reach *him*. He surged forward and grabbed the merchant by the shirtfront in both hands, shoving him back and releasing him, watching him smash into his shelves. The fat man slipped and barely managed to keep from going to the floor among tumbling cans and sacks.

Without a backward glance, Kittery stepped outside and onto the porch, halfway regretting his rashness in the store but not so much that he would ever admit it to anyone. He was glad he had shut the storekeeper's mouth, even if it did mean he had proven the weaker of the two. Pushing these thoughts from his mind, he loosened Cotton Baine's Colt in its holster, which rode his right hip today. He moved on to the sheriff's office.

The man they called Slicker Sam Malone stomped his foot to get the blood moving. He paced the dank shed, puffing on his third cigar. He kept sticking his left hand deep into the pocket of his coat, bringing it back out, over and over again. He went to the crack in the wall and peered out. The gloominess out there weighed heavy on him. The sky was dreary and hung low over the gray mountains, blue curtains of moisture draping out of it to quench the desert. The sight made him shiver. He drew long and hard on the cigar between his teeth and wrapped the thin brown poncho tighter about him, swearing under his breath.

He was a slender man of average height. A crooked nose, shaggy mustache and pale eyes made him a homely man, not a man who would bring any undue notice in most western towns. He dropped the cigar stub to the floor and twisted the toe of his boot on it, then blew the last of the soothing smoke out his nostrils. He drew an 1875 Remington pistol from its holster, opening the cylinder gate and revolving the cylinder to check the loads for what must have been the tenth time in two hours. Six .44 shells. More than enough for any man. He stared unseeing at the dim words engraved on the barrel and shivered again. *God bless "E. Remington and Son,"* he thought.

The door swung open. Cocking the Remington, he leveled it at the torso of the man who stepped inside carrying a '76 Winchester. "Your man just headed for Kingsley's Café," said the new man. "You'll never have a better chance. And you'd best make sure he's dead."

Sam Malone nodded and rubbed a hand down over his drooping mustache. He knew he didn't smell good. He hadn't bathed in he didn't know how long. But this man was worse. He smelled like death.

Holstering the Remington, he reached out and took the big rifle. "All right . . . Mouse. Now get outta here."

Mouse threw Malone an angry glance, then turned and disappeared outside. For a moment Malone stood stroking the dark walnut stock of the beautiful rifle. He worked the lever until the rifle had deposited its twelve .45-75 cartridges in his palm, then he reloaded them. He nodded, satisfied, then gritted his teeth against the cold and stepped into the rain.

He stayed at the rear of the buildings to make his way to the head of the street, and here, from the south corner of Greene's Dry Goods, he peered down the long street toward the café. It wasn't really a long street. But under the circumstances it seemed many hundreds of feet away. There was no one in sight. The slanting dark rain had driven them all into shelter. It puddled the streets, made mire around the hooves of the few hitchracked horses. One lone dog lay under the shelter of the blacksmith's awning, its hair in clinging muddy clumps.

Malone strode across the street, went down an alley and moved along behind the businesses until he reached the café. Stopping there, he lit another cigar, then moved up its side back to Main Street, where he stationed himself to wait—the worst part of all.

Inside the sheriff's office, Miles Tarandon grimly cleaned his Schofield. Tappan Kittery had left him not ten minutes earlier with the news that Malone and Doolin had circled around the posse and back to town. As Tarandon wiped the pistol clean of oil and thumbed in six shells, his mind turned to Tappan Kittery, his only backup should trouble arise. They had had their differences in the past, but in this time of danger he conceded there was no one he would rather have behind him.

Pushing up from his chair, he placed his hat just right on his head and clutched the door handle. He stepped outside, seeing little movement in the town. A freight wagon stood tarped and ready to roll in front of the warehouse. A buckskin cowhorse with its back humped and tail between its legs stood forlorn in front of the dry goods store, rain running off its hips and sides and saddle. Three more horses lined the hotel's hitching rail. One of them stamped, splashing mud, then shook itself, spraying water through the air and making its stirrups flap against its sides.

Abruptly, the rain ceased. It took Tarandon a few moments to notice it, and then he looked to the south, where the sky was bluing up again. He nodded with satisfaction and stepped from the porch into the mud, moving toward the mercantile.

Seeing no one on the street, and with the knowledge that Kittery was making his rounds of the establishments, he decided to work the backstreets and alleyways, looking for the two outlaws' mounts. If they were in town, their animals would be around somewhere. So he walked around behind the mercantile and began to work his way through the Mexican quarter.

Standing beside the café, Sam Malone smiled when he felt the rain cease and heard it stop pattering on his poncho. Having heard nothing from the café, he rested his rifle against the frame of the building and pulled out another cigar. He had just started to light it when he heard the café door open and then women's voices. After his start, he began to relax again and was touching the match to his cigar when a man laughed. He whipped his eyes to the street.

There was his man! He dropped the cigar and match and reached for his rifle. The man was nearly to the middle of the street, the two women ahead of him and to the right.

He stepped out from the corner enough to clear the rifle barrel and called, "Hey, Red!" He brought the rifle chest high. "Soldier boy! Turn around!"

The words caught army investigator Lieutenant Bab Peppridge midstreet, headed for the livery stable. Recognizing the threat in the voice, he snatched at the flap of his holster as he whirled. The Winchester boomed and slammed against Malone's bony shoulder, and the .45-75 slug ripped through the left side of Peppridge's chest, lifting him from the ground and tossing him back a couple of feet from where he had stood. He landed spread-eagled on his back.

Not waiting to see if his task was complete, trusting to the power of the big rifle, Malone whirled toward the Mexican quarter of town, where his and Doolin's horses waited.

Instead of the horses, he saw a man with a badge standing there.

Miles Tarandon drew his pistol and fired twice at the outlaw, missing both shots as Malone wheeled around the corner of the café onto Main Street.

Tarandon came splashing through the mud close behind him. The outlaw's first instinct was to run, and hearing the footsteps, he dashed for the horses tied in front of the hotel. A bullet whizzed past him, and he turned and threw a haphazard shot with the rifle, missing his assailant.

Tarandon was too near. Malone had to run right past the horses at the hotel and chose to try instead for the buckskin at Greene's. The two women with the umbrella, frozen by the sudden outburst of violence, stood in the street. Malone, running with no thought as to who got in his way, crashed into them, sending them careening to either side. Stumbling, Malone cast the heavy rifle aside and drew his pistol. He could hear the lawman's steps close behind him.

The outlaw spun, wasting another shot toward the deputy. He slid in the mud right past the buckskin and nearly slammed into Greene's porch. Grabbing for the buckskin's reins, he saw Tarandon was too close. So he turned and broke for the door of the dry goods store.

On the porch, he grasped the doorknob with his left hand and heard the sound of a bar sliding across the door from inside. A blast came from behind, and the first slug slammed through his ribs and planted in the door. Whipping about, he saw Tarandon standing spraddle-legged in the street, turned to the side and using his raised left forearm to support the pistol in his right fist.

Malone cocked and raised his gun, but Tarandon was ready and fired two more rounds into the outlaw's body. Malone staggered back against the doorframe then slid along it, landing with a dull thunk on his left shoulder. His revolver hit the porch and skidded across, teetered on the edge for half a second, then dropped into the mud with a soft splash.

As the first shot from Malone's rifle beat through the rain-washed streets, Tappan Kittery whirled away from café owner Bartholomew Kingsley and rushed for the door. Blind to the room because of his urgency, he stumbled over a table and nearly fell headlong, only managing to catch himself as he went to one knee. Pain shot through his leg, and he lunged up, clutching the knee. He groaned but pushed the pain aside and made his way to the door.

By the time he stepped outside, he could see Miles Tarandon standing in the middle of the street thirty yards away. The deputy's pistol exploded, and gunsmoke leaped out before him.

Suddenly, a big, black-haired man rushed from behind the livery stable. When he saw Kittery step from the doorway of the café, he slid to a halt ten yards away, grimacing. He looked from Kittery toward the horses tied a hundred feet away behind the café. The burly man straightened out of his crouch and turned fully to Kittery. He speared the lawman with his eyes.

Kittery stood half out of the doorway, holding the other man's gaze, knowing without question who he was. He remembered him from his poker game in the cantina and from his fistfight with Bison Sabala. The wanted posters' description of the man was coldly plain. Bloody Walt Doolin. Blackie. The Bear. The man who had helped kill Joe Raines, then carved his initials into his flesh, like a notch on a pistol.

This end of the street was nearly empty. Kittery heard the last shots of Tarandon's exchange fade away but never let his eyes leave Doolin's hard-cut black-whiskered face.

He watched with no expression and allowed Doolin to move within five yards of him. He saw nothing but those brutal hands, those eyes of an angry hawk.

"Doolin, it's time for you to die."

Chapter Thirty-Six
More Gifts Than One

A wolfish look washed over Walt Doolin's face. His lip corners curled. "The mighty Captain Kittery. You're supposed to be so damn tough. Prove it."

Without warning, Doolin lunged forward and grabbed Kittery by the coat with both hands. He smashed him against the wall of the café, where a protruding nail head penetrated the lawman's coat and shirt and punctured his skin, sinking half an inch into his back. Kittery jerked away from the nail, feeling flesh tear. He sprang at Doolin, and the outlaw lashed him across the face with the back of his hand.

But when Kittery came in a second time, his fighting blood was aroused. His bitter hatred resumed the fight. He knew he should draw Cotton Baine's Colt and shoot this man down. He knew it with every grain of common sense inside him. But, like Doolin, he listened only to the overwhelming drive to destroy this opponent with his bare hands, to feel flesh and bones sag beneath his pounding fists.

He smashed a knee into Doolin's groin, weakening the outlaw's simultaneous blow to his stomach. When Doolin doubled over, Kittery smashed him an uppercut to the face with both hands. Doolin reeled back. But with a fiendish grin showing his blood-pinkened teeth he came in and threw a left hook, then a right cross. The first connected—the second did not. Kittery staggered out of range of another cross, then lowered his head and sank a left of his own into Doolin's abdomen, then a right. Another right fist drove the outlaw to his knees in the mud, and Doolin came up swinging and missed with both fists.

A crowd had gathered, but neither man seemed to notice. Doolin would die here—that was plain. But he only saw this hated enemy before him. Kittery might die, too, but even then he would win, for the crowd would eat Walt Doolin alive.

Kittery struck out again, no longer cognizant of the weight of the .45 on his hip. His knuckles bit into Doolin's cheek and knocked him back, and he closed again. From out of nowhere a fist connected with his own face. He felt his head snap back but felt no pain. He hammered both fists into Doolin's stomach, then one into his chest where his heart would be—if he had one. Doolin struck him again, and he didn't try to dodge or block it. There was no finesse in this fight, no use of a pugilist's tactics or even awareness of their existence. Brute strength ruled, and both men's faces ran with blood.

Doolin threw all his weight into a right cross that missed, and Kittery hurt him first with a solid blow to the ribcage high under that arm, then another to the belly. He broke his nose with the flat of his palm.

Doolin roared so it seemed to shake the street. His face, besides the blood, was red with fury. He cursed and growled again and reached for the Spiller and Burr .44 on his hip.

Kittery had no doubt he was about to taste a bullet. But something whooshed through the air over his left shoulder and smashed a hard blow to the shoulder of Doolin's gun arm. The outlaw nearly lost his grip on his pistol. Kittery had the briefest of glimpses at the meat cleaver that went tumbling to the mud. Then Doolin, grimacing, once again brought his revolver to bear.

Taking advantage of the outlaw's temporary weakness, Kittery stepped to Doolin's right side. He drew open his coat, his hands numb from the bruising they had taken against Doolin's face, and fumbled out the Colt Peacemaker. Today it would make no peace.

He cocked the revolver as it came up and fired as Doolin's weakened arm tried to react to his change in position. The slug smashed through Doolin's midriff.

Doolin groaned and shut his eyes, dropping the pistol. Kittery shot him again in the same place before the pistol hit the ground. Doolin stayed on his feet. His left hand came up to clutch the bone handle of his knife. His right clutched at his midriff, where the blood flooded between his fingers and down his shirt onto his trousers. Doolin glanced around at the onlookers but seemed not to notice anyone but Kittery.

He lifted his right hand and looked at his bloody fingers, then at the blood all over his shirt and pants. Kittery waited.

Straightening, Doolin slid the knife from its sheath and managed one weak step forward. Behind his fierce gaze was a look of begging. Kittery raised the Colt and shot him in the exact center of his corrugated forehead. Bloody Walt Doolin toppled backward, falling to earth like a great oak. The mud splashed in all directions, then flooded back and seemed to engulf him.

When the smoke drifted away, Bloody Walt Doolin lay dead, tiny rivers of red swirling away from his head and body to join the muddy brown that made the street look like a sea infiltrated by soggy islands.

The crowd pushed forward, oblivious of the mud and water. Somehow among the throng Kittery saw Tania. The look of diffused terror dissipated from her face as she made her way toward him, and he waited where he was. At last, she reached him and threw her arms around him, squeezing as tightly as her swollen abdomen would allow. He held her and whispered, "I'm home, girl. I'll not be leavin' you again." Tania wept and clutched him tighter still, burying her face in the folds of his coat.

Behind him, Doctor Hale spoke his name. Kittery turned to see the bearded doctor kneeling beside Doolin in the mud. "Come over here, Captain."

Kittery stepped away from Tania and wiped a smear of blood from her cheek. He glanced at Doolin, then back at the woman. He didn't want her to walk over there, to see what he had done. But he could tell by the look on her face he couldn't stop her, so he took her hand and they walked to the doctor and the dead man.

The doctor had opened Doolin's shirt to reveal more than just the outlaw's wounds. There had been a jagged white scar on the outlaw's chest Kittery had noted during the fight with Bison Sabala—everyone had noticed it, and talked about it later. The doctor now touched it with his finger. It was hardly visible, even up close, under the matted black hair. But a close inspection revealed two crude letters that had been carved there—M. D.

Kittery looked over at Hale, and the doctor shrugged, saying, "It looks like at least one of the legends is true." Kittery nodded, thinking of the popular story that Walt Doolin's father, Matthew Doolin, in a drunken rage had held his son down after some childhood misdeed and carved his initials in his chest.

"Sad way to start out in life," said Kittery. He noticed the meat cleaver lying in the mud. Picking it up, he turned around, finding Bartholomew Kingsley in the crowd. The café owner stood close by, seemingly only half-interested in the proceedings. Kittery walked up to him. "Thanks, Kingsley. Looked like he might've had me, till you stepped in."

Bart Kingsley looked over Kittery's battered, bloody face. "Looks like he had you anyway." Taking the cleaver, he slung some of the mud from it, then looked back up at Kittery. "You're welcome, Kittery. I'm glad I was here." He turned back into the establishment, the cleaver dangling from his fingers.

Kittery stepped onto Main Street two days later a different man, at least in outward appearances. He had washed away the trail dust and grime and had his hair cut. His neck and ears were uncovered for the first time in months. The ruffled three-inch-long beard was also gone. Its only vestige was the mustache he had kept, its tips slightly curled to lend him a sophisticated air. It was the same style of mustache Joe Raines had worn.

Warmth had returned to Arizona—not yet stifling heat, but predicting that to come. Kittery stood on the porch of the cantina and tamped his black pipe and lit it, enjoying the sunshine and the fresh smell of newly watered vegetation the breeze brought down from the ridge. A puff of pipe smoke obliterated the other smells of the spring, and he looked down with sudden distaste at the thing protruding from his mouth. He studied it for a moment, thinking of his friend, Joe Raines. Kittery had never been a smoker. He would never admit it to anyone, but the only reason he had taken up the pipe was as a sort of tribute to his friend. But now he had the mustache, just as good a memento to remember his old partner by. He took the pipe in both hands and broke it in half, dropping it into the dirt at the edge of the boardwalk.

Across the thoroughfare, Miles Tarandon busied himself tacking notices to the outer wall of the jail. Kittery let his curiosity carry him over there, and when the deputy had gone inside he bent over to work his way through the print of the latest additions.

One of them in particular caught his eye, and he peered closer.

To whom it may concern:

Albert Oscar Hagar was pronounced dead by Doctor Hale at the scene of his shooting on Monday, the second of April. The new proprietors of the warehouse, Dabney Trull and Solomon Hart, of Castor, and B. M. Lackton and Henry Bloom, of Tucson, are interested in opening up the business to the sale of shares. Any of them might be contacted prior to the first of May.

Furthermore, Solomon Hart, new owner of Hagar's Mercantile, wishes to sell out at the earliest date possible for the best offer. Interested parties should see Mr. Hart at the store.

Kittery straightened up, nodding. He wanted to stay in Castor. And Tania would give anything to remain here where she had grown up. He was not a merchant, but there had to be something more to life than chasing outlaws and drifting from town to town, especially now that he had a wife and approaching baby to think of. He pondered as he walked along the porch, pinching his lower lip and nodding to himself. Why not? Sure, it was sudden. But at least he could give it a try.

Tania was in the kitchen baking biscuits, and she smiled at him when he walked in. She started to say something, then tilted her head. "What's happened? What are you up to, grinning like that?"

"What do you think about the idea of me standin' behind a counter wearin' an apron?"

She placed her spoon on the countertop and turned to face him fully, wiping her hands on her own apron. "I think you're a silly man."

"May be. But do I look like a merchant to you?"

He came near and kissed her on the mouth, and she giggled like a little girl, wrinkling up her nose. "Well, right now you look like Frankenstein's monster, with those bruises all over your face. What are you talking about, Tappan?"

"I'm talkin' about us stayin' here in Castor. Findin' a house here. The mercantile is up for sale. I'm thinkin' about buyin' it."

Tania's mouth dropped open, and she covered it with the back of a flour-dusted hand. "You can't be serious! Do you mean it?"

He shrugged. "I have five hundred dollars left from the money I got for Rico Wells, and I have six hundred coming for Denton and another thousand from Doolin. It may not buy the place, but it's a beginning. What do you say about that?"

Tania stood shaking her head. "I think I love you, Tappan." And she held him tight.

The following morning, Kittery rode the black stallion down into Main Street. He reined in at the mercantile, where he dismounted and tied the black and stepped up on the porch. Before walking in, he removed his hat and smoothed back his hair.

Solomon Hart stood behind the counter, sliding a dust rag across it with his one hand. "Why hello, Mr. Kit'ry." White teeth flashed across his face, in stark contrast with his dark skin. "What can I do for you?"

"Well, Mr. Hart, I came to talk to you about buyin' the place."

Hart stopped dusting. "You? Ah, go on! You don't say. Well, if you're serious, have a chair." He smiled again, indicating a chair against the wall. He strode out from behind the counter, and removing his dusty apron he seated himself across from Kittery, a little table with a checkerboard sitting between them.

"What sort of figures are we talkin' about?" began Kittery.

Hart began to scratch the top of his frizzy head, then chuckled. "Shoot, Mr. Kittery, I'm still tryin' to picture you behind this here counter. You sure you wanna do that?"

"I'm sure, Mr. Hart. I'm tired of livin' on death. I want to raise a family here, and this is the only way I can figure to do it right now." Kittery waved a hand to indicate the street outside. "Castor's a good place to begin life."

"You really are serious. Well, I'll be hornswoggled. In that case, I'm goin' t' give you the best deal you ever gonna get."

The black man stood up and began to walk the room, peering at the most insignificant of items as if to him they were of utmost importance. "Mr. Kittery, I can go through an entire inventory for you, or you can pay a flat rate. I got ten new Colt pistols anywhere from seventeen to twenty-nine dollars; I got .44 Winchester rifles for twenty dollars and up, '76 Winchesters for twenty-five dollars; I got a whole lotta shirts, a whole lotta trousers, a whole lotta victuals . . . Well, shoot, Mr. Kittery, you c'n see for yourself. Fact is, I gotta whole lotta junk, and I don't wanna add nothin' up." He grinned and stepped closer. "I like you, Mr. Kit'ry, so I'm gonna tell you somethin' I don't want gettin' around. When me an' Mr. Hagar did inventory last January we was sittin' on twenty-three hundred dollars, merchandise alone! The lot itself is worth two hundred dollars, and the building? I figger a thousand. If my calculations is correct, we're lookin' at a total worth of three thousand five hundred dollars. I'll give it to you, Mr. Kit'ry, for three thousand flat."

Kittery chuckled. "Sounds like you've done a lot of figurin'. You got it down to the penny, almost." He walked about the room. He made a show of inspecting the woodwork, the walls, the frames and glass and scratched counter, the braid rugs and wall hooks. He knew the goods were quality. He had bought some of them himself. And he figured the entire enterprise had to be worth at least four thousand dollars, if he was any judge. Still, even three thousand was more than he had seen together in one place before. It wasn't something to throw around.

"Mr. Hart, I'd like to sleep on this a day or two and talk to my wife about it. But before I go, is there any way I could put a thousand down and pay the rest on time?"

Hart cocked his head and gave Kittery the evil eye. "Now, Mr. Kit'ry. You wouldn't be tryin' t' swindle me, would yuh? I don't know a whole lot about these here matters, you know."

Kittery grinned in response to the joking look on Hart's face. "Mr. Hart, you and I both know you probably have more business savvy than most people in this town could have after ten years in business. Don't you play dumb with me. I'll be back to talk to you in—let's say three days. Thursday morning."

As Kittery had promised, Thursday morning, at fifteen minutes past ten, he stepped into the mercantile. Hart sat behind the counter drinking coffee. He looked up and smiled broadly.

"Mr. Kit'ry! You actually came back!"

Kittery smiled back at him and nodded. "I said I would. I came to make you an offer."

"Well, blame it, I'm sorry. They done bought me out already. This is the last day I'll be the owner of the store."

Kittery felt like he had been punched in the stomach. But his face remained calm. "They bought you out? Who's they?"

"Well, it says here a . . . a . . . Listen here, Mr. Kit'ry, I don't have my spectacles. Could you read that for me?" He slid a bill of sale across the counter.

Perplexed and disappointed, Kittery picked up the piece of paper and read:

I, Solomon Hart, on this day, April 5, 1877, sold one mercantile store, all goods and grounds included, to the undersigned for the amount of three thousand dollars, paid in full.

The words meant only that the place was gone. But when Kittery reached the bottom of the page and stared down at the name, although it was written clearly in black ink he had to peer closer and read it again. The "undersigned" was Tappan Kittery, written in his own hand.

Incredulous, Kittery looked up. "I don't understand."

Solomon Hart laughed with delight and wiped his eyes. "Jus' read the other side, Mr. Kittery."

With knitted brow, Kittery flipped the paper over. On the opposite side was a note addressed to him, signed by no less than fifty people.

Our dearest United States Deputy Marshal Captain Tappan Kittery— he chuckled at that heading— *I know this will come as a shock to you, but I can assure you it is all real. I, in behalf of Governor Safford, Marshal Standefer, Sheriff Barden, the Tucson Stockmen's Club, and half the town of Castor, who have signed below, wish to thank you for what you've done for Castor and indeed for the entire territory of Arizona.*

We hope you will accept this gift, as you will be forced to anyway. Besides, we used your reward money from Denton and Doolin for part of it. The money you have saved from Rico Wells is yours to keep. Thank you for your undaunted stubbornness and drive.

There is one more thing: there should be a wooden box on the counter before you. Open it.

Best wishes,
James Price

P.S. It was Tania who forged your name. Not bad, I thought.

As the letter indicated, a polished, handmade box sat on the counter, tied in a cheerful scarlet bow. Kittery took it, slid off the ribbon, lifted the latch, and laid back the lid. There, snuggled deep in a bed of red velvet, a legend shone out at him. Two ivory-handled Colt pistols lay there in luxury, their gunfire stilled forever. An inscription on the underside of the lid read: In honor of a good friend, Captain Tappan Kittery. The guns of Noble "Silverbeard" Sloan. By James Danning Price.

Kittery closed the box, took it under his arm, and smiled up at Hart. "I guess words won't do to thank you, Mr. Hart. I couldn't say enough."

"No need, Mr. Kitt'ry. Jus' take care of my store."

Business at the Kittery Mercantile went well for a man starting out with little experience in that line of work. Buyers came from as far away as Tucson to patronize the establishment of the man who was becoming one of Arizona's largest living folk heroes.

He hired the little Mexican boy named Pepe to sweep the floor and porch and dust the merchandise at two dollars a week. He also took on Doctor Hale's eighteen-year-old son, Jeremy, to tend to financial affairs in his absence. This left him free to take an occasional hunting expedition into the mountains or spend some time searching for a suitable place to build a home.

By the end of the first week in June, Tania's time was drawing close. It was obvious, for as Kittery jokingly told her she looked like she was trying to steal a calf. Kittery stayed at the store more often, looking up the hill toward the house, wondering if today would be the day. Nine months seemed so long. When would his child come into the world?

One afternoon, when the day dragged on long and hot, Kittery sat on the porch, watching the quiet dusty street. The sun was a blurry ball in the sky, making sweat run down his sides and down the backs of his legs. Only the flies were stupid enough to be out at this time of day; he had no rest swatting them away from his face while he tried to enjoy the peace of siesta.

Glancing north, he saw two riders come over the rise leading into town. He recognized them as they rode under the shadow of the sign welcoming visitors to Castor as Jake Long and Lafayette Bacon. The pair had taken on ownership of the Double F Ranch. As they neared the store, they smiled and touched their hats, dismounting before him.

"Afternoon, boys." Kittery smiled up at them. "Rough day to be out ridin'."

The two nodded and came to the porch. Long sat down, and Bacon rested a dusty boot on the edge. Bacon removed his hat with his right hand and wiped his sweaty forehead with the same sleeve, throwing back his long, yellow hair and replacing the hat.

He shook his head wearily. "Yeah, Cap'n, it's hotter'n Texas t'day."

"What brings you to town at this time of day?"

"Me and Jake have an offer for you."

Kittery looked over at Long, who sat a few feet away from him with his head lowered, breathing shallowly against the blistering heat that came up off the white dust. He looked back at Bacon. "An offer. What of?"

"You said you used to buy horses for the army an' such. Well, you know how it is this time of year, what with roundin' up and takin' herds to Tucson and Lowell and all that."

Kittery nodded. "Gets pretty busy."

"Right." Bacon looked over at Long as if to petition his aid. Long was still studying the dust, impassive, so Bacon returned his gaze to Kittery. "Well anyhow, you've owned this here store for over a month now, an' it seems you spend an awful lot of time gettin' away from it. As for us, we been lookin' all around Tucson for a certain type of feller—someone that knows horses and brandin' an' foalin' an' all that, but also managin' financial affairs. We wondered if you might be interested in goin' in with us on the place."

"You're offerin' me a job?"

Jake Long looked up for the first time. "Wouldn't call it a job. More like a partnership. We ain't been doin' so good since old Zeff's bin gone. That's the sorry truth."

"I'm flattered, boys. It's a rough time to ask though. Tania's gonna have her baby soon, an' I'm kind of lookin' around for a place to live."

"We thought about that too," said Bacon, coming up another step. "We got Zeff's big house all cleaned up. Kinda figgered you might stay there. He built that place for a family, you know, except his wife died of the dysentery before they really got a start."

Kittery mulled over the offer. The more he thought about it, the more he realized he would be a fool to pass it up. Opportunity like this was rare.

He stood and looked up and down the street, then at Long and Bacon. "Boys, I'm gonna consider your offer. What if I come out to the place this evening or sometime later this week and we talk?"

"Sounds fine, Cap'n." Bacon smiled, and then his face went serious. "There was one other thing."

"What's that?"

"Well sir, when we were cleanin' up the big house we run across a locked strongbox. Couldn't find a key, so we busted it open. It had Zeff's personal papers in it, includin' this letter." Bacon reached into his pocket and drew out a damp, folded piece of paper and offered it to Kittery. "Sorry about the sweat."

As Kittery shrugged and started to unfold the letter, an odd chill came over him. He began to read in silence.

To the people of Castor:

Because of recent events in town, I take pen in hand in hopes of clearing my name of the grave charges brought against me. I don't expect this to be found until I am old, gray and deceased, but hope it may do something toward bringing me back into the good grace of your memories.

Six years ago I was a broken man without a penny and without a family. My wife died of dysentery. Times were hard. What animals the Apaches didn't run away with starved or died of thirst. The stock that did manage to survive was invaluable. Pete Bandy was my foreman at the time, a conniving thief. He convinced me to allow him to use my ranch as a gathering ground for other ranchers' unbranded stock, in exchange for a percentage of profits he made by selling them off. It is something I will always be ashamed of, for I worked with Ned Crawford and several other notorious outlaws. Once, I even came close to working with Baraga, though nothing came of that. But for over a year it went that way, and I rebuilt the spread.

Recently, Captain Kittery accused me of working with Baraga—of being a traitor. I swear that being a traitor is something I would never do, and I am innocent of those charges. But I can't set myself in the clear now because it would bring out my previous mistakes.

The night I was charged with treason, Silverbeard Sloan and Crow Denton came to me, aware of my past dishonesty and hoping to begin a partnership with me, using my land much the same way Bandy had done. I refused them. I would not go back to my old ways. I pray that whoever reads this will believe that. I have no reason to lie.

It eases my feelings to write this, even if no one ever reads it. I pray to God I can be forgiven my misdeeds.

Signed,
Zeff Perry

Kittery finished reading and closed his eyes, letting out a long breath he had been holding. "Wish I'd known this back then." He looked from Bacon to Long. "It's a hard way to be buried, everyone thinkin' yer a traitor. But this clinches one thing: Perry wasn't with Baraga, and as far as we know whoever was is still alive."

When Long and Bacon had gone, Kittery sent Pepe up the ridge to the doctor's house to fetch Jeremy. Even before they returned, he left a note telling where he was going, took Denton's chestnut out of the stable and rode north.

He took close to twenty minutes to reach the cemetery, and there he threw the reins over the fence and passed through the gate. His heart felt heavy, and his throat tight. He stopped before a grave, now grown over in grass, bearing a cross dappled in shade from the leaves overhead.

The inscription was branded into the wood: ZEFFANIAH PERRY, DIED VALIANTLY, 1876. He removed his hat, letting it dangle in his hands, folded before him.

"It's a real nice day, Perry," he said. "I kinda miss seein' yer friendly face around town on Saturday nights. I don't know if you can hear me, but I came to say I'm sorry. I read you wrong. You only tried to kill Crawford to prove you were worthy to ride with us, not because you thought he'd give you away as Baraga's inside man. I know that now. There's not a lot more I can say, friend. But I hope they sat you next to Cotton and Joe and the rest, because you deserve it."

Turning, he clamped on his hat and mounted and rode back toward town. Half a mile from Castor, he saw dust ahead, marking the passage of a lone horse coming fast. He rode forward warily, remembering the last time he had seen such a cloud of dust coming out of town. When the rider drew near he recognized Pepe. The boy sawed his pony to a halt, trying to catch his breath.

"S—Señor!" He kicked the excited horse in a circle, trying to calm it down. "Come quick! You' wife—she have baby!"

Kittery's heart leaped, and he laid heels to the gelding, leaving the boy in his dust. The animal had the speed of a racehorse, and in no time they were charging into town and up the ridge road.

In the yard, he jumped from the horse before bringing it to a full stop and rushed into the house. Before he could reach the bedroom, Rand McBride stood before him, placing restraining hands on his shoulders.

"You wouldn't want to go in there just yet, Tap. They're having some kind of trouble."

"What's happened?" The big man looked toward the door, eyes widening. He could hear Doc Hale's muffled voice and Beth Vancouver's soothing one.

"I'm not sure," responded McBride. "The baby's caught somehow."

"What?" Kittery threw his father-in-law a concerned glance. Then he heard Tania cry out in pain. "I'm goin' in there."

As he was pushing the door open, Beth met him, and placing a hand on his chest, she pushed him back into the main room with her, shutting the door.

"You'd better not go in, Tappan. Tania has had some complications. I think it's best you didn't see her. But there is some wonderful news as well—you have a beautiful baby girl! Come with me."

Beth reached out her hand, and Tappan took it against his better judgment, feeling like a helpless child as she led him back out into the bright sunshine. She turned toward him, taking his other hand and looking into his eyes.

"Tappan, I know you want to be in there with your wife, but it's best you stay outside this time. Right now the best thing you can do is pray." She smiled. "I know it hurts, but you'll have to get used to it from now on. Not being all-powerful, that is. All your life you've been strong and able to handle anything. But once you have a wife in labor, and from then on, it's like becoming a child all over again. All those things you never had need to know before you now must learn."

"What's wrong in there?"

"I don't know, exactly. Somehow one of . . . well, the baby managed to get its arm or leg tangled up in the cord. It's happened to the doctor before. You know, he didn't seem too concerned. I'm sure she'll be all right."

"Beth." Kittery stepped toward the door, making a decision. "If Tania's hurtin', I wanna be by her side—she'd be by mine. Sure, I'm new to this kind of thing, and I don't know much about it, but I'll never learn sittin' out here. If I can't do anything else, at least I can hold her hand."

Beth sighed and let go of his hands. "I'm sorry. I should have known how you'd feel."

As Kittery stepped through the door, the sound of crying arose from the bedroom, and Hale burst out, wiping his hands on a towel. The moment he saw Kittery, he grabbed him by the upper arms. Before Kittery realized the look in the doctor's eyes was one of excitement, he was ready to shove him aside. He had been stopped enough.

Then it dawned on him that Hale's eyes were lit up like torches. "Captain!" he yelled, shaking him. "He's all right. He's out of danger and healthy as can be. And Tania's out of danger, too!"

"It's a boy?" asked Kittery, unbelieving. "But I thought—" He looked over at Beth. "I thought it was a girl."

Hale threw a questioning look at Beth, who was squeezing her hands together, staring anxiously. "You— You didn't tell him?"

Beth let out an involuntary laugh, a sound of relief. "Well, I tried! I just . . . No, I guess I didn't tell him."

Hale looked back up at Tappan Kittery and held out his hand to shake. "The first one *is* a girl, Tappan. Number two is the boy!"

Dazed, Kittery turned and pushed through the door, followed by Hale, Beth and Rand. Tania looked up at him with a weak smile. "Hello, Daddy." She lowered the top of the blanket, revealing two tiny red heads covered with black hair. They both lay contentedly against Tania's soft cotton gown, their crying stilled. "Meet Jacob Derek and Rachel Ann Kittery."

All the helpless father could do was stand there and stare. "I'll be chopped up in a grinder," he said with a huge grin spreading across his face. "I thought a store full of merchandise was a dandy gift. This one beats all."

Chapter Thirty-Seven
Into the Den

    Summer heat had returned to the Baboquivaris after the short cool winter and the respite it brought from the sun. Vegetation withered beneath that sun, bleaching under the onslaught of its fire. The earth absorbed the heat and baked white, but the rocks parried, reflecting the heat toward the sky.
    Baboquivari's dome, gray-blue on the horizon, jutted bold and raw and indomitable into the sky. Its proud bald head was submerged in a wandering thunderhead whose looming belly glowed with energy.
    Baboquivari Peak. There was magic in the name. Mystery. Power. Baboquivari, lair of the Papago god, I'itoi, where none but the brave or foolish dared wander.
    The entire range was a massive, ragged playground of jumbled boulders and broken backbones of ocher-colored stone. Deep, winding canyons veered this way and that, disappeared, reappeared, butted against sheer rock faces that vaulted toward the cloud-studded sky.
    Spreading oaks, patches of prickly pear, and yucca with long green sword blades and towering stalks dotted the hills that climbed to the peak. Yellow-trunked paloverde, curve-thorned catclaw and velvet mesquite, whose name belied its harsh character and long, wicked spikes, sewed the dry washes together in a thicket that resembled a blanket for the devil. This made passage anywhere but on the beaten path nearly impossible to a horseman except at the cost of much blood.
    In the canyon north of looming Baboquivari Peak, the so-called Desperado Den secluded ten men from the sun's brutal assault and the hawk's ever-searching gaze. The ice-blue eyes of Savage Diablo Baraga regarded each of the nine men in his company, his impassive expression masking his thoughts. He assessed the newcomers thoroughly.
    Baraga's faithful henchmen, Morgan Dixon and Colt Bishop, had placed themselves at strategic points across the floor. He nodded at them in turn, watching the Major run an oiled rag down one barrel of his shotgun, while Bishop sipped a cup of coffee. They, too, watched through veiled eyes the newcomers and the men who had ridden with them at times in the past. Who could they trust? Who would fight, and who would turn and run or surrender in a tight spot?
    These ten men would ride hard today. And some might die. It was a chance one took to ride with Baraga. An army payroll wagon was on its way from Yuma to

Tucson and Camp Lowell. The wagon carried thirty thousand dollars scheduled to reach Tucson by nightfall, but it would never reach its destination. The road along which it rolled cut across the northern tip of the Coyote Mountains, an arm of the Baboquivaris, and it was there that Baraga's Desperados would lie in wait.

The morning wore on. Baraga sat against the wall and listened to the song of a canyon wren, a series of clear, descending whistles from somewhere out in the brush. He closed his eyes for a moment and let his mind drift back over the past ten years, over his triumphs and victories, then to the sudden decline of the Desperados Eight. It had begun with the death of Rico Wells, moved to the loss of Denton and Sloan, and now recently to the killings of Doolin and Malone. And all was due in large part to the entrance on the scene of one man: Captain Tappan Kittery.

His eyes turned hard at the thought of the name. Tappan Kittery. Kittery . . . The name brought back memories from long ago. Once, it had been no more than a name, like any other. Now he detested the sound of it. He had cursed that awful name countless times, wishing death upon the one who bore it. Kittery had led the vigilantes. He had killed Ned Crawford and others, and now Doolin and Malone. Denton and Sloan, too, but that was of little consequence, for they had betrayed him.

And now, mostly due to Kittery, the last of the Desperados made ready for their last raid. This was it—the final coup. Regardless of the outcome, the Desperados would disband, and only Mexico, its food, tequila and pretty, dark-skinned ladies, awaited them . . . forever.

Captain Tappan Kittery. Baraga sat and pondered the name. A Union man. A Yankee officer. They were the worst of all. Yet he had known the name of Tappan Kittery long before those bloody, awful days of the war. He remembered a good boy, loyal to his family, loving and full of the desire to succeed. How could such a boy have so disgraced his family? How could he have turned on his own folks?

Baraga shook his head. That was past, those long ago days in the Great Smoky Mountains of Carolina. Only memories remained. Memories of a young, dark-haired boy playing in the creek, walking in the woods, wearing a buckskin shirt and growing tall and strong. Would he ever meet Tap Kittery again? And if so, which one would die?

Pushing these thoughts aside, Baraga lunged up and thrust his short-barreled Colt into its holster on his left hip. The others all ornamented his body like metal hands, along with the two concealed in his boots. His veiled eyes skimmed the room, and he reached down to seize a Winchester carbine. All eyes watched him, and he nodded.

"The time has come."

In an instant, the room came to life as men holstered pistols, stuffed on hats and took up rifles, moving into the adjacent cavern, where their saddled horses stood waiting. Baraga, Bishop and Dixon moved out last and swung into their saddles.

Bishop spoke suddenly. "Where's the Indian's horse?"

Baraga peered back into the shadows, searching along the wall, but the Indian's bay pony was gone. Riding outside, he glared up into the rocks, where Paddlon, the Papago, had stood guard every day for the past eight years. He swore, turning to his comrades.

Dixon spoke. "He must've ridden out during the night. I didn't hear a thing."

"I reckon it was time," Bishop mused. "Leavin's been in his eyes for weeks. He knew we were through."

Baraga shrugged. "One less for the split."

With a last look up at the empty rocks where Paddlon had always sat, the trio turned and rode in single file after the other horsemen, strung out along one hundred yards of the trail.

The Papago, Paddles-On-The-River, made the trail east to the lonesome sound of his bay horse's hooves on rocky soil. Insects were silent, birds didn't move, there was no breeze. There was only the sun and the Indian and his horse. Ghostly echoes in the canyon.

Paddlon had pondered long and hard how he would leave Baraga. Should he just leave, a loyal servant who saw the end in sight and acted accordingly? Or should he lash out against the mistreatment he had received for the past eight years? Paddlon was only a Tohono O'odham, an outlaw in the eyes of white man and his own tribe alike. Yet he had been faithful in his service to Baraga's gang, and he should have been rewarded in kind. What did he have for the years of service? Nothing, while the Desperados had everything. And Paddlon would always have nothing, but he would not be alone in his misfortune. He would bring down Baraga with him.

At eleven o'clock that morning, Paddles on the River loped onto Caster's hard-packed Main Street. He eased the bay to a saunter as he drew past the warehouse, and everyone who noticed him turned their eyes to stare at the odd sight of a bandoleroed Papago Indian riding so boldly up their street. His eyes gazed straight ahead, and his highborn blood showed in his proud bearing and square shoulders. He climbed down at the sheriff's office and stepped inside.

Luke Vancouver swung his feet to the floor, startled. "What can I do for you?"

Paddlon looked on stoically. "I come bring you warning. You are starman?"

Vancouver reached into his pocket, withdrew his badge, and held it up. "I'm the sheriff, yes."

"Then I come tell you. The long-knife gold that goes Tucson, it will not arrive. Diablo Baraga attacks the wagons near Coyote Mountains."

Vancouver lunged up from the desk, upsetting his coffee over a sheath of wanted posters. "How do you know this?"

Paddlon tapped his chest. "I was with Baraga. I guard his camp much years."

As Vancouver tried to clean up his desk, he fired questions at the Indian, learning as much of Baraga's plans as possible. Then he strapped on his gunbelt and grabbed a rifle and canteen. "Thanks, friend." He handed the Indian two silver dollars. "Go have yourself something to eat. They'll let you in at the café across the street."

With that, he rushed out the door and ran to Kittery's store. Kittery stood behind the counter. He caught the sheriff's urgent look the moment he cleared the doorway.

"What's up?"

"A raid. Baraga's going to hit the army payroll headed for Lowell. I need you, Tappan."

"You sure about it?" asked Kittery.

"I'm sure of nothing. But I don't want to gamble against it."

Kittery picked his marshal's badge from a shelf and pinned it to his vest, then came around the counter, snatching his gun belt and holstered Remington from a nail on the wall. He slapped on his hat and grabbed a Winchester and box of shells, then turned to Pepe, who stood nearby.

"Pepe, run and get Jeremy, then come back here and have him get me out a canteen and bedroll. You fill the canteen and bring them both to the stable—all right?"

Turning, he placed a hand on Vancouver's shoulder. "Let's go!"

He rushed him out the door. On the porch, he faced him. "Go home and get your gear, then try and round up Tarandon and a posse. I'll go get Price and Beck and see if Sabala will go."

Kittery whirled about and trotted toward the blacksmith shop, while Vancouver darted through the alley and headed for the hill.

Fifteen minutes later, Kittery, having recruited Beck, Price and Sabala, was saddling his chestnut in the stable. In the excitement, no one had seen Paddlon disappear into the dry goods store, and no one seemed to notice as he threw the door open and strode from the porch, his face set in hard lines. He carried a brand new Winchester.

Greene rushed out behind him, halting on the porch. His right hand held a pistol that dangled, already cocked, along his leg. "Hold up there, Injun! Thief!" The bold tone of his voice became urgent as the Indian walked on toward the jail, and he yelled again. "Come back here, old man! Now!"

Still, the Indian did not stop, and Greene came down the steps into the street. Oblivious to the stares of several who had gathered along the thoroughfare to watch the disturbance, he raised the pistol and aimed it at Paddlon, who now had almost reached the jail's door. The .44 barked twice, the slugs slamming into Paddlon's upper back. A woman screamed, and her husband gasped as Paddlon slumped to his knees, then fell on his face in the dirt. Greene stood near his steps, the pistol smoking in his quivering hand.

The posse members, hearing the shots, came on the run. Luke Vancouver reached the scene first. People had begun to approach, crowding around near the jail, and when he looked down he beheld the old Indian. Glancing about, he saw several people staring toward Greene, and he shifted his eyes that way. The merchant glared at him.

Vancouver walked toward him, vaguely aware of Tarandon and Kittery arriving nearby in the street. Three feet from Greene, he pulled up short. "What happened, Greene? Why'd you shoot the Indian?"

Greene lowered his eyes to the pistol, then looked back up, forcing himself to meet the accusing gaze. "He stole a rifle from me. Can't you see?"

"I would have arrested him," retorted Vancouver.

"I stopped him. I knew he'd try to kill someone if I didn't prevent it."

"What?" barked Vancouver. "How could you know that?"

"He was going to kill someone."

Vancouver's neck began to redden. Greene had shot a man in the back, a man who was not running, a man who apparently had not been making any kind of aggressive move toward anyone. Yet even so, the sheriff's reaction was surprising. Everyone considered him a cool-headed man who never lost his temper and handled any situation with tact. But suddenly his right fist lashed out, catching

Greene on the bridge of the nose. The fat man staggered back, then sat down hard on the street, dropping his pistol. He looked up and about him at the staring faces, then reached again for the weapon. Before he could pick it up, he heard a rush of feet and the click of pistol hammers being cocked. He looked up into the bores of Vancouver's, Tarandon's and Kittery's pistols. Pulling his hand away from the gun, he bent and buried his bleeding face in his hands.

Vancouver dropped his pistol into its holster and rubbed his bruised knuckles, looking over at Tarandon, who still had his weapon trained on Greene. "Miles, I'm afraid you'll have to sit this posse out. I need you to take this murderin' trash to Tucson and turn him over to Sheriff Barden for first degree murder and the attempted murder of an officer of the law. And stop by the Double F to let Tania Kittery know Tappan won't be home tonight."

"Sure thing, Luke." The deputy turned to see Kittery looking down at Greene. He cleared his throat. "Kittery." The big man looked up and met his eyes. "Good luck out there. Bring Luke back safe . . . and yourself, too."

Kittery was surprised to see the lawman's hand come out toward him. The handshake was firm, and warm. "Thanks, Miles," Kittery said. And he smiled at the man he knew would become his friend.

Turning, Tarandon waited for Greene to rise, and Vancouver and the rest of them went to their horses.

Pepe arrived at the stable with the canteen and bedroll Kittery had requested. Kittery put a hand on his shoulder. "Pepe, I'm gonna ask you to do something for me that'll make a hero out of you. I wouldn't ask you to do this, except I know my horse likes you, and I know you can ride like the wind." He reached into his pocket and came up with a five-dollar gold piece, which he handed to the boy. "Take Satan and ride as fast as you can to Tucson—just don't kill the horse. Go to the camp where the soldiers are and tell 'em their payroll wagon is going to be attacked by Baraga in the Coyote Mountains. I'm counting on you to get there long before the deputy and his prisoner. When you're done, use that money to get a room and food for the night, and you can keep the rest."

"Si, Captain," Pepe burst out. "I will do it." With that he dashed toward the back of the store, where the black was tied.

Kittery turned to the posse, who stood beside their horses, watching him. He nodded, and they swung into their saddles and rode north out of town. He explained his strategy as they rode.

They would cut cross-country, past Twin Buttes and around the Sierritas, Gunsight Mountain and Soto Peak, to the Coyote Mountains, where the ambush was to take place. The chances of catching the outlaws before the commission of the crime were slim indeed; it was a good thirty-five mile ride through hostile country. But they rode in hopes of coming upon them making their escape toward the Den. It was their only chance.

At eleven o'clock the following morning, Diablo Baraga sat in the Den nursing a flesh wound in his side. He watched the trail south with longing in his eyes. The last of the dust had long since sifted back to the brush and rocks, the dust of Dixon, Bishop and three other outlaws who had survived the raid. And it was over. The infamous Desperados Eight had ridden into bloody oblivion. Out of the three remaining members, two now rode south to Mexico. Their leader sat licking his wounds and reminiscing, wondering how life would have been had his last big job come through for him.

The raid was a disaster. Three wagons formed the payroll train; two of them were loaded with twenty of the best soldiers an army could muster. That was the Desperados' greeting in the Coyote Mountains. For the outlaws it was a rout. Four of them had gone down in the first wave of gunfire, and the attack broke up. One of the fallen men reached his mount and ran. Another of the wounded men had also been able to rise, and with a friend he pulled back from the line of fire, a bullet in his side.

The firing continued sporadic from the two wagons and from the outlaws' hiding places in the rocks until one of the army sharpshooters put a .45-70 bullet through an outlaw's neck at two hundred and fifty yards. After that Baraga, too, pulled out. Carrying Dixon, Bishop and the others with him, he fell back to the horses.

The three Desperados and the remaining outlaws made their dash for safety, but as they rode into the brush a scattered volley broke out from the rocks behind them. One bullet lashed at Baraga's side.

Ten minutes later, they came upon the horse of the outlaw who had run. It stood in the trail with a broken foreleg, and Colt Bishop put it out of its misery with a shot to the head.

They found its rider a half mile farther, sitting in a sweat on the rocks, clutching a bloody leg. As he saw them, he lowered a canteen and stood up, waiting for them to draw near.

"What happened? I thought you were all behind me. Did you fetch me a horse?"

"No horse, you turntail, but we'll leave you better than you left your horse," said Baraga with great disdain. He drew his short-barreled .45 and shot the outlaw twice in the abdomen, leaving him whimpering in the sand as the others rode on by without looking back.

Upon reaching the Den, the Desperados' plans had gone unchanged. They slept one last night there, then broke apart. Dixon and Bishop left talking of starting a ranch in Mexico, and Baraga, who had begun it all, watched the last of his fighting force ride away. He had molded his mighty seven, and now had watched them vanish. He was alone—Baraga, the final Desperado. The newspapers' *Desperado One* took on a whole new meaning now.

His mind wandered to Old Mexico, where he should be by this time tomorrow if fate smiled on him at all. There he would live the good life. He had sisters somewhere. If ever the yen came upon him, perhaps he would visit them. He didn't have the riches he had desired, but enough to live comfortably. His had been a profitable operation over the years. He should have perhaps been there even now, but he had wanted the extra day to say goodbye to this canyon that had been home to him for the past ten years.

He thought once more of Tappan Kittery. So the boy from Carolina had grown to a man. He had made himself a Confederate lieutenant at a very young age—not bad. And then a captain for the Union. The side was despicable, but the title . . . Now he was a lawman—a federal deputy marshal. It was something to speak of. The boy had made something of himself, and despite his bitterness, Baraga felt a glow of pride. Yet now the boy he had watched grow up was out there hunting him, wanting him dead, tearing his plans apart. If Kittery had only known it was he, perhaps then . . . No. Tappan Kittery was too honest; he always had been. And knowing who Baraga was would have changed nothing.

He tried to think of other things, yet that name and face kept coming back. Tappan Kittery must have changed considerably. Would they recognize each other now, sixteen years later? Would they want to?

"Damn that war," Baraga spoke aloud to himself. He reached over and massaged the stub of his arm. It had been a good arm. He and the other boys used to see who could throw a stone the farthest off a ridge, and he always won. He had been the best at everything he did. The best shot, the best hunter, swimmer, wrestler, runner. A colonel in the Confederate army. But what had become of him? With his arm went his will to live, to succeed in life. Left was only the urge to kill, to destroy others the way he had been destroyed. What a fool he was to live that way! What a fool, now that all was said and done and Tappan had surpassed him in everything they had ever wanted when they were young.

Tappan had a wife now, and a son and daughter. He had his own store and was managing a ranch, and all his dreams were coming true. While here sat Baraga, broken and alone.

"Damn this world," he snarled, and then a shot rang out, and a bullet smacked the wall and ricocheted across the room. A barrage of gunfire followed, and Baraga lunged for the wall behind his big wooden chest and lay flat, his pistol drawn. Ricochets lashed the room, whining death. Smoke drifted past from outside, where there must have been twenty guns firing at once. Rock chips flew everywhere, broken from the walls by the maddened chunks of lead. He tried to will away the terrifying sounds of the bullets as his mind brought him back to Cemetery Ridge. He saw the gray uniforms falling around him, the blue smoke, the blood. He heard a hundred cannons roar, saw the beloved guidon fall. He picked it up and ran like a madman up the hill. Toward the endless blue line muddied by smoke. He saw a man sitting by a fallen comrade, crying as he reloaded his musket. Then there was a blinding flash and a horrible sound of someone screaming. He realized it was he. And now he lay silent, covering his left ear, feeling the stub of his arm throb with the memories, knowing he would die.

As suddenly as it had begun, the shooting ceased. It took him a moment, but when he realized it had gone quiet he stood and let out a roar. Ears ringing so he could hardly hear a thing, he stood with braced legs and cried out in anger to the gunmen concealed in the rocks around the cavern.

"Don't shoot! You've done enough!"

"Come on out!" a voice boomed. "Drop your weapons and come out touching the sky!"

Outside, crouched against the wall of the cavern that had served as a stable, Tappan Kittery waited for the man who had spoken to appear. Was it Baraga, or had he headed out already? There was just one horse left in this room, and it was obvious the others had left that morning. This couldn't be Baraga. He couldn't imagine the outlaw leader giving in so easily. But no one was invincible, and the threat of ricocheting bullets was an awesome thing.

He glanced over at Luke Vancouver, who nodded at him calmly. "It's about over, Tap."

All around the cavern, men hid in the rocks, crouched low, clutching their rifles, their faces drawn taut. A handful of them were soldiers, the ones from the payroll train. They had followed Baraga from the scene and joined Kittery's posse along the way.

Kittery waited, wanting to shoot Baraga the moment he stepped into sight, but knowing it was no longer in him to kill in cold blood.

"I'm coming out. I have no gun." The words came from a deep, steady voice.

Kittery stepped out from the cavern wall into the trail, the Colt cocked in his hand. His thoughts raced as he waited for the man to appear. If it was Baraga, what would he look like? Would he look as fierce, as diabolical, as they said? Would he be the towering, bloody giant of legend?

Then the outlaw appeared in the sunshine, his hand held toward the sky, bent at the elbow. His unkempt hair hung loose on his forehead, his beard untrimmed. He wore a flowered gray shirt and brown canvas pants, and a bandoleer full of shells crossed his torso. Kittery saw the stump of the right arm and knew this was the mighty Baraga, and he saw the ice-blue eyes and sharp nose, the hard-set mouth and wide forehead. And then he saw none of this, and the Colt pistol lowered to his thigh without his realization. His jaw went slack, and he stared.

It couldn't be . . . Kittery's mind raced. He felt the hairs rise up on his neck. His skin broke out in goose bumps as Savage Diablo Baraga took a couple of tentative steps toward him.

"Hello, Tappan," the outlaw said, nodding gravely. "They didn't lie. You *did* grow a little."

Chapter Thirty-Eight
The Demise of the Mouse

Kittery stared at the terrible Savage Diablo Baraga about whom he had heard so much. He wanted to speak, but his tongue wouldn't work. All he kept seeing was Baraga as he had been a long time ago, a lanky young man teaching Tappan to shoot a rifle, throw a rock, stalk a buck or plow a field. Tappan's mind reeled back over all the distant years, and Joe Raines's and Cotton Baine's deaths and all the violence of the previous year washed away. Washed away in the thick blood of family.

Somehow he could see only his older brother, his one-time idol, Jacob Atticus Kittery.

Tappan Kittery took a step closer. He couldn't feel his feet, didn't know they had moved. He couldn't see a thing beyond the narrow tunnel of vision before him, where an odd blond apparition hung, with blue smoke for its eyes and dark smoke for its beard.

"Jake?"

Baraga nodded. There was no expression on his face. "Hello, little brother."

Kittery looked down on Baraga, too stunned to catch the irony in the fact that the little brother now stood six inches taller than the big brother. "So . . . you're him." His voice was shaking and very quiet. "You're Baraga."

"I'm Baraga. And what's left of your brother. What the war didn't take."

A rush of all-consuming grief exploded inside Kittery. He let out a roar that was half cry. He could hardly see his surroundings, but he could see enough to stagger toward the lip of the canyon. He stepped over and somehow managed to stumble down the slope through the heavy brush and rocks and oak trees, leaving Baraga and the posse behind him. Again he cried out, firing his pistol into the depths of the canyon. He triggered it again and again until it was empty, and then he fell sideways into the dirt and rocks and prickly plants and lay still, hidden from view of the posse by the trees. He wanted to cry, to wash his soul, but he couldn't. He couldn't think, he couldn't breathe, he couldn't see. He wished with everything in him he could have died rather than face the truth of this day.

God or angels or a powerful conscience came to Kittery as he lay there breathing in the choking dust. He couldn't lie here. He couldn't pretend this one away. He had to get back up to the others and try to be the man Jake would have expected. His chest was swollen to exploding, swollen with the millions of tears that begged escape. Sitting up in a daze, he tried half-heartedly to dust himself off. He reloaded his pistol mechanically and made his way back up to the posse and Baraga.

Luke Vancouver was the first one to speak, and he came over before Kittery could get near the others and put a hand on his broad shoulder. "You all right, Tap?"

"I'm fine." Anyone watching would know that was a lie.

Vancouver's eyes flickered over at Baraga. "Tap, is this true? He's your . . . brother?"

Kittery nodded, unable to meet his friend's eyes. He turned and walked back over in front of Baraga, who stood watching him with a mask over his face.

"What did you do to yourself, Jake?" Kittery heard these words, but it took a moment to realize he had spoken them.

Still, Baraga stood placid. The hate he had borne for so many years had drained from his eyes, leaving only those blue holes full of distant smoke. "What did I do to myself? *They* did it." His loose wave indicated the troopers, dressed in their gray woolen pullover shirts and navy blue trousers. "Your all-powerful Union did all that needed to be done to me. They took what they wanted and threw the rest to the wolves. But I came back."

Kittery shook his head, sickness filling the pit of his stomach anew, embarrassed that Luke Vancouver and Adam Beck had to hear this and know this was the beloved brother he was always bragging about. Jacob had the upper hand here. He had apparently known about his younger brother for some time. Kittery was still struggling with the shock of it, and his words stumbled out.

"Yeah, you really came back, Jake. I guess you had reason to change your name; you went against everything Mama ever taught us."

Anger flashed like lightning in Baraga's eyes. "Mama taught us loyalty, too. I guess you forgot that."

Kittery grunted but couldn't respond. He turned numbly to Sheriff Vancouver. "You were right, Luke; it *was* about over."

Vancouver, trying to feel the shock Kittery must feel, reached out and laid a hand on his arm. "Let's go home, Tap."

As the posse wound its way back toward Castor, the soldiers loped out before them until their backs and the rumps of their dark mounts faded into the distance and their dust settled back over the earth. Adam Beck brought up the rear of the cavalcade. Kittery and Baraga rode side by side in silence ahead of him, the sheriff and the other posse members strung out ahead in single file.

Baraga turned to look at his younger brother, scratching his chin. "Well, I guess you do remember something about loyalty."

Shaken by the sound of the voice, Kittery glanced over. "Why?"

"You're riding beside me. I wouldn't expect that from someone who wanted me dead as bad as you seem to have."

Kittery's eyes returned straight ahead, concentrating on the trail he could see between the chestnut's pricked ears. He was still numb, unable to convince himself beyond any doubt this was not a dream. He wished Jake had died in the war.

"Well," he replied at last, "you may be a murderer and a thief, but Jake must still be in you somewhere, or you wouldn't *want* me at your side. I hate what you became—but I can't forget what we were before."

Baraga nodded. "I want you to know something, because I think after all they say you did for Arizona you deserve it. I wouldn't tell this if I didn't believe the man a worthless coward, but there's one in your town who's sold me much information for the past couple years. In fact, he told me about this payroll and nearly ended *all* our lives. He also turned over to me a list of every man involved in your little vigilante crusade. We call him Mouse. You interested?"

Kittery's body had gone taut. His vision blurred in apprehension as he looked over at Baraga, then ahead at Luke Vancouver. "What about the sheriff? Can he hear this?"

Baraga shrugged. "Suit yourself."

Main Street was quiet when the posse drew into it, and Luke Vancouver rode in the lead. He dismissed the posse, and most of them broke out of the cluster of riders and headed for their homes or for the cantina or café, where the soldiers' horses were already tied. That left Kittery, Vancouver and Baraga, who dismounted at the jail and went inside.

Locking Baraga in and leaving him with Kittery, Luke Vancouver walked back out to the main office in time to see Miles Tarandon clear the front door. The deputy's look of anger did not escape him.

"I'm glad to see at least *you* were successful."

Vancouver's eyes narrowed. "What are you talking about?"

"Thaddeus Greene's back in town. That worthless judge in Tucson said he was lettin' 'im out because his arrest was on shaky grounds. Said he was just protecting his property against savages, and you had no right to hit him, either. I think that prejudiced old—" He stopped as Vancouver whirled away.

The sheriff slammed through the door and strode from the porch. Tarandon followed him outside, smiling to himself. He unbuttoned his coat as he stepped into the street, twenty feet behind Vancouver.

Greene must have seen them coming. At that moment he burst from the door of his store and out on the porch, his eyes full of fire. Vancouver kept coming, staring at the merchant, his jaw muscles bunched.

The lawman stopped, still forty feet from Greene. "You should have ridden away when they let you go, Greene."

The merchant looked from Vancouver to Tarandon, who hung several yards back. He rubbed the bridge of his nose, where the sheriff had struck him. He looked at the ground where he had fallen.

"And you'll wish you'd stayed out in the desert till I was gone," Greene said. "Judge Hart gave me permission to walk all over you for what you did to me. He's calling for your badge."

Vancouver didn't respond to that. He just said, "I brought Baraga in."

The heavy-set merchant's face went chalk-white, and he turned his head to glance back at the door of his store. His chest rose and fell with a great breath, and he turned his eyes on the sheriff. "Well, good for you. It's about time you worthless lawmen stopped wasting taxpayers' money and did your job."

Vancouver again ignored him. "The soldiers broke Baraga's raid. I guess you'll get nothing out of that one."

Again, the color went out of Greene's face, and his eyes darted toward the front of his store. "What are you talking about?"

"Shut up, Greene. Be thankful this is me, not Kittery. I only intend to take you back to your crooked Judge Hart and the governor. Kittery would kill you. He still will if he finds out you're in town."

The crazy look went out of Greene's eyes, and he dropped his eyes and sighed. "Let me get my bags."

"Stop!" Vancouver barked when the merchant began to turn.

The fat man stopped and turned back toward him. His eyes were dark with storm clouds of hate. The gray sacks underneath them appeared dark purple against his pallid cheeks. The left corner of his mouth twitched.

"All right . . . *Sheriff.*" He spat the title contemptuously. His lips and chin quivered as if he were cold. "I'll go with you to Tucson. You don't have a thing on me except the word of a murderer." He stepped off the porch and into the street. There he stopped again, facing Vancouver only twenty-five feet away.

Vancouver stood in the exact center of the street. His face settled into hard lines, and he took a deep breath. There was something about the way Greene was talking, about his stance, about the intensity in his eyes. Nothing the fat man said could be trusted. His actions and his words had no connection. He motioned Tarandon farther away from him as heads began to appear in doorways and windows.

"Walk to me, Greene." Vancouver's eyes searched the fat man for a weapon.

"You come over here and get me," Greene shot back. His right hand shivered like his lips and chin.

A chuckle escaped Vancouver's lips—a reaction of nervousness. "Baraga told us everything, Greene—just so you know. We always knew there was an inside man here. Even your crooked judge can't protect you anymore . . . Mouse."

The word snapped along the street like a whip. Greene's hand quit shaking, and his arms flexed and went stiff at his sides. His eyes widened. He squared himself, and his neck seemed to sink into his shoulders. His face settled into lines of resignation, and the color flowed back into it.

"You should have let it go," said Greene.

Vancouver grunted. "What does it take for a man to sell out the people he lives with, Greene? What'd they give you?"

"You should have let it go," Greene repeated with a growl in his throat. A drop of sweat fell from the tip of his nose, and he reached behind him with his right hand and pulled a Smith and Wesson that had been hidden in his waistband.

As the fat man started to bring the pistol up, Vancouver jerked his own Colt from its holster and fired into his chest. But Greene was a driven soul. He finished his lift and fired. The slug pierced Vancouver's left shoulder, spinning him halfway around. Vancouver steadied himself and shot Greene again. The bullet slammed in above the fat man's hip.

For a man who has never been in a gunfight, it is hard to imagine how one man can miss another at a range as close as twenty-five feet. But the stress of someone shooting back can send even an experienced shooter's bullet yards away from its target. One breath drawn at the wrong second, a flinch at the barking of the foe's gun—many factors can throw a shot wide. Only a man who has been there can attest to the power of these forces.

Greene grunted with the force of the bullet in his hip, but he snapped off two quick shots. The first threw up dust from the street, the second sank into the flesh of the sheriff's thigh, bringing him to his knee. Greene leveled the pistol for his killing shot.

From behind Vancouver, Miles Tarandon's .45 cracked, and his bullet missed. He fired again, and luck drove the bullet into the big man's left breast. He staggered to the side, smashed against the hitching rail. He bounced away, still clutching the pistol and gritting his teeth while he tried with both hands to bring it to bear.

The sudden, exaggerated blast of a shotgun boomed in the afternoon, and Greene took the full load in his chest from thirty yards. His legs kicked upward spasmodically as his shoulders rammed to earth. Before his blood-soaked body hit the street, he was dead. His nerves shuddered, settling him into the dust.

Tappan Kittery stood in front of the jailhouse. He clutched a twelve-gauge shotgun he had drawn out of Vancouver's rack. Blue smoke curled out of both its barrels. Maria McBride stood silent at the door of her hotel, her hand over her mouth and her eyes wide on the still, bloody form of the man she had married long ago in Mexico—the man known as the Mouse.

Chapter Thirty-Nine
The Devil Goes Home

Savage Diablo Baraga spent two weeks in the Tucson jail. When they were ready to hear his plea, the courthouse drew people from all walks of life, more than overflowing the blue-trimmed adobe building. When people couldn't get inside, they milled around the big blue doors, waiting for news of the proceedings. The colonel from Camp Lowell represented the Army, and United States Marshal Wiley Standefer was there for the civilian government, as well as Governor Anson Safford and Sheriff William Barden. A troop of smartly dressed soldiers waited outside the building, ready in case of any rescue attempt.

Justice did not linger in the territorial days of Arizona, and for some men it was faster than others. Especially when the man on trial had no friends. But Baraga's "trial" never occurred. Savage Diablo Baraga tried himself. He was found guilty.

Standing in front of the judge, with a horde at his back who hated and feared and stood in awe of him, and one man who was his brother and didn't yet know where he stood, Baraga pled guilty to a list of charges as long as his leg.

They took a short recess where disappointed men retired to the saloons and cantinas to talk about the trial that wasn't to be. When they returned, it was for sentencing. The judge had dealings in another town and wanted to end his business here that day.

Tappan Kittery stood at the back of the courtroom and listened to the judge declare that his brother would be hung by the neck until dead, and then he turned with a cold feeling in his heart and walked away.

Baraga sat in his cell that night and thumbed through his journal. He made sketches of guns and horses and nooses. He drank some water as he had every day since his capture, but he would eat nothing. He had no hunger, but more importantly he was not going to let the crowd see him soil his drawers. No, not Baraga.

They brought his tray then took it away several hours later, as full as when it arrived. Finally, they had just stopped bringing it. His lack of nourishment had given him a gaunt, haggard aspect. His cheeks were sunken in, and his eyes looked big and haunted. His hair had been trimmed, as well as his beard, and this coupled with his hawk-beaked nose gave him the air of a fierce aristocrat.

That was how he looked when Kittery came to visit, bringing with him a plate brimming with navy beans. He had not seen his brother up close since the day of his arrest, and his appearance now made Kittery's stomach tighten. His brother was handsome, and he was robust. He had the perfect facial structure and powerful neck and jaw, the big strong hands, the wide shoulders and deep chest of a Roman warrior. But in spite of it all Jacob Kittery looked like he was standing with one foot in the grave.

They sat together inside the shadowed cell and talked in low tones. "You oughtta eat, Jake," Kittery said. "You're startin' to look like a ghoul."

Baraga scoffed. "I *am* a ghoul." He nodded toward the door. "Ask them out there. And what's more, I don't *feel* so good, either. The noose is already tight."

"How does it feel?" asked Kittery. "Or do you feel anything?"

"I feel regret. I want myself dead more than I wanted you dead three weeks ago. And, Tappan, that's a hell of a lot. It just took being captured to find out. I should've died on Cemetery Ridge with the rest of those gallant boys of mine. Couldn't have chosen a better place, and I'd have died for a good cause. Not for this."

"Other people have come back from war prisons, Jake. And amputated limbs. You could have tried to go right."

"Don't preach to *me*, boy," Baraga said, looking through bitter eyes at the window that seemed so distant above him. "You take a turn lying for never-ending weeks in a pit of human wastes, eating nothing but rice flavored with maggots, watching your friends starve to death around you. You lie there in the cold and take off your shirt, then wrap it around you just so you can pretend it's a blanket. You sweat for three days in a metal box like an oven only to be pulled out and peed on. You look at your friends and wait for them to die, hoping you'll be first to know so you can steal their shirt and have an extra to keep you warm on those bitter northern nights, and so you can eat what's left of their maggot soup. They rode us hard up to Gettysburg, boy. I couldn't have weighed over one hundred seventy-five pounds when they took me to that prison to chop off my arm—and I'm a big man. I weighed one hundred twenty-five that April day they carried me out. Until you've been there, Tappan, keep your preaching to yourself."

Kittery, a little hurt, stood up to leave. Baraga waved him back. He would not apologize out loud, but Kittery accepted the gesture as such.

Baraga looked way up at the window again, his eyes wistful. "You know all those stories Ma would tell us about Heaven? How if we were good that's where we'd go?"

Kittery smiled. "Sure."

"You know, I used to try real hard to be good—before the war. I wanted to go there—to Heaven. Ma talked so nice about it, about the green fields and the clover and the fat horses and cold water. And to tell you the truth, I always kind of thought the man I grew up to be died there in that prison camp and went to Heaven. He didn't deserve what they gave him. He was just fighting for what he knew was right—to keep the rich man from running over the poor, to keep the powerful men from telling everyone else what to do. Yeah, old Jake's up there with Ma. And me, I'm here."

Baraga's predicament seemed to pry the deepest ponderings from him, thoughts he had pushed aside for many years. It was gradual, but he opened up entirely. He talked to Kittery as he hadn't talked to anyone in years, not even to his henchmen, Dixon and Bishop. *Especially* not to them.

"That's why I changed my name," he offered. "I knew I was a different man. And why Savage Diablo? To make people afraid. Everyone's afraid of the devil." He chuckled without humor. "And by hell it worked. I was the most feared man in Arizona. And now I'm going to pay. Jake went to Heaven, Tap, and I'm going right where Ma always said we'd go if we went bad. It's where the devil belongs."

Tappan stood up again, saddened by this talk. His image of the blond young man throwing rocks in the river was fading too fast. "I'll bury you as Jake Kittery."

"No." The word sounded almost angry as Baraga lunged up, and his eyes met and held his brother's. "Jake had two arms, remember? Good arms. Diablo has only one. As far as Ma knew—or anyone else—they buried Jake a long time ago."

"Then I'll do it your way. Savage Diablo Baraga." Tappan hesitantly turned toward the cell door, fighting a rush of tears he knew was too close to the surface. He wanted to reach out, to touch his brother, the man who had been his hero. He wanted to shake his hand, to put his arms around him one last time. But did a man embrace the devil? And yet if he thought about it he was *kin* to the devil. The Devil's blood.

"Hey, little brother." He turned back to Baraga, blinking against his dimming vision. "If you ever see any of the girls, tell them I died in the war. Tell them I died with honors."

"I'll tell 'em Jake did," he gruffed. "Sure. Is there anything I can get for you? Or do?"

"Yes. Talk them into letting me swing there in Castor, would you? I hear you're a powerful man around here now, and they're still debating letting me have that as my last request. I don't want to be here in Tucson where the governor and everybody is just a step away from the gallows."

"Sure, Jake. That's fair." He could hardly see his brother, who swam behind the wall of his tears.

"Hey, little brother. There's one other thing." There was a look of near pleading in Baraga's eyes.

"Yeah?"

"I'm a proud man. But I know I won't see you again—and I know that's only right. Still, since you're here . . . I'd be honored if I could shake your hand."

Tappan Kittery couldn't help himself. His tears spilled over and rolled down his cheeks, and there was no stopping them. Blindly, he put out his big left hand to shake that of his brother. Flesh met flesh, and he could feel the bones in Jake's hand, the power of his grip. He didn't think about doing it, it just happened: in a moment he had his arms around Jake and was squeezing him tight, afraid to let go. Jake just stood there for several seconds, but then, hesitantly, his arm came up to hold onto his little brother. It shocked Baraga to realize it was the first time he had felt loved by anyone in more than twelve years.

"You've made me proud, little brother," he whispered.

Kittery's throat ached and was far too tight to reply. Baraga pushed away from him and cleared his throat. "Now stay away from here on. All this talk is ruining my image. I haven't talked so much since the day they cut off my arm."

Kittery nodded, finding his voice. "I know; you kept it all inside. It was the worst thing you could've done."

Baraga shrugged loosely. "So I'm stupid. Shut your mouth and go. And take this foul plate of beans with you. I'm not in hell yet, am I?"

One week later, to the day, they escorted Savage Diablo Baraga back to Castor. His last request, with Kittery's persuasion, had been granted. Onlookers swarmed Castor as they had filled Tucson for the sentencing. They had traveled from as far as a hundred miles away to see the infamous Baraga swing.

Kittery had ridden Satan to town, leaving Tania home. And the saddle continued to be his seat at the gallows. He didn't want to mingle with the crowd. He wanted to feel alone. He watched as the cavalry troop rode into the street in columns of two. Along with the colonel from the military post, an imported hangman rode at their head, for James Price hadn't been able to hang his friend's brother. The outlaw rode in the center of the cavalcade, protected there from the many people he had hurt in one way or another who the Army thought might try to kill him before he reached the gallows.

Anyone could pick Baraga out, not only by his civilian clothing and the missing arm, but also by the straight-backed, arrogant way he sat the saddle and gazed down from his pinto. Among the horsemen, he was only slightly over average height yet seemed to sit high above them. Even unarmed—in more ways than one—he appeared a man of authority and importance, a man whose features and bearing and fiery gaze set him apart from the others, no matter how straight they might manage to sit their mounts. This was the way Kittery would always remember him.

Everyone gathered around the gallows when Baraga dismounted and followed two soldiers up the steps that would carry him to eternity. The hangman followed him twenty feet behind, carrying a black hood beneath his arm.

Kittery watched in silence. This had been Savage Diablo Baraga, bandit, murderer, one-time leader of the most ruthless and daring band of outlaws in Arizona. Now he was simply Jacob Atticus Kittery, the leader of no one—a man condemned to die. As they placed the noose around his neck, he stood calm and defiant, the way he had lived. No one would ever hear Baraga cry for mercy. He was invincible! It would be written, he died as he had lived, unafraid.

But then his eyes changed, perhaps imperceptible to anyone but Kittery—perhaps imagined even by him. Baraga looked out at his brother sitting his horse at the edge of the crowd, and somehow the sounds of the people disappeared, and it was only he and Tappan alone there, facing each other.

Baraga's face showed a momentary sorrow. Kittery choked down a lump in his throat, and tears brimmed his eyes, but they didn't spill. The outlaw brother clenched his teeth. "Tappan!" he said, just loud enough for Kittery to hear him above the din. Then he mouthed his next words, and the peace-keeping brother understood. "I'm going home now."

The hangman stepped forward, and Baraga looked down at the black hood. He started to raise it, and Baraga started to object. Then he looked toward Kittery. The hangman hesitated, as unsure of himself as if he were hanging the devil himself. At last, Baraga turned back toward him and gave a solemn nod. The hangman slipped the hood over his head, and Tappan Kittery looked into those blue eyes for the last time. The hangman drew the hood's latchstrings tight and took two steps back. He placed the palm of his hand on the big metal lever.

A lump rose again in Kittery's throat, and he couldn't swallow it. He hoped his mother's spirit wasn't here to watch this. His vision blurred, and his ears closed over until he was barely aware of Marshal Standefer reading something from the Bible. With harsh suddenness, the hangman thrust the lever forward, springing the trap door with no more emotion on his face than a man killing a fly.

Time seemed to stall. Would the door open? Could Savage Diablo Baraga really die? Or was he invincible, as they had said? Perhaps he really *was* the devil.

But no. He was only a man. The hinges cracked. The door came down. It slammed against the underside of the flooring, and Baraga's feet and legs disappeared, then his chest. The rope cracked tight, and the weight of the man slammed with final brutality. Thus ended a life.

Tappan Kittery turned the stallion away from the gallows, not wanting to see Baraga hanging dead at last. Somewhere in the crowd a baby, startled by the sudden *crack* of the gallows door, began to wail—Baraga's solitary mourner.

But that baby *isn't* your only mourner, Jake. A flood of tears came up and streamed down Kittery's suntanned cheeks, and he put heels to Satan's ribs, making him jump forward in surprise and jog between the water tower and Greene's Dry Goods.

"Savage Diablo Baraga is dead," were the last words he heard someone utter as he rode away.

And it was true. Like Jacob Kittery ascending to Heaven, El Diablo had gone to his home.

Chapter Forty
Kittery and His Mustache

There is that time deep in the summer and out in the midst of the day when the sun is like a furnace and the heat beats clear through a man, when nothing, not even a hawk, is on the move, and the occasional grasshopper in the brush chatters out the only natural sound. This day was like that, except that the sound of the colts dancing and blowing and whinnying in their big long corral kept reminding Tappan Kittery that he was at work. That he was alive.

It was dinnertime on the Double F ranch, and Tania had just called to Kittery, Bacon and Long, busy green-breaking the young stock. They walked toward the house, massaging elbows and hands and passing around a gritty once-white towel to wipe the sweat from their faces and necks.

Before they reached the shady porch, Tania opened the door and stood looking down the road. Lafayette Bacon turned to follow her gaze, then touched Kittery's arm to halt him. "Somebody comin', Cap." He pointed toward the winding lane and the tiny, dust-trailing spot that was a wagon.

When it rolled closer, Kittery recognized the Vancouvers seated on the front seat of the high-piled rig. A twinge of sadness came over him. He motioned for Tania to come, and the door swinging shut, clapping against the doorframe, was all that broke the silence until the wagon rumbled near.

Vancouver drew up in the yard far enough away to keep the dust from drifting to the house. He stepped weakly to the ground, then managed to help Beth from the seat, despite his still-mending wounds. The others smiled at his courage. Kittery walked up and held out a hand to Luke, who shook it. "I'm happy to see you movin' around, Luke."

Luke chuckled. "Well, two months gives a man a long time to heal."

"Yeah, I reckon so. But those were some nasty wounds."

Tania walked up now beside her husband and the two punchers. She smiled and greeted the Vancouvers. "I'm glad you two didn't try to sneak off without coming over. Dinner's on," she added, by way of invitation.

"Well, it's a long road ahead of us, Tania," Luke said with regret. "We'd better not stay that long."

Kittery pointed to the wagon, loaded three feet over its sideboards with furniture and other household goods. "Think all that'll make it to Prescott?"

"I sure hope it does," Vancouver said with a laugh. "I plan to buy some mules in Tucson."

"Sure you won't change your mind about goin', Luke? You're leavin' me with a big job."

"Tappan, with your reputation outlaws are going to walk through town on pussy feet—or go around. Besides, Miles will take care of the county duties. Your biggest problem is trying to stay behind the counter of that store, when everyone knows you want to be outside."

Kittery shrugged with a lopsided smile. "Well, Jeremy's a darn good clerk. He takes care of things." He paused. "I guess we can't keep you then."

"No. In fact, we'd probably better be on the road. Tucson's a long way by wagon."

Kittery smiled, looking apologetically at Beth. He knew she wanted to stay. He looked back at Luke. "I sure hate to see you go. I hope you decide Castor's better than Prescott. This is your town, my friend. You made it."

"Yeah. Who knows? Well, old man, I hope our trails cross again. Maybe we can clean up a town together some day."

"Maybe, but don't wait too long. We're not gettin' any younger, you know."

"Nope."

Beth's eyes rimmed with tears, and she stepped close to Tania, gripping her shoulders. "You take care of yourself, you pretty little thing." She wiped her eyes with her fingertips. "And watch out for this man of yours. He thinks he can take on the world."

She giggled and poked Kittery in the ribs. Then she and Tania embraced, both letting the tears run down their cheeks with no shame. Beth came over and put her arms up so Kittery had to bend down and let her kiss him on the cheek. Then he gave her a bearhug and held onto her for a long time.

Luke Vancouver walked over and threw an arm around Tania, kissing her on the forehead. "You know, I think you can tame this big brute after all."

Tania laughed, and the tears ran down her cheeks anew as she threw her arms around his neck. "Say hi to the children for me, okay? Tell them everyone misses them."

"We'll do that," Luke assured her.

He turned then to Bacon and Long. "Well, boys, it's been good to know you.

You'll always be welcome in any town where I'm sheriff."

"Thanks." Bacon smiled with embarrassment, shaking Vancouver's outstretched hand.

Jake Long shook the hand next. "G'bye, Sheriff. G'bye, Mrs. Vancouver." He tipped his hat.

Turning, wiping tears from her eyes, Beth allowed Luke to boost her onto the wagon seat. Then Luke shook Kittery's hand once more with deep warmth, and climbed up himself, pulling off the brake. "So long, folks." Vancouver nodded. They waved, and the wagon rolled around and bumped off down the rutted lane.

It was October sixth, over a month since the Kitterys' first anniversary. Stars hung like diamonds across the autumn sky, and a poor-will's flute-like music drifted one last time over the desert. Kittery, hearing the commotion of the horses in the corral, snatched his Winchester from behind the door and stood waiting just out of the open doorway, ears tuned to any sound.

"It's me, Captain. Jed Polk," came the call from the dark yard.

Kittery stepped into sight, filling the doorway, yellow light flooding out past his silhouette. Light reflected off his rifle barrel, touching Polk dimly as he dismounted and moved partway up the steps.

"I got some real bad news," said Polk.

Kittery's heart quickened, and he stepped to the side, glancing involuntarily around to see if Linus Beehan was somewhere out in the shadows. "Come in, Jed. We have coffee on."

Jed Polk stepped inside, removing his hat upon seeing Tania, and smoothed back his gray hair with the palm of his hand. "Howdy, Tania. How're the children?"

"They're real fine, Mr. Polk," she assured him

He stood twisting his hat in his hands, biting the inside of his lower lip as Tania poured three cups of coffee. He accepted his then glanced uneasily at the big man across from him, who spoke.

"I reckon you'd best say what you came for, Jed."

"Captain . . . I'm real sorry. Uh, Linus . . . They killed him."

Kittery let out a captured breath, his face losing color, and put his cup down on the table. He glanced at Tania and walked across the room, resting one hand on the wall. He dropped his head and closed his eyes. Then he turned and stepped back to Polk, whose sad gray eyes stared at him.

"I'm real sorry," Polk repeated, shaking his head.

"Who killed him?"

"Morgan Dixon."

Kittery slumped into a chair, staring at his feet. "When is this ever gonna end?" he said, more to himself than anyone.

"He left this." Polk handed Kittery a small, folded piece of paper.

Kittery unfolded it with dread creeping up inside his chest.

Kittery:

You don't know me, but you know who I am. You have wanted me dead for over a year, and I have you as well. Now I am offering you the opportunity to do something about it.

I have with me a list of over fifty men who will die within the ensuing month. On this list, you are last. Baraga paid a merchant in your town five hundred dollars for this list of vigilantes who failed in an attempt to "regulate" Baraga's men.

My proposal is simple. I will kill these men as I find them, and if you care about them whatsoever, you will meet me and stop me. If you choose not to, it is just as well. In the end, I will find you, when you are all that is left. That, you can take as God's truth.

Major Morgan Dixon

"Jed," Kittery said quietly. "You said 'they.' Who's with him?"

"Colt Bishop. Cap, get away from Castor!" he said, raising his voice. His eyes flickered to Tania, then back. "He won't quit till you're dead."

"Or until he is. You know I can't quit, Jed. You read the note. I'm left without a choice. Unless I want to see a lot of good men die because of my cowardice. And that includes you."

Polk nodded with resignation. "I knew you wouldn't. Cap? Don't let Dixon get close. Use a rifle and take him from behind, and Bishop, too. Dixon's usin' a sawed-off shotgun. It's what he killed Linus boy with."

Kittery's eyes took on a faraway look at the mention of that name. Linus would have made a great man. Now he was dead . . . his killer would pay with his life.

After Jed Polk had retired to the bunkhouse, Tappan sat out on the front porch swing. Its chains made no noise, for he sat still. Tania came out and stood at his side, gazing at the star-dazzled sky. Finally, she stepped around and sat down beside him, making the chains complain. She slid an arm behind his back and looked up at his bitter face.

"What are you thinking?"

He squinted his eyes out across the dark, silent yard. "I was thinkin' that not too long ago, the day I killed Doolin, I told you I wouldn't leave you again. I already broke that promise once, to go after Baraga. It was because of my duty and my promise to Joe to see the Desperados all dead. Now it looks like the promise I made to Joe is the one I have to keep again—not yours. Dixon and Bishop won't stop while I'm still alive."

Tania was silent for a moment, stroking his back and watching the stars wink in the sky. "You know, Tappan, these are our stars. I watch them at night and know I wouldn't want to watch them anymore if I didn't have you here to see them with me."

Kittery made no reply. Her comment broke his heart.

Tania sat for a long minute, then went on. "But there is a star up there somewhere that you're following, I've decided. You have to find it. So I want to give you another one. See that star up there, the biggest one?" She pointed. "I want you to think of me when you see that star because I'm giving it to you. Wherever you are, no matter where you roam, when you see that star know that I am here waiting for you and praying you come home.

"I love you, Tappan. And I know what you must do. Perhaps it's better this way. Now it will all be over, and we'll have peace. But come back when you're through. Without you there's no life left for me."

Tappan Kittery looked into those large, dark eyes, thinking of all the love they had shared together, and the hard times they had conquered. She meant the world to him. He thought of little Jacob and Rachel Ann, his two sweet babies. He could never leave them alone. Now that he had found what life was all about, he would not give it up.

He kissed Tania gently on the mouth. "They ain't made the bullet that can keep me from comin' back to you."

"So where do we begin?"

Manhunter Adam Beck sat across the table from Kittery in the cantina, puffing on a fat cigar, his beer mug half full.

Kittery flipped a card over on the table, then shrugged. "The Desperado Den. Where else?"

Beck screwed up his eyes. "Would they go back there?"

"They lived there for years. What better place?"

"Yeah, I guess they might at that. But if not there, then what?"

Kittery lifted his hands from the table in a half shrug. "We take it as it comes. But I'll bet money our best lead is the Den. They may not stay there all the time, but I bet they go back now and then to rest up."

Beck lifted a finger in mild admonition. "Are you sure you want to do this without a posse? I mean, there's a dozen men I know would take the trail with you at the raise of a finger."

"I know there are. But enough of my boys have died. Besides, there's five thousand dollars bounty on each of these men now. You wouldn't want to divide *that* up, would you?"

Beck chuckled. "You must think my life is run by the all-powerful dollar."

"You mean it isn't?"

"Well, maybe it is," the manhunter said with a laugh. "By the way, when are you gonna shave that silly thing off?" He pointed at Kittery's mustache.

Kittery ran his fingers over the thick dark tuft of hair and felt its curved ends. Of course Beck had no way of knowing he had kept the mustache because of Joe. He hadn't meant any harm in bringing it up. But it hurt to think it had been so long since his promise to Joe, and still it was unfulfilled.

"Looks like Joe Raines', doesn't it? That's why it's there, to remind me of him. Well, Adam, I told myself when I left it I'd shave when the last of the Desperados was dead. So the day they die, it's gone."

Dawn broke clear and cold. Above, the blinking stars shone like distant campfires, too far away to shed any heat but lending a guiding light. There among them glowed the one Tania had given Tappan. It was like she was up there looking down to keep him safe.

In the darkness, Kittery spotted a big buckskin gelding on the far side of the corral. He climbed inside and used the hooley ann throw to snake a loop between the other horses and over the buckskin's head, dragging him away from his corral mates.

Satan had appointed himself monarch of the enclosure, as always, and he stood in the corner alone. He trotted eagerly over and stopped to sniff Kittery's chest, ignoring the buckskin.

"Get out of here, boy. Can't you tell I'm restin' you? I don't wanna lose you. When I finish what I'm settin' out to do, we'll ride again."

The stallion pushed his head against Kittery's chest, not liking this rejection. It was something Kittery didn't often let his horses do, and he was surprised Satan would try it, but this time he didn't try to stop him. He scratched his friend's head and patted the muscular neck. "We'll ride again when I'm back. I promise." *If I ever make it back* . . .

He led the buckskin out into the yard, closing the gate behind. He saddled up and lashed a bedroll and saddlebags on at the rear. He had Cotton Baine's Colt on his right hip, his Remington in his waistband. On a sudden whim, he drew the Remington and studied it for a moment. Then, reaching into his saddlebag, he drew out the Remington's holster and thrust the pistol into it. He removed Baine's belt and pistol and stuffed them into the saddlebag, then strapped the old Remington around his waist. In spite of his many tries, the old familiar things that were his friends still held him captive. He guessed he was a relic himself.

Soft hooffalls caused Kittery to whirl around, Remington in hand. Adam Beck sat silhouetted against the dawn sky. "James Price tells me that's how angels are born," Kittery said with a chuckle of relief.

Beck shrugged, staring down at the pistol. "I thought you were going modern—with Baine's pistol. What's with the Remington?"

Kittery turned the Remington on its side for a moment and studied it in silence. "You and me will never change, Adam. We've been usin' these old things too long. It's in the blood. This old shooter's never failed me yet." With that, he returned the pistol to its holster. He turned and checked the cinch one last time and swung into the saddle. "Let's go."

As the pair rode toward the Baboquivaris, light grew in the east, and Beck looked over and noticed for the first time the absence of Kittery's mustache. "I thought you said your mustache stayed till Dixon and Bishop were dead."

"No, no. I said the day they died it was gone. That day is today."

Beck studied him, and he seemed to sense the same foreboding Kittery had had all morning, the feeling that by the end of this day their quest would be ended, one way or the other.

"Well, speaking on a personal basis, I liked your old self, Tap," said Beck with a smile. "Glad to see you back."

They touched heels to their horses' flanks and loped out across the creosote flats.

Chapter Forty-One
The Season's End

The air was cool in the Desperado Den, at least compared to the heat that rolled across the desert under the onslaught of the October sun. The pervading aroma of coffee and frying bacon filled the space, eddying around in invisible swirls. Outside, the sky was brilliant blue, and deep in the maw of the canyon the spring trickled through the rocks. A cactus wren flung its harsh cry from a catclaw bush.

Colt Bishop crouched beside a tiny fire, nursing a cup of coffee, his eyes riveted to the floor. Morgan Dixon, face drawn in tight lines, paced the cave's entrance, clenching and unclenching his fists. Of a sudden, he turned and glared at his partner.

"I can't understand your apathy. Do you have any loyalty?"

Bishop glanced up. There was no sign of any inner struggle on his placid cheeks, just whiskers and the grime of the campfire and of the trail. His hair made its uneven track across the wrinkles of his forehead. "Morgan, I've tried to explain, but you don't have it in your blood to understand. I have a reason for wanting Kittery left alive. It's over—don't you understand that? I shouldn't have to explain anything. Your common sense should be enough."

"It's a fool idea you have—no matter what it is! If we ride away now, you make a liar of me. I promised I'd come for him."

Bishop shrugged and stood up, his slow steps taking him to the cave's entrance, just out of the glaring sunshine that beat against the outer wall. After a few minutes of feeling Dixon's anger drive against his back, he returned to the fire, stopping several feet away from his partner.

"Morgan, Baraga's dead. And the rest of them didn't mean a thing to you. I know you thought a lot of Baraga. So did I. But remember, it wasn't Kittery who killed him."

"But"—Dixon jabbed a finger at Bishop's chest — "he was the one who brought his death on. Open your eyes!"

Bishop looked calmly away, then back.

"My eyes have been open all the time, Morgan. For once in your life, forget your hate. We've got a stake now. We can make something out of that place in Mexico. Don't throw that away. Would it help if I told you I know you'd let Kittery live if you knew my reasons?"

"Why should it help?" Dixon growled. "I *don't* know them."

"But do you trust my judgment?"

"Less and less," Dixon growled.

With sudden distaste, Bishop flung his cup and its dregs to the floor. He squared himself with his partner, the anger of frustration glowing for the first time in his eyes. "I'll promise you one thing, Morgan. I'm headed for Mexico and that little ranch at noon today. With you or without. I'm hoping it'll be with."

Major Morgan Dixon stared at his partner with something akin to hate in his eyes. He turned and scanned the smoke-blackened walls, then looked outside. At last, his gaze met Bishop's again, and there was a softness there, a resignation. "Damn your stubborn hide," he said, and he closed his eyes.

Tappan Kittery stood and stared at the empty cavern. The scent of bacon and coffee hung strong in the air. He turned knowing eyes on Beck, who stood wary beside him.

On a sudden impulse, he strode across the floor and tested the fire's ashes with the flat of his hand. He jerked his hand away from the heat and turned to look up at Beck. A chill climbed his spine.

"They couldn't have left long ago, Adam. 'Fraid there's no time for dinner."

"Nope."

In a rush, they left the cave, searching the ground for fresh sign. The most recent tracks trailed south . . . toward Mexico.

"What do you suppose they're up to?" Beck wondered. "I don't think Dixon would leave without seein' to you."

Kittery swung onto the buckskin's back. "He might try, but he won't get far. He killed Linus, and before long he'll kill somebody else . . . unless we stop him."

Followed by Beck, he pushed the buckskin as hard as he dared down the trail, his Winchester in hand. They drove brutally for six miles, then left the oak-clad mountains and moved into scrub country, the foothills west of the Baboquivaris. Fifteen miles to the northwest, the tiny crest of the Artesa Mountains lifted above the horizon, the bulky mound of Las Animas Mountain rearing from the desert eight miles to the west. Before them, bleached volcanic soil buried the plain that stretched into the distance, disappearing at length into a forest of mesquite and desert scrub about a mile in width and still two miles away.

Like a serpent the Desperados' trail wore on, winding its way around through the brush until it faded from sight. Kittery and Beck followed it from the backs of their mounts and bore on at a butt-grinding trot.

Soon, a half mile-long spur of the Baboquivaris loomed up ahead, forcing them to turn their mounts west to skirt it. As they pulled out of a gully, Beck jerked his horse in with one hand as he pointed ahead with the other.

Kittery eased the buckskin up beside him and stared straight ahead. A thin but unmistakable veil of dust drifted off the point of the hill yet half a mile distant. He looked over at Beck. "That's it, partner. Tie yourself down. We're goin' for a ride."

With these words, Kittery yelled at the buckskin and kicked him into a gallop so the dust rose up behind. Unfortunately, they still hadn't caught sight of their quarry by the time they reached the first of the mesquites, and here the trail became more difficult to follow.

They made good time through the heavy cover of the trees. Once, they jumped a herd of javelinas that charged up the side of the wash, startling them. Both horses skidded to a halt to keep from running the little animals over, and Kittery sat staring in disgust after the disappearing horde, his pistol in hand. He turned to Beck then, and they laughed with relief. But when they rode on it was more slowly, and they more wary.

As Kittery rode, nightmarish thoughts kept racing through his mind. He didn't want to die. He wanted to go home to Tania. But worse, he didn't want to see his friend die. There had been too much of that. Yet in his mind he kept seeing Beck die, over and over again. One flash of such a thought would have meant nothing. A mind under stress draws up unpleasant images. But the fact that it kept coming over and over seemed like a warning to Kittery, a premonition he must heed. But how could he keep his friend safe now? Beck was here. There was no way out of that.

And then there was a way. It came down like a blessing sent from the sky when Beck called a halt to answer the call of nature. Kittery stepped off the buckskin to let him blow while Beck wandered off into the mesquite. He had lifted up one of the buckskin's feet to check the hoof for stones when a plan struck him. It was one of those ideas that comes crashing through the consciousness to strike like lightning. It had to be acted on in the instant, or it would be too late.

Taking the jackknife he had been scraping around the frog of the buckskin's hoof with, he hurried to Beck's sorrel. He coaxed it to raise its left front foot, and without any hesitation he drove the blade of the knife an inch deep through the soft part of the frog. The horse let out a grunt and pulled away, nearly knocking Kittery down before he could back off. He closed the knife and dropped it into the pocket of his trousers and looked up at the horse, saying, "Sorry, boy. It's for your own good."

The foot would heal in time. Adam Beck was good to his horses, and he would see to it the sorrel rested well until the little wound healed. But that wound was all it would take to lame the animal for now—and all it would take to set Tappan Kittery on his own.

Adam Beck returned, and they climbed onto their mounts and rode on. It wasn't more than another two miles before Beck's horse started to favor its left foreleg. Beck wanted to push it on a little, but Kittery kept suggesting he take a look at it. After three more miles of slow going, when they could see that not much farther the trees began once more to turn to scrub and cactus, Beck had to rein in.

The manhunter called out to Kittery and climbed down. Kittery pulled in and came around to sit his horse there above his friend. A feeling of guilt and fear crept up inside him, squeezing his chest. He didn't know if he had done the right thing, and he knew he was about to ride on alone. But he didn't want Adam Beck to die. He had lost too many friends.

Beck leaned over his horse's hoof, his head blocking Kittery's view. He was there a long time, wordless, and the buckskin fidgeted, wanting to move. At last, Beck looked up at Kittery. Kittery had never seen such a look of despair in the manhunter's face. "Tap, we ain't goin' on." He dropped the horse's leg. "It looks like I'm walkin' back to Castor. Stone must've cut 'im."

Kittery turned to look south, the dread swimming circles around his heart. He glanced back at Beck in time to catch a strange expression coming over his face.

He was studying the landscape around him, recognition seeping into his eyes.

"I know this place, Tap! This is Coyote Canyon. We're not five miles from the border!"

Kittery cursed, pivoting his gaze in the direction the outlaws had taken. "I'm sorry, Tap," he heard Beck say. "They're gonna get away this time."

"They're not gettin' away, Adam. Not after we went through all this. I shaved off my mustache, remember?" Kittery turned his horse and took one last long look at Beck. "Sorry, my friend. You look after Tania for me."

With that, he turned again and laid spurs to the horse, pushing the pain and fear out of his heart. He wanted Beck with him in the worst way, but he knew his friend was safe. Maybe now, even if he died, it would satisfy Morgan Dixon. Maybe the bloodshed would end.

At a reckless gallop Kittery plunged down the brush-choked wash. The dust ahead of him was gone, but the trail remained, and Kittery plunged along it at a breakneck run. His own funnel of dust rose up behind him long and heavy. His mind was almost feverish. He couldn't let Dixon and Bishop escape now. If he had to follow them across the border he would. They had killed Linus—in cold blood!

Ahead, four miles to the southeast, rose the tiny Pozo Verde Mountains, straddling the border. The trail led their way, and he pushed for them. The outlaws had gone far, for still he saw no dust. But he would not stop or even pause at the border. He would not have stopped at the border to hell.

The sudden neigh of a horse from the brushed-over hill to the right caused him to saw back on the buckskin's reins. One glance showed him where the outlaws' trail veered off and turned up the sidehill. And he had been so intent on the mountains ahead!

A deafening roar from thirty or forty yards up the hill sent shockwaves through his body. Something struck him in the leg like a club. The buckskin gave a terrified squeal and lunged to the side. It all seemed to happen so slowly, yet with no time to react. Kittery grabbed at the saddle horn as the horse started to go down. It seemed like perhaps he would only stumble and then catch his balance, but slowly, slowly, he closed the gap between them and the rocky earth. Kittery tried to throw himself from the saddle, and at the same time he pulled at the Remington. The horse came down on top of his leg, grunting with pain.

Above, the sound of a shotgun crashed again. Kittery winced in anticipation as the buckshot swept him. He felt weight slam against his right arm, against his leg. The rest of the lead made a dull *whump* against the side of the horse, and its weak struggling ceased. It laid down its head, and its breaths came in labored gasps.

In shock, Kittery had no doubt he was about to die. He was helpless as a fledgling dove once the bobcat has found its nest, trapped under thirteen hundred solid pounds of horse. His numb right hand tried to close on the Remington, only to find an empty palm, an empty holster. He swore, and the earth rolled and spun around him.

Far back in his consciousness, he could hear angry voices, voices arguing. He fought to clear his head, to rid himself of the dizziness and nausea. He heard the sound of loose shale sliding down the hillside and the sound of his horse sucking for air. With the last of his strength he lifted his head and caught a glimpse of two men coming down the slope on the run, dust and gravel spraying out at each stride.

One of the two, a mustached man wearing a long denim vest, carried a pistol in his right hand. He reached Kittery several yards ahead of a black-haired, bearded man, who bore a shotgun and was clad in the long gray coat of a Confederate officer. This would be Morgan Dixon, so the first must be Colt Bishop.

"Let me finish him," Dixon growled. Kittery assumed they were arguing over the honor of the killing shot. "He's nothing to you. Craven Union trash! Turncoat to boot!"

"Put away the shotgun, Morgan. I won't have it this way." Bishop's voice came to Kittery as that of a man forcing himself to remain cool while anger boiled inside.

Morgan Dixon broke open the shotgun and began to reload, not saying a word. Bishop disappeared behind Kittery's head. In another five seconds, he stepped back around. He had holstered his own Colt and picked up Kittery's Remington, which dangled from his fingers. He stood looking down at Kittery's bloody shirt and pant leg.

Again, Kittery forced his dizzy, reeling mind to hold onto consciousness. The pain of his wounds began to thread its way to his brain, and he was thankful; it helped keep him alert.

He looked up at Dixon's hooded eyes, watched him thumb the two shells into the chambers of his shotgun and shut it with a crisp, solid snap. "Worthless trash," Dixon said with a sneer. His thumb drew back the hammers.

Kittery knew he was about to die. He had broken his last promise to Tania. He saw the weapon lift. His fingers grasped the buckskin's mane. But his eyes remained open, staring at his killer.

Bishop spoke, his voice urgent. "Dixon! Don't you do it!" But Dixon didn't hear him. The shotgun continued to come up.

Bishop raised Kittery's Remington. It bucked. A foot-long flag of orange flame and a cloud of smoke burst out the bore. Dixon staggered backward, eyes wide with shock. His jaw went slack, and he looked down at the hole in the middle of his chest. Then his eyes turned to Bishop. The gunman looked on with an expression almost as surprised as Dixon's. His jaw muscles were tightened into knots.

"It's over, Morgan," said Bishop in a soft voice, the way a man might speak to a brother he loved. He lowered Kittery's Remington to hang along his thigh, exposing himself to the dark twin barrels of the shotgun. Bishop sucked in and blew out a deep breath in the same second, his shoulders sagging. "Go ahead, my friend. I make a big target."

Dixon swayed on his wide-spread feet but managed to keep from falling. The dark spot on his shirtfront was spreading with incredible swiftness. His fingers rested on the big gun's twin triggers. His eyes were clear. He stared unbelieving at his partner. "So you killed me," he croaked.

"You left me no choice," replied Bishop with as much sadness as Kittery decided it was possible for his stony face to express. "Go ahead, Morgan. Kill me, too, if you're of a mind. Pull the trigger. Civilization has killed us all."

Seeming to draw on some inner strength, Dixon pulled himself up tall and straight for one last time, and his fingers tightened on the triggers. He lowered the weapon and blasted a hole in the ground at his feet. The kickback jolted him just enough, and he went to his knees, gritting his teeth against the pain. He dropped his empty weapon, his upper body beginning to rock back and forth.

"Colt . . . you are part . . . of civilization now."

With a sigh, Dixon slumped onto his side, his hat rolling away. A cloud formed over his eyes.

For a long, silent moment, Colt Bishop stood still, unblinking, staring at Kittery's pistol in his fist. He raised his head and walked over to his dead partner. Squatting down, he placed the man's hat over his sightless eyes. Then he stood and moved to the fallen buckskin. It had struggled to rise again at the sound of the shot, but it was too weak. Taking it by the bridle, near the mouth, Bishop tugged at it and spoke encouraging words.

"When he moves, try to roll out," he ordered Kittery.

The buckskin took strength from Bishop's calm voice, and between its weak strugglings and his steady, hard pressure on the bridle, the animal was able to move far enough for Kittery to roll away from it.

Kittery's blood seemed to have stopped flowing out of his wounds. But there was little feeling in his left leg. Still he felt strong enough to talk.

"You're Colt Bishop."

Startled by the voice, Bishop spun to look away from the horse to Kittery. He straightened up, stuffing the Remington behind his belt. He walked over and crouched beside the fallen lawman.

"I'm Bishop," he affirmed as he reached for a button of Kittery's shirt. "And you're Captain Kittery—Baraga's little brother."

Tappan nodded. He followed the movements of Bishop with his eyes as the outlaw unbuttoned the shirt then laid it open. With gentle fingers he probed around the area of the bullet hole. Kittery gritted his teeth.

"You're lucky. Most of the load missed you. But the one that didn't went in pretty deep. It looks like your arm's broke by the others."

Kittery grunted without surprise and watched his benefactor remove a bandanna and press it against his wound while he glanced down at his left leg.

"It looks like your leg might be broke, too. Sorry, Kittery—I'm not much good at this doctor stuff. If it helps, I've lived through worse."

"So have I. How's the horse?"

Colt Bishop stood and looked again at the animal. "Sorry, Kittery. He's give up."

Kittery nodded, unable to keep his thoughts from going back to the moment he decided to leave Satan home. He was sorry to lose the buckskin, but his eyes filled with tears of gladness. He tried not to show the pain he was in as he studied Bishop's face for a length of time before speaking.

"Why'd you do it, Bishop?"

The man glanced up at him, then back at the wounds in his leg. He had pulled a jackknife from his pocket and began to cut several inches of material away from the bullet holes.

"Why'd I do what?"

"You killed your partner so he wouldn't kill me." A wave of weakness swept over Kittery, and he paused and closed his eyes. He gathered his strength and went on. "Now you're doctorin' me. If you'd let him kill me, you'd both be in Mexico by now."

"Don't talk," Bishop ordered.

The outlaw didn't speak again until he had wrapped a bandage around the big man's leg and tied it. Now he stood and walked away, and for a moment Kittery thought he meant to ride off. He closed his eyes and prepared to finish dying here alone.

Bishop had ascended the brush-choked hill, and several minutes later he came back leading his blood bay and Dixon's gray. He walked toward Kittery, who forced his eyes open once again. Bishop bent over without comment and handed him a slip of paper, and Kittery took it and read. It was a letter from Baraga to Bishop, sent from the Tucson jail to Sonora, Mexico.

Bishop,
They have captured me. I am sitting in a cell in Tucson, and I have been sentenced to hang. I pled guilty to everything, for we cannot fight destiny, can we?
But with me dead I know you and Dixon will be on a vengeance trail against Tappan Kittery and the rest of the vigilantes. That is the reason I take pen in hand to ask you to go another way. I know Dixon well enough to know his mind won't be changed by a letter, so I address this to you. Knowing you have some control over him, I beg you to keep him from taking vengeance on Kittery. Now, you will wonder my reasons, and to explain I must introduce myself as I was known before I became Baraga. My name was Jacob Atticus Kittery. Tappan Kittery is my little brother. With me gone, he is all that remains of my family name, and he has made a life for himself and begun to raise a family of his own.
So you will see why I ask you, in my memory, to spare his life, and to protect him from Dixon if you can. But do not betray my secret to Dixon. To him it is better that I remain and die Savage Diablo Baraga. I don't wish him to ever know I had a brother who left the cause.
I wish you the best of luck always, with your life and with that ranch. May you find it fruitful.

Yours truly,
Baraga

When Kittery finished reading the letter, he lowered it with a sigh. The irony was staggering. He had wished death upon the Desperados for so long and so fiercely, and here at the end two of them had saved him—from one of their own. One of them had done it from the grave. Jacob was still looking out for him.

"He must have meant a lot to you," Kittery judged, "to make you kill a friend over me."

Bishop nodded. "It wasn't just him. This might sound funny coming from a man like me, but I chose you over my partner because he was a killer, and you're not. You're a lawman, and there are too few good ones left. And he had no loved ones, which you do. If you didn't . . ." He shrugged and left the rest unsaid.

"And now?" Kittery gave him a questioning look. "If you leave me here, I'll die. Will you ride to Castor to carry me in, or do you ride south?"

Bishop sat down on his heels and met Kittery's gaze. He seemed about to smile, then he just nodded his head and looked out across the desert. "I pondered that myself. But judging by the dust cloud I saw from the top of the hill a few minutes ago, I'd say you have a good number of friends. And they're no idlers, either, judging from this past year. I'd say you'll be in good hands."

He stood again and pulled Kittery's Remington from behind his waistband. He pondered it for a second, turning it over once in his hands. In his eyes was the closest thing to sadness Kittery could imagine. "Funny," he mused. "It was like it was meant to kill my friend." He stooped, and in a gesture of complete trust or perhaps fatalism he placed it near Kittery's hand, then turned to his horse.

In a moment he returned, this time carrying a book. "Somethin' I've been readin'," he said, opening the front cover of the book and handing it to Kittery. "I figured it'd mean more to you than me. I found it in your friend's saddlebag—your friend the marshal. The note is for you." He touched his hat, turned, and, after struggling to heft Dixon's body and tying it to the gray horse, took its reins in his hand and climbed onto his own mount.

He moved the horse closer and looked down at Kittery. "It's my bet those wounds aren't fatal, Kittery. So make use of your life and that name my friend Baraga was so set on preserving. Raise your family well. Your friends are a little too close now, so I'll be on my way."

He gave a long, searching look to Kittery, then let a great breath swell his chest. Then Samuel Colt Bishop, the last of the Desperados Eight, yelled at his horse and kicked him into a gallop, towing the gray behind. And Old Mexico was a dirty gray line on the near horizon.

With the beating of the two horses' hooves fading into the distance, Kittery looked down at the book the gunman had given him. It was the copy of *The Book of Mormon* he had given Joe Raines. A simple note was scrawled across the inside cover.

Tappan,
If you read this I am probably dead. It may be too late, but knowing your nature and the depth of our friendship, I think you would avenge my death. I give you one last request. Say a prayer over my grave, then ride away. Don't waste any of your life avenging mine.

Best regards forever,
Joseph P. Raines

Kittery closed his eyes. So his promise at the marshal's grave had fallen on unwilling ears. But he could find in his heart no regret for the crusade he had undertaken, for he and his vigilantes had made Arizona Territory a safer place to live.

When he opened his eyes again, the sound of many pounding hooves rose out of the brush of the desert, and he watched a dozen men break from the trees, then many more in the swirling dust behind. Adam Beck rode double behind the saddle of one of the horses, and Kittery knew at a glance that Beth Vancouver's dream had come true. Sheriff Luke Vancouver was the man in the saddle.

Epilogue

    Luke Vancouver rode to Tappan Kittery's rescue that fateful day after returning from Prescott with his family to reassume his duties as under-sheriff of the town of Castor. *His* town. He came with Beck, James Price, Lafayette Bacon, Jake Long and many others who had shared a score of campfires with Tappan Kittery, the man who came to be known as "the Vigilante." Even Miles Tarandon was in the posse.
    There came to be a friendly rivalry when Luke Vancouver bought Greene's Dry Goods store from Maria McBride, changed its name to Vancouver's Hardware, and ran it with much success across the street from Kittery's Mercantile. The rest of the town remained as unchanging as the hills.
    Tappan Kittery, the Vigilante, lived for many years after his narrow escape from death at the hands of Major Morgan Dixon. He raised a strong family of four sons and three daughters, and the Kittery name carried on.
    Arizona achieved statehood in 1912. She may never relive the kind of violence that was part of her territorial days, her legacy in the nineteenth and early twentieth centuries. Never again may she see the death and bloodshed that once symbolized her dark and bloody ground, impeding her acceptance as a state for so many years. The laws of Arizona now are strong, as are its lawmen. They have the power to protect and uphold the law. But this was not always so.
    In time, those who called themselves the Castor Vigilantes faded into the pages of history. Men like Adam Beck, the manhunter, James Price, the hangman, and Miles Tarandon, deputy sheriff. They are labeled heroes, by some, and others call them outlaws. But they accomplished what they set out to do: they purged the range and cleansed their homeland.
    The mines eventually filled with water and had to be closed. The streams dried up, even the Santa Cruz river that brought the valley to life, and the last of the beaver died with them. Now the town of Castor, like its namesake the beaver, is gone, vanished with the winds of change, buried in the desert sands. The bones of those who built her lie with her, the men who once helped her grow and thrive.
    Captain Tappan Kittery, too, is gone, but his legend lives on. A man traveling through Arizona, stopping at some small town café, may hear his story and his name. He may hear of the Castor Vigilantes, the men who went outside the law because they had no other choice. He may hear the name of Jacob Atticus Kittery,

the man who called himself Savage Diablo Baraga. And he may listen to the mystery of the lone gunman, Samuel Colt Bishop, the last of the Desperados Eight. Castor will be remembered, with her vigilantes, and her outlaws. But when is told the tale of this Arizona town, the vigilantes, and the Desperado Den, they will hear before all else the name of Captain Tappan Kittery . . . the Vigilante.

Song of the Vigilante

There is blood on the land,
Spilled there by the killer band;
Soon they'll lie beneath the sand,
Laid there by the Captain's hand.

Like a breath of desert wind,
From the north, he rode in,
Seeking justice for eight men;
He'll make them pay for their sins.

They say he won't make it
'Cause he's only one man,
But his pride is strong as his callused hands.
He doesn't care about what they say;
He'll bury the outlaw band one day.
His name is Captain Kittery;
He'll go down in history.

He learned to kill in the war;
He didn't want to anymore;
But the wolf howled at his door;
Now he had a bloody chore.

So he headed out of town;
Toward eight killers he was bound;
Now across the bloody ground
Lie the men that he put down.

They said he won't make it
'Cause he's only one man,
But his pride was strong as his callused hands.
He didn't care about what they said;
Now the killers lie behind him dead.
His name is Captain Kittery;
He's gone down in history.

About the Author

Kirby Frank Jonas was born in Bozeman, Montana. He lived along a once-remote road in a rambling vee in the mountains known as Bear Canyon, where cattle range gave way to spruce and fir, and the wild country was forever ingrained in him. It was there he gained his love of the Old West, listening late at night to his daddy tell stories and sing western ballads, and watching television Westerns such as *Gunsmoke, The Virginian* and *The Big Valley,* and listening to a well-worn long playing record of Davy Crockett.

Jonas next lived on a remote farm in the middle of Civil War battlefield country near Broad Run, Virginia. That was followed by a move to Shelley, Idaho, where he completed all of his school years, wrote his first book *(The Tumbleweed)* in the sixth grade and his second *(The Vigilante)* as a senior in high school. He has since written six published novels and two which are forthcoming, one of which was co-authored by his older brother, Jamie. He is currently co-authoring a novel entitled *Yaqui Gold.* His partner is none other than his hero, actor Clint Walker.

Besides writing novels, Jonas also paints wildlife and life in the West. He has done all of his cover art and hundreds of other pieces. He is a songwriter and guitar player and singer of old Western ballads and trail songs. Jonas enjoys the joking title given to him by his friends, "The Renaissance Cowboy."

After living in Arizona to research his first two books, and traveling through nine countries in Europe, to get his glimpse of the world, Jonas settled in Pocatello, Idaho. He has made a living fighting forest fires for the Bureau of Land Management in five western states; worked for the Idaho Fish and Game Department; been a security guard and a guard for Wells Fargo in Phoenix, Arizona. He was employed as an officer for the Pocatello City Police and currently works as a city firefighter. He and his wife, Debbie, have four children, Cheyenne Kaycee, Jacob Talon, Clay Logan and Matthew Morgan.